Epigenetics

Epigenetics

Lyle Armstrong

GS Garland Science
Taylor & Francis Group

NEW YORK AND LONDON

Garland Science
Vice President: Denise Schanck
Senior Editor: Elizabeth Owen
Development Editor: Elizabeth Zayatz
Editorial Assistant: Louise Dawnay
Production Editor and Page Design: Ioana Moldovan
Typesetter: Georgina Lucas
Illustrator and Cover Designer: Drippy Cat Software Limited
Copyeditor: Bruce Goatly
Proofreader: Sally Livitt
Indexer: Bill Johncocks

About the author:

Lyle Armstrong is a Reader in Cellular Reprogramming
at the Institute of Genetic Medicine in Newcastle
University where his research program contributed to
the derivation of some of the UK's first human embryonic
stem cell lines and the development of the world's first
cloned human embryos. He is now working on new
methods to reprogram cells into medically useful cells,
focusing on the possibility of reversing aging during
reprogramming and how this might be valuable for
repairing organ damage or treating human diseases
using induced pluripotent stem cells. Of particular note
are investigations aimed at treating cardiovascular
disease and age-related hearing loss. Epigenetics is the
cornerstone of this reprogramming process.

ISBN 978-0-8153-6511-2

Library of Congress Cataloging-in-Publication Data
Armstrong, Lyle, author.
 Epigenetics / Lyle Armstrong.
   p. ; cm.
 Includes bibliographical references.
 ISBN 978-0-8153-6511-2 (alk. paper)
 I. Title.
 [DNLM: 1. Epigenomics. 2. Epigenesis, Genetic. 3. Genetic
Predisposition to Disease. QU 460]
 RB155.5
 616'.042--dc23
                    2013040242

Published by Garland Science, Taylor & Francis Group, LLC,
an informa business,
711 Third Avenue, New York, NY, 10017, USA, and 3 Park
Square, Milton Park, Abingdon, OX14 4RN, UK.

Printed in the United States of America

15 14 13 12 11 10 9 8 7 6 5 4 3 2 1

 **Garland Science**
Taylor & Francis Group
Visit our website at http://www.garlandscience.com

Preface

When this book was first conceived, the motivation to write was simple. My research program focused on analyzing the mechanisms through which epigenetic reprogramming functioned during somatic cell nuclear transfer. Several graduate students were interested in this project but there were few easily digestible texts that could provide an introduction to the subject of epigenetics. That is not to say that excellent textbooks did not exist on the subject, but these were perhaps less suitable for students whose projects were concerned with the impact of epigenetics on cellular functions rather than understanding the basis of epigenetics in its own right.

The book is the result of a long and iterative process. The field of epigenetics has advanced rapidly during the assembly of the text so that several updates have been necessary. I hope that the result is a general introduction that will enable students who need to appreciate how epigenetics can influence cell biology and the onset of human disease. **Epigenctics** is divided into three sections. Section 1 describes chromatin architecture before going on to the details of DNA methylation and histone acetylation. The second section explains how normal cellular functions (gene expression, the cell cycle, gene imprinting, and differentiation) are affected by epigenetics and how epigenetic modification patterns can be reversed to create pluripotent stem cells. The final section discusses the evidence for epigenetic involvement in disease, with emphasis on cognitive dysfunction and cancer.

For these reasons the book is aimed at a wider audience than those involved in understanding of purely epigenetic phenomena; my hope is that researchers in all areas of cell biology and medicine will find the text useful and informative, and if by reading this they are encouraged to add to the body of knowledge of epigenetics, I will consider the book to have been a success.

Acknowledgments

I would like to acknowledge the support of my long-suffering family during the writing of this book.

The author and publisher would like to thank the reviewers who provided helpful comments on the original proposal and gave detailed feedback on draft chapters:

Keith Brown (University of Bristol); James Catto (University of Sheffield); Frances Champagne (Columbia University); Hugh Dickinson (University of Oxford); Steven Gray (Trinity College Dublin); Jay L. Hess (University of Michigan); Megan Hitchins (University of New South Wales); Louise Jones (University of York); Diane Lees-Murdock (University of Ulster); Amanda McCann (University College Dublin); Jonathan Mill (Institute of Psychiatry); Peter Meyer (University of Leeds); Fernando Pardo-Manuel de Villena (University of North Carolina at Chapel Hill); Gerald Schatten (University of Pittsburgh); Doris Wagner (University of Pennsylvania); Colum Walsh (University of Ulster).

ONLINE RESOURCES

Accessible from www.garlandscience.com, the Student and Instructor Resource Websites provide learning and teaching tools created for **Epigenetics**. The Student Resource Site is open to everyone, and users have the option to register in order to use book-marking and note-taking tools. The Instructor Resource Site requires registration and access is available only to qualified instructors. To access the Instructor Resource Site, please contact your local sales representative or email science@garland.com. Below is an overview of the resources available for this book. On the Website, the resources may be browsed by individual chapters and there is a search engine. You can also access the resources available for other Garland Science titles.

For students

Flashcards
Each chapter contains a set of flashcards that allow students to review key terms from the text.

Glossary
The complete glossary from the book can be searched and browsed as a whole or sorted by chapter.

For instructors

Figures
The images from the book are available in two convenient formats: PowerPoint® and JPEG. They have been optimized for display on a computer.

PowerPoint is a registered trademark of Microsoft Corporation in the United States and/or other countries.

Contents

BASIC CONCEPTS

SECTION 1

1

INTRODUCTION TO THE STUDY OF EPIGENETICS

1.1 THE CORE ISSUE: CONTROLLING THE EXPRESSION OF SPECIFIC GENES

The inheritance of the genetic information contained in our DNA has been studied for the past 60 years, with the research efforts culminating in the Human Genome Project. The Human Genome Project aimed to map and sequence every gene in the human organism and to understand how these were positioned in our chromosomes. The successful completion of the project required an internationally coordinated effort by many research groups and has provided the basic information we need to understand how genes work at the molecular level. This latter effort has been described as being part of the "post-genomic" era, and its aims are to understand how information contained in the sequence of DNA bases can be turned into proteins that create the structures of the cell, and how this process is controlled.

The mechanism by which genes are controlled is probably the most active and fascinating area of research in science today. How does one type of cell, for example a fibroblast, "know" that it is different from a neuron or a muscle cell, given that all these cells have essentially the same information about protein synthesis contained in their genomes? At least some of this control of cell type comes from specific transcription factor proteins that instruct some genes to express while others remain silent, but this cannot explain how the cell remembers to only produce other cells of the same type when it divides. Maintaining cellular identity and function is most probably effected by using so-called "epigenetic" mechanisms.

1.2 DEFINING EPIGENETICS

The term **epigenetics** seems to be used to explain a wide range of biological observations, so it is useful to have a precise definition of its meaning. The word "epigenetics" was first used by Conrad Waddington in 1942, and his definition of the subject was as follows: "a branch of biology which studies the causal interactions between genes and their products which bring the phenotype into being." With the benefit of modern hindsight, we can see that this is quite a broad description, as it covers most of the mechanisms by which cellular identity and function can be maintained. However, Waddington's concept of a "gene" did not benefit from the investigations of the 1950 and 1960s. In spite of this, his original definition is remarkably close to the epigenetic control of gene transcription that we will cover in later chapters of this book.

Our increasing knowledge of genome functions has refined the definition of epigenetics, and today the term is generally accepted as meaning *the study of changes in gene function that are mitotically and/or meiotically heritable and that do not entail a change in the sequence of DNA*. We cannot assume that this definition will be final, of course; even with our current knowledge we can see that restricting the focus of epigenetics to "gene" function alone might be viewed as erroneous because there is increasing evidence that epigenetic mechanisms can control the functions of noncoding sequences of DNA (that is, those sections of the genome that do not contain sequences formally identified as genes). However, for the purposes of this broad discussion of the topic of epigenetics, the definition is good enough.

1.3 THE NATURE OF EPIGENETIC MARKS

The literal meaning of "epigenetics" is that it is something that is outside the traditional study of genetics. Although it is unlikely that Waddington's original definition would have been conceived with this in mind, the concept of an additional or external control system (that is, one that does not directly originate from the sequence of DNA bases) that is imposed upon genes fits very well with our modern knowledge of epigenetic modifications. Gene control is achieved by semi-reversible covalent chemical modifications of DNA bases and the proteins with which DNA is associated in the cell's nucleus. The description of these modifications and their effects on gene and cell function forms a large part of the topics described in this book.

1.4 THE IMPORTANCE OF EPIGENETICS

A system capable of controlling gene function is interesting to most biologists, but some of the possible consequences of epigenetic control of gene expression have also attracted interest from nonscientists because of the impact that our lifestyles may have on the health or behavior of future generations. At a first glance this would seem to go against the "traditional" principles of genetics, because the bulk of the scientific data in the first 50 years of the twentieth century seemed to suggest that the phenotypes of animals and their offspring can be determined solely on the basis of the genes they possess and pass on to future generations. Any changes that occur in phenotype would thus be due only to alterations in the DNA sequences of the genes and would therefore fall under the heading of "evolution" in the Darwinian sense of that word.

The acceptance of Darwin's theories took years, and Darwin's concept of evolution had to compete with several nineteenth-century ideas. The most prevalent of these competing hypotheses was probably the theory of inheritance of acquired characteristics, published by Jean Baptiste Lamarck in 1801. Lamarck suggested that if an organism changes its phenotype to adapt to its environment, those changes could be passed on to its offspring. Darwin suggested that this would not occur and that the only mechanism by which phenotypic change could occur stemmed from some organisms' having variations that help them to survive in their environment and be more successful at producing offspring. Because these useful traits arise from the parent's genes, the only way that the offspring will acquire such a survival advantage is by receiving copies of the advantageous genes from their parents. Ultimately, the evidence generated by the study of genetics supported Darwin, and Lamarck's theory was discredited, leading to a central dogma of twentieth-century biology that the only way for traits to be passed on was through the inheritance of genes and that genes could not be affected by events in the outside world. This belief seemed to stand to reason; after all, it seemed to be common sense that something bad that happened during the life of one's grandfather could not have any effect on one's own health. However, some of the worst events of the twentieth century seem to suggest otherwise.

Epigeneticists often refer to the "Dutch Hunger Winter" of 1944, and we will be making our own references to this World War II event in the chapters describing the impact of epigenetic mechanisms on a variety of diseases. The Hunger Winter resulted from a German blockade of food and fuel shipments into western Holland from September 1944 until the liberation of the country from Nazi German rule in May 1945. The blockade, coupled with a harsh winter, caused widespread famine and resulted in a large number of deaths. Because the Hunger Winter was well documented, it has allowed us to measure the effects of famine on human health. The results of many of these studies suggest a link between starvation of pregnant women and the health of their offspring. Several epidemiological investigations found that the resulting children were more susceptible to diseases such as diabetes, heart disease, and obesity than the children of normally fed mothers. Furthermore, mental illnesses such as schizophrenia seemed to be more prevalent in the children of Hunger Winter mothers. More surprisingly, similar propensities to develop diseases were eventually observed in the grandchildren of Hunger Winter mothers, thereby countering a possible argument that starvation in the mother could simply lead to alteration in the development of her unborn fetus. These data seem to suggest that a grandmother's diet could affect the health of several generations, which further implies that an adaptation to her environment has produced a heritable trait—something that is not supposed to happen if we accept that Darwinian evolution is the only cause of phenotypic change.

One might still be tempted to argue that it was the mother's malnutrition that damaged the developing fetus and therefore caused some form of damage to the child's genes by introducing mutations in the DNA sequences whose harmful effects became evident only later, in adult life; however, recent research indicates otherwise. Prompted by the findings from the Dutch Hunger Winter, more recent studies have focused on the problems arising in the children of men who either are obese or have suffered starvation. It is known that there is a correlation between the pattern of epigenetic marks (mostly DNA methylations) on the insulin-dependent growth factor gene *IGF2* and the body mass index of the father, with hypomethylation of this gene being observed in newborns arising from couples in which the father is obese. These data imply that the nutritional state of the father may also contribute to the health of the children; this could only have been transmitted to the child via the father's spermatozoa. Although it is still possible that the starvation or obesity of the father could have introduced mutations into his own *IGF2* gene before the child's conception, it seems unlikely that similar mutations would occur in all obese individuals to confer similar *IGF2* hypomethylation in the offspring. It is noteworthy that hypomethylation of *IGF2* was also observed in the children of Hunger Winter mothers six decades after the children's birth in 1944–1945. There is therefore a considerable body of evidence supporting the acquisition of a heritable trait as a consequence of the nutritional state of the parents.

Other diseases can also be considered to have an epigenetic basis, regardless of how well a child's parents ate. Only 10% of breast cancers that run in families can be linked to known genetic mutations. Exposing pregnant rats to chemicals known to influence breast cancer risk in humans (such as a high-fat diet or the synthetic estrogen ethinyl estradiol) caused female offspring to develop a higher incidence of mammary tumors than the offspring of rats that were not exposed to these risk agents. Interestingly, the increased risk of breast cancer does not seem permanent, at least for some of the risk agents. For example, the great-granddaughters of rats exposed to a high-fat diet had no greater tumor incidence than those of control pregnant rats. The effects of ethinyl estradiol, however, seemed to be more durable: the great-granddaughters also had an enhanced risk of developing cancer.

We do not know the exact mechanisms by which chemical exposure can alter the pattern of epigenetic modifications in the genomes of affected animals, and this is one of the reasons why epigenetics is such a fascinating and potentially fruitful area of investigation. After all, if certain environmental exposures or behavior patterns in previous generations can influence the health of the current generation, altering those exposures or behaviors in the current generation might improve the health of the next. Better still, if we understand the mechanisms that caused events such as the Dutch Hunger Winter to predispose the descendants of those affected to disease, we might be able to develop methods to prevent the accumulation of harmful epigenetic changes and reduce the disease risk, regardless of environmental influences. The influence of this on human health and well-being could be great indeed.

FURTHER READING

Bird A (2013) Epigenetics: discovery. *New Sci* **217**:ii–iii (doi:10.1016/S0262-4079(13)60030-5).

Burgess DJ (2013) Epigenetics: mechanistic insight into epigenetic inheritance. *Nat Rev Genet* **14**:442 (doi:10.1038/nrg3525).

Eichten S & Borevitz J (2013) Epigenomics: methylation's mark on inheritance. *Nature* **495**:181–182 (doi:10.1038/nature11960).

Ferguson-Smith AC & Patti ME (2011) You are what your dad ate. *Cell Metab* **13**:115–117 (doi:10.1016/j.cmet.2011.01.011).

Greer EL & Shi Y (2012) Histone methylation: a dynamic mark in health, disease and inheritance. *Nat Rev Genet* **13**:343–357 (doi:10.1038/nrg3173).

Grossniklaus U, Kelly B, Ferguson-Smith AC et al. (2013) Transgenerational epigenetic inheritance: how important is it? *Nat Rev Genet* **14**:228–235 (doi:10.1038/nrg3435).

Gruenert DC & Cozens AL (1991) Inheritance of phenotype in mammalian cells: genetic vs. epigenetic mechanisms. *Am J Physiol* **260**:L386–L394.

Hackett JA & Surani MA (2013) Beyond DNA: programming and inheritance of parental methylomes. *Cell* **153**:737–739 (doi:10.1016/j.cell.2013.04.044).

Hackett JA & Surani MA (2013) DNA methylation dynamics during the mammalian life cycle. *Phil Trans R Soc B* **368**:20110328 (doi:10.1098/rstb.2011.0328).

Hanson MA, Low FM & Gluckman PD (2011) Epigenetic epidemiology: the rebirth of soft inheritance. *Ann Nutr Metab* **58** (Suppl 2):8–15 (doi:10.1159/000328033).

Hardison RC (2012) Genome-wide epigenetic data facilitate understanding of disease susceptibility association studies. *J Biol Chem* **287**:30932–30940 (doi:10.1074/jbc.R112.352427).

Heijmans BT, Tobi EW, Stein AD et al. (2008) Persistent epigenetic differences associated with prenatal exposure to famine in humans. *Proc Natl Acad Sci USA* **105**:17046–17049 (doi:10.1073/pnas.0806560105).

Hesman Saey T (2013) From great grandma to you. *Science News* **183**:18–21 (doi:10.1002/scin.5591830718).

Ho MW & Saunders PT (1979) Beyond neo-Darwinism—an epigenetic approach to evolution. *J Theor Biol* **78**:573–591 (doi:10.1016/0022-5193(79)90191-7).

Maynard Smith J (1990) Models of a dual inheritance system. *J Theor Biol* **143**:41–53 (doi:10.1016/S0022-5193(05)80287-5).

Moazed D (2011) Mechanisms for the inheritance of chromatin states. *Cell* **146**:510–518 (doi:10.1016/j.cell.2011.07.013).

Monk M (1990) Variation in epigenetic inheritance. *Trends Genet* **6**:110–114.

Roberts AR, Huang E, Jones L et al. (2013) Non-telomeric epigenetic and genetic changes are associated with the inheritance of shorter telomeres in mice. *Chromosoma* Jul 18 [Epub ahead of print] doi:10.1007/s00412-013-0427-8).

Sarkies P & Sale JE (2012) Cellular epigenetic stability and cancer. *Trends Genet* **28**:118–127 (doi:10.1016/j.tig.2011.11.005).

Shea N, Pen I & Uller T (2011) Three epigenetic information channels and their different roles in evolution. *J Evol Biol* **24**:1178–1187 (doi:10.1111/j.1420-9101.2011.02235.x).

Shiota H, Goudarzi A, Rousseaux S & Khochbin S (2013) Transgenerational inheritance of chromatin states. *Epigenomics* **5**:121–122 (doi:10.2217/epi.13.12).

Skinner MK, Haque CG, Nilsson E et al. (2013) Environmentally induced transgenerational epigenetic reprogramming of primordial germ cells and the subsequent germ line. *PLoS One* **8**:e66318 (doi:10.1371/journal.pone.0066318).

Soubry A, Schildkraut JM, Murtha A et al. (2013) Paternal obesity is associated with IGF2 hypomethylation in newborns: results from a Newborn Epigenetics Study (NEST) cohort. *BMC Med* **11**:29 (doi:10.1186/1741-7015-11-29).

Thayer ZM & Kuzawa CW (2011) Biological memories of past environments: epigenetic pathways to health disparities. *Epigenetics* **6**:798–803 (doi:10.4161/epi.6.7.16222).

Xu M, Wang W, Chen S & Zhu B (2011) A model for mitotic inheritance of histone lysine methylation. *EMBO Rep* **13**:60–67 (doi:10.1038/embor.2011.206).

THE BASIS OF THE TRANSCRIPTION PROCESS

The genomes of virtually every eukaryotic organism contain thousands of genes, but only a fraction of these will be actively transcribed at any one time. Exactly which genes are active depends on the type of cell they are in, its position in the mitotic (or meiotic) cell cycle, environmental influences on individual cells, and even the age of the cell relative to its neighbors. Maintenance of this gene expression pattern is complicated enough in a single-celled eukaryote such as *Saccharomyces cerevisiae*, but it is exponentially more complex in multicellular organisms, where most of the cells have exactly the same genome yet need to express very different genes to perform their specific function. Deciding which information to use from the genome is a complex task and is the principal reason why epigenetic modification of the genome has evolved. Many of these epigenetic modifications impinge on the ability of gene promoters to initiate transcription, so it is useful to review the transcription process briefly before launching into the molecular basis of epigenetic control of gene expression.

2.1 THE NEED FOR SPECIFICITY

The objective of gene transcription is to make ribonucleic acid (RNA), but the exact type depends on the nature of the gene being transcribed. Many RNAs fulfill the role of transmitting information from the genome (DNA) to allow the synthesis of proteins that are needed to perform biochemical or structural functions in the cell (hence the rather obvious name of messenger RNA or mRNA), but there are other types of RNA molecules with very different jobs. Transfer RNAs (tRNAs), for example, spend their days transporting amino acids to the ribosomes for incorporation into growing polypeptides, and the ribosome itself is composed largely of ribosomal RNA (rRNA). There is also increasing evidence that still other RNAs (the so-called noncoding RNAs (ncRNAs)) have a role in modulating the control of gene expression.

The various types of RNAs are made using three different RNA polymerase enzymes that consist of multiple polypeptide subunits. Our knowledge of how all these subunits interact and function is still incomplete. However, we do know that all the polymerases possess a similar catalytic domain (which is not surprising, because they all make RNA), so the other domains are probably involved in recognizing the elements (such as enhancers, promoters, and gene specific sequences) within the underlying DNA sequences that identify the specific gene that a particular type of polymerase is meant to transcribe. By this we mean that there are features encoded into the DNA sequence that distinguish genes meant to produce mRNAs from those intended for tRNA and rRNA synthesis.

There is some degree of control imposed on transcription by the subcellular location of the respective polymerase. For example, RNA polymerase I is localized largely in the nucleolus, which is a small vesicle-like structure within the cell's nucleus whose job is to assemble ribosomes. This is a demanding task because all the other transcriptional activity of the cell depends on successful ribosome synthesis, and rRNA transcription by RNA polymerase I within the nucleolus therefore accounts for most of the cellular RNA synthesis. In contrast, RNA polymerase II is located in the nucleoplasm (the part of the nucleus that is not the nucleolus); its job is to make heterogeneous nuclear RNA (hnRNA), which is the precursor of mRNA. The activity of RNA polymerase II probably accounts for much of the remaining RNA synthesis in the cell, including that of microRNA, which probably arises by cleavage of much longer polymerase II-transcribed RNAs. Transfer RNAs and other small RNAs are made by RNA polymerase III, which is also located in the nucleoplasm. The latter implies that polymerase location alone is not sufficient to differentiate the activities of the three enzymes in RNA synthesis.

2.2 PROMOTERS AND THEIR TATA BOXES

The specificity of RNA polymerases rests on their ability to recognize sequence elements within the gene known as promoters. In essence the promoter is a specialized region of DNA sequence that is distinct from the arrangement of DNA base pairs that encodes the sequence of amino acids needed to make the gene product. It can be defined more precisely as a regulatory region of DNA, probably (but not necessarily) located upstream of the coding sequence, that is essential for the transcription of the gene's polypeptide-encoding regions. Promoters are often located quite close to the DNA base pairs that comprise the first codon of the coding sequence (the transcription start site), but this does not always apply, and we also see some genes that have promoter elements located at considerable distances from the transcription start site. In eukaryotes, there are often several regions of DNA in or near the promoters whose function is to regulate the rate of transcription up or down, or to stop it altogether if a specific set of circumstances occurs within the cell. This complexity and diversity makes it hard to produce a common definition of the eukaryotic promoter. What we can state unequivocally is that promoters are really all about getting transcription to start at the right place. Thus many promoters (but by no means all) have a sequence called the TATA element (or TATA box) situated about 30–50 base pairs (bp) away from the start site (**Figure 2.1**).

The TATA box serves to recruit a specific TATA box binding protein (TBP) that is part of the complex of proteins generally referred to as transcription factor IID (TFIID). As shown in **Figure 2.2**, the interaction between the TATA box and TFIID becomes stabilized by the binding of transcription

Figure 2.1 Typical configuration of a eukaryotic mRNA gene. Shown is the generally accepted position of the TATA box and possible locations of other control elements, such as promoters and enhancers. The TATA box is usually found within the core promoter and functions as a binding site for the TATA box binding protein that permits assembly of the transcriptional machinery in the correct position to begin transcription. The direction in which the gene is transcribed is indicated by the black arrow at the start of the first exon.

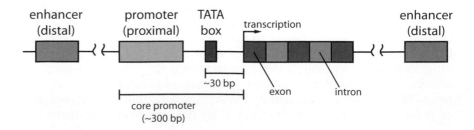

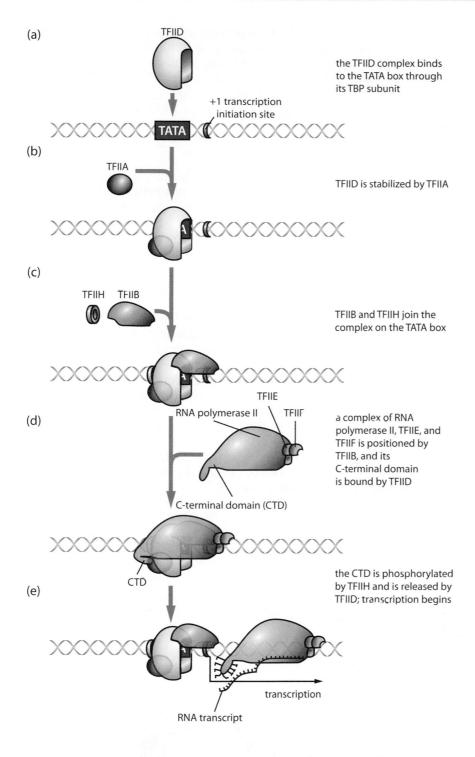

(a)

TFIID

the TFIID complex binds
to the TATA box through
its TBP subunit

+1 transcription
initiation site

TATA

(b)

TFIIA

TFIID is stabilized by TFIIA

(c)

TFIIH   TFIIB

TFIIB and TFIIH join the
complex on the TATA box

TFIIE

RNA polymerase II   TFIIF

(d)

a complex of RNA
polymerase II, TFIIE, and
TFIIF is positioned by
TFIIB, and its
C-terminal domain
is bound by TFIID

C-terminal domain (CTD)

CTD

the CTD is phosphorylated
by TFIIH and is released by
TFIID; transcription begins

(e)

transcription

RNA transcript

Figure 2.2 Assembly of the pre-initiation complex (PIC). Transcription requires the formation of a pre-initiation complex of proteins that includes the polymerase and a variety of TFII proteins. This occurs at the TATA box. (a) The assembly of PIC begins with the TATA box attracting a specific binding protein (TBP) that resides within a TFIID protein complex. (b, c)The TATA box binding protein recruits TFIIB to join TFIID (and TFIIA if present). (d) This growing complex of proteins provides an attachment site for the polymerase. However, before entering PIC, RNA polymerase II and TFIIF are bound together and subsequently recruited by TFIIB. (e) Finally, RNA polymerase II recruits TFIIE, which further recruits TFIIH to complete the PIC assembly.

factor IIA (TFIIA), which seems to encourage other transcription factors to join in. Ultimately this block of proteins sitting on top of the TATA box provides a site to which RNA polymerase II can attach. This initiation mechanism involving the construction of an RNA-polymerase-containing complex on any gene promoter (ribosomal RNA, noncoding and mRNA-encoding) with the help of transcription factors is common to all RNA polymerases, although the recruitment of proteins to a TATA box is unique to the protein-encoding genes. After phosphorylation of its C-terminal domain (which is a bit like unhooking it from the attachment site), RNA polymerase II can move off down the DNA, making the RNA transcript as it goes.

To describe the activity of the RNA polymerases in this manner is, however, an oversimplification. There are probably hundreds, if not thousands, of genes with TATA boxes in broadly similar positions with respect to the transcription start site, so how does the cell know which genes to transcribe? A lot of this specificity seems to rely at least in part upon interactions between the TATA box protein itself and TATA-binding-associated factors, or TAFs, although probably not because of any direct binding of the TAFs themselves to important and gene-specific regions of the promoter sequence.

The functions of the TAFs seem to be to stabilize interactions between other transcription factors (so called for want of a better description) that do bind to highly specific sites in the promoter region. If all of these factors are available in the correct positions on the gene promoter, they may be able to interact to ensure that the TATA box binding protein stays in its place long enough to be able to recruit RNA polymerase.

2.3 ASSEMBLY OF THE PRE-INITIATION COMPLEX

Since individual gene specific transcription-factor-binding sites are often located quite far from the TATA box, pre-initiation complex assembly often requires bending and looping of the promoter DNA, as shown in **Figure 2.3**. The distortion of promoter DNA by the pre-initiation complex is presumably not straightforward, because it is energetically unfavorable and therefore must involve some energy expenditure on the part of the cell. This implies that the cell has good reason to partake in this activity. For example, if it is one of the mechanisms through which specific gene expression is brought about, then the energy investment is probably worthwhile, because specific and precisely controlled patterns of gene expression are the cornerstones of cellular identity that enable them to carry out their functions in the organism.

The binding sites for the transcription factors cannot all be located in the same stretch of the DNA, because the looping or bending of DNA would then be far too extreme to be easily achievable. It can also be

Figure 2.3 Transcription-factor-binding sites (TFBSs) are often located far apart, and the DNA must form loops to bring these elements into proximity. CRM represents Cis-Regulatory Module. [Adapted from Wasserman WW & Sandelin A (2004) *Nat. Rev. Genet.* 5, 276. With permission from Macmillan Publishers Ltd.]

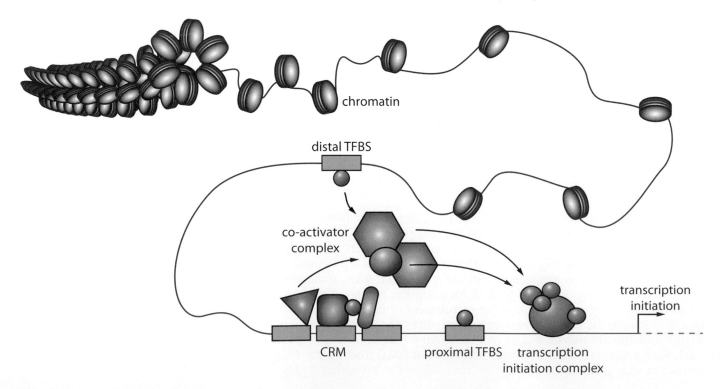

inferred that, once formed, the loop stabilizes the interaction of the RNA polymerase in some way. An interesting bit of recent evidence seems to support this possibility: it has been shown that, for a few genes, treatment with antisense oligonucleotides can actually activate transcription. Antisense techniques are a well recognized means of reducing the synthesis of gene products because the introduction of a sequence of RNA into the cell can interfere with the processing of its mRNA homologue at the ribosome. This is so because the antisense oligonucleotides bind to the normal "sense" mRNA through Watson–Crick base pairing to produce a three-stranded "triplex" that is no longer recognized by the ribosome. The activation of transcription that occurs in some instances seems to result when there is a degree of sequence similarity between the antisense oligonucleotide and parts of the promoter sequence of the gene, because this can permit the oligonucleotide to "invade" the double-stranded DNA at that position to form a triplex. This causes a "bulge" in the promoter (more properly called a D-loop), which seems to be interpreted in a similar manner to the loop created during formation of the initiation complex and results in recruitment of RNA polymerase II and consequent synthesis of the RNA transcript.

It should be noted that many genes do not have TATA boxes, although most of these seem to be housekeeping genes whose transcription occurs in all cell types at all times. This does not seem to present a problem to the cells, because they simply use some other protein in place of the TATA box binding protein that allows TFIID to bind and begin the sequence of events culminating in the recruitment of RNA polymerase II. Why this alternative method of initiation evolved is very much open to question. It may be that TATA-less promoters are much easier to initiate, which fits well with the use of this system on permanently expressed types of genes for which we do not need to establish and control a complex spatio-temporal expression pattern during development.

For transcription of most other genes, the specificity of the three RNA polymerases depends, as mentioned earlier, on their ability to distinguish promoters. RNA polymerase I recognizes promoters comprising two distinct sequence elements; one element defines the core promoter region and the other is a control element that lies a precise distance from the promoter core and is always upstream of the transcription start site. The presence of multiple promoter elements is not unique to the rRNA genes transcribed by RNA polymerase I, but the relative positions of the individual elements along the DNA define the specific interaction with the polymerase and ensure that only this enzyme can transcribe these genes. RNA polymerase III looks for promoters that are nested within the coding sequences of the genes. This arrangement is typical for tRNAs and structural RNAs, and makes sure that their interaction with polymerase III is specific. Promoters that are recognized by RNA polymerase II tend to be more complex, which probably reflects their developmental and cell-type-dependent specificity.

2.4 INITIATION OF TRANSCRIPTION

It is important to realize that the RNA polymerases do not initiate transcription by themselves. They require the simultaneous presence of a wide range of other proteins and factors at the gene to make sure that transcription can begin, and the loss of even a single member of this aggregation of proteins generally halts the process. At this point transcription becomes really complex because not only are there thousands of potential transcription factors that can bind to precise sites within the coding and/or promoter sequences of the gene, but there are also numerous additional proteins—such as the transcription factors—that form part of the pre-initiation complex. If you add to this the possibility (indeed the probability) that many of these pre-initiation complex elements are able to bind and interact with proteins responsible for performing epigenetic modification of the genome, it becomes evident that a precise description of the mechanisms through which transcription operates is too large and complex a topic to incorporate into this relatively modest text.

The key to understanding the enormous impact of non-sequence-based modification in controlling genome activity is the appreciation that DNA does not exist in isolation within the cell's nucleus but is associated with chromatin. The latter is a very general term for a range of proteins whose primary function is to package the enormous volume of DNA into the relatively small volume occupied by the nucleus. Of special interest to us is that these proteins also serve to control access of the transcriptional machinery to the information encoded by DNA. Chromatin is the topic of the next chapter.

KEY CONCEPTS

- Genes can be divided into those that encode ribosomal, transfer, noncoding, and messenger RNA. The last category includes the bulk of the protein coding genes.

- Synthesis of RNA from each of these classes of gene is referred to as transcription. A group of enzymes called RNA polymerases perform the RNA synthesis steps but they cannot act alone: a range of other proteins are needed to recruit the polymerases to the correct location on the gene. The combined assembly of proteins plus the RNA polymerase is known as the pre-initiation complex.

- Genes have noncoding regions that contain the sequence information needed to direct the assembly of the pre-initiation complex. Most of these sequences are contained within a region known as the promoter close to the transcription start site, but there are other control sequences in more distal locations (sometimes known as enhancers or repressors). Both the promoter and these distal regions have binding sites for transcription factors.

FURTHER READING

Archambault J & Friesen JD (1993) Genetics of eukaryotic RNA polymerases I, II, and III. *Microbiol Rev* **57**:703–724.

Aso T, Conaway JW & Conaway RC (1994) Role of core promoter structure in assembly of the RNA polymerase II preinitiation complex. A common pathway for formation of preinitiation intermediates at many TATA and TATA-less promoters. *J Biol Chem* **269**:26575–26583.

Asturias FJ (2004) RNA polymerase II structure, and organization of the preinitiation complex. *Curr Opin Struct Biol* **14**:121–129 (doi:10.1016/j.sbi.2004.03.007).

Berg JM, Tymoczko JL & Stryer L (2002) Biochemistry, 5th ed. W.H. Freeman & Co Ltd.

Boeger H, Bushnell DA, Davis R et al. (2005) Structural basis of eukaryotic gene transcription. *FEBS Lett* **579**:899–903 (doi:10.1016/j.febslet.2004.11.027).

Carter R & Drouin G (2009) Structural differentiation of the three eukaryotic RNA polymerases. *Genomics* **94**:388–396 (doi:10.1016/j.ygeno.2009.08.011).

Carter R & Drouin G (2010) The increase in the number of subunits in eukaryotic RNA polymerase III relative to RNA polymerase II is due to the permanent recruitment of general transcription factors. *Mol Biol Evol* **27**:1035–1043 (doi:10.1093/molbev/msp316).

Chung WH, Craighead JL, Chang WH et al. (2003) RNA polymerase II/TFIIF structure and conserved organization of the initiation complex. *Mol Cell* **12**:1003–1013 (doi:10.1016/S1097-2765(03)00387-3).

Flanagan PM, Kelleher RJ, Feaver WJ et al. (1990) Resolution of factors required for the initiation of transcription by yeast RNA polymerase II. *J Biol Chem* **265**:11105–11107.

Ginsburg HS & Vogel HJ (eds) (1984) Transfer and expression of eukaryotic genes. Elsevier.

Hashem Y, des Georges A, Dhote V et al. (2013) Structure of the mammalian ribosomal 43S preinitiation complex bound to the scanning factor DHX29. *Cell* **153**:1108–1119 (doi:10.1016/j.cell.2013.04.036).

Kornberg RD (2001) The eukaryotic gene transcription machinery. *Biol Chem* **382**:1103–1107 (doi: 10.1515/BC.2001.140).

Kornberg RD (2007) The molecular basis of eucaryotic transcription. *Cell Death Differ* **14**:1989–1997 (doi:10.1038/sj.cdd.4402251).

Kornberg RD (2007) The molecular basis of eukaryotic transcription. *Proc Natl Acad Sci USA* **104**:12955–12961 (doi:10.1073/pnas.0704138104).

Krumm A, Meulia T & Groudine M (1993) Common mechanisms for the control of eukaryotic transcriptional elongation. *Bioessays* **15**:659–665.

Li Y, Flanagan PM, Tschochner H & Kornberg RD (1994) RNA polymerase II initiation factor interactions and transcription start site selection. *Science* **263**:805–807 (doi:10.1126/science.8303296).

Nielsen S, Yuzenkova Y & Zenkin N (2013) Mechanism of eukaryotic RNA polymerase III transcription termination. *Science* **340**:1577–1580 (doi:10.1126/science.1237934).

Nogales E (2000) Recent structural insights into transcription preinitiation complexes. *J Cell Sci* **113**:4391–4397.

Nonet M, Sweetser D & Young RA (1987) Functional redundancy and structural polymorphism in the large subunit of RNA polymerase II. *Cell* **50**:909–915 (doi:10.1016/0092-8674(87)90517-4).

Rhee HS & Pugh BF (2012) Genome-wide structure and organization of eukaryotic pre-initiation complexes. *Nature* **483**:295–301 (doi:10.1038/nature10799).

Sayre MH & Kornberg RD (1993) Mechanism and regulation of yeast RNA polymerase II transcription. *Cell Mol Biol Res* **39**:349–354.

Strachan T & Reid A (2010) Human molecular genetics, 4th ed. Garland Science.

Sweetser D, Nonet M & Young RA (1987) Prokaryotic and eukaryotic RNA polymerases have homologous core subunits. *Proc Natl Acad Sci USA* **84**:1192–1196.

Tang H, Sun X, Reinberg D & Ebright RH (1996) Protein-protein interactions in eukaryotic transcription initiation: structure of the preinitiation complex. *Proc Natl Acad Sci USA* **93**:1119–1124.

Tsai FF & Sigler PB (2000) Structural basis of preinitiation complex assembly on human pol II promoters. *EMBO J* **19**:25–36 (doi:10.1093/emboj/19.1.25).

3

DNA PACKAGING AND CHROMATIN ARCHITECTURE

Many cell types have quite high levels of transcriptional activity, particularly if they are in a state of active division or growth. However, even cells that are less active need to retain their entire complement of genes and other, noncoding, sections of DNA, even if the DNA is not in regular use. This represents something of a problem because it means that for most cell types there is actually quite a lot of DNA in relation to the cell's overall size. In most eukaryotic cell nuclei, the DNA macromolecule would extend for approximately two meters if it were stretched out end to end, so to get this inside a nucleus of micrometer dimensions is quite an impressive feat of packaging. The main strategy of eukaryotic organisms for fitting the DNA into the cell is to wrap the DNA into tight spheres called **nucleosomes** (a bit like beads on a string) and then to form higher-order packaged structures in what may be thought of as folding up the string of beads. **Figure 3.1** gives an overview of this packaging process.

3.1 NUCLEOSOME STRUCTURE AND CHROMATIN

Chromatin consists of DNA plus many proteins

Chromosomes have an iridescent coloration, which may have been the reason for naming them chromosomes in the first place. This word is derived from the Greek words for color (*chroma*) and body (*soma*) and sounds a lot more erudite than 'colored bodies.' Extending this naming system, the material that comprised the chromosomes was referred to as **chromatin**.

Although the chemical nature of DNA and its presence in chromosomes had long been known, there was no specific information about the function of chromatin until Francis Crick and James Watson in 1953 determined the contribution of nucleic acids to heredity, but by then it was also evident that there was more than just DNA in the chromatin structure. In fact, by far the greater part of a chromosome is not DNA at all but a wide range of proteins whose functions were for many years assumed to be simply structural in nature, there only to support the DNA in the nucleus. We now know that this is partly true: proteins do form the basis of the nucleosome structure, which does indeed support the DNA and present it to the transcriptional machinery of the cell. However, also present are many other proteins whose functions are more concerned with controlling the way in which a cell can access the information contained within its DNA. It should be noted that these proteins also seem to have some degree of influence on the chromatin structure, so it can be difficult to distinguish between the functions of individual proteins.

Figure 3.1 Overview of the packaging of DNA. Each DNA molecule is packaged into a mitotic chromosome that is 50,000 times shorter than its extended length.

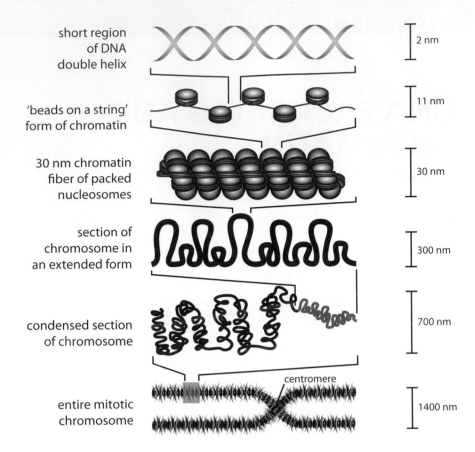

short region
of DNA
double helix
— 2 nm

'beads on a string'
form of chromatin
— 11 nm

30 nm chromatin
fiber of packed
nucleosomes
— 30 nm

section of
chromosome in
an extended form
— 300 nm

condensed section
of chromosome
— 700 nm

centromere

entire mitotic
chromosome
— 1400 nm

The nucleosome is the basic unit of chromatin

The fundamental unit of chromatin, the **nucleosome core particle (NCP)**, organizes 147 base pairs (bp) of DNA in a 1.7 left-handed super-helical turn around an octamer of four core histones: H2A, H2B, H3, and H4 (**Figure 3.2**). Small sections of DNA called linker DNA serve to join the nucleosomes together, and these are associated with the linker histone H1.

The structures of the individual histones determine how they bind together to assemble the octamer. X-ray crystallography shows the octamer to be a tripartite arrangement in which two (H2A–H2B) dimers flank a centrally located (H3–H4) tetramer. Comparison of the amino acid sequences of histone proteins across a wide range of species shows quite large differences, which led to initial speculations that the histones do not share any functional relationships. However, they do share a common structural motif known as the **histone fold** (**Box 3.1**) that seems to be essential to the dimerization of histones during octamer formation.

This structural element has been conserved throughout evolution, having first emerged in the archaebacteria where its purpose was to assist the interaction of proteins that do not assemble nucleosomes in the same way as in eukaryotic cells. Despite several hundred million years of evolution since the emergence of the eukaryotes, the histone fold motif has changed little, indicating the enormous importance of this protein structure to eukaryotic cell function. The other sections of the histone protein are probably necessary for DNA binding. The latter may depend on gene, cell type, and species, so it is perhaps not surprising that there is a much greater degree of variability in these regions.

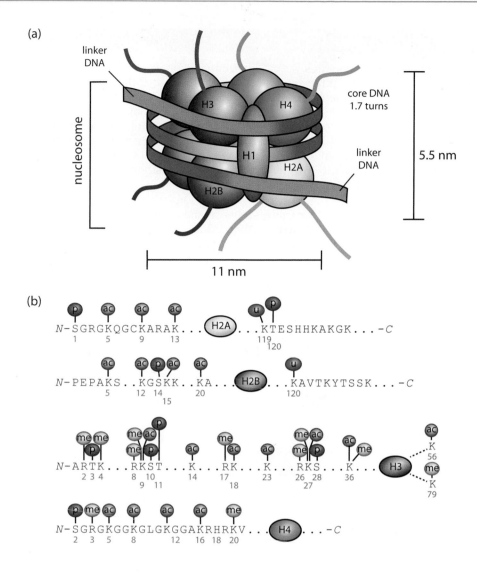

(a)

Figure 3.2 The nucleosome core particle and the histone octamer. (a) The structure of the histone octamer, with the sequences of the N-terminal tails shown stretched out. (b) The locations of the amino acid residue sites where the histone modifications occur are indicated above the sequences. The various types of modification include acetylation (ac), methylation (me), phosphorylation (p), and ubiquitination (u).

(b)

DNA binds to the histone octamer

The importance of the histone fold motif lies not only in its ability to drive histone-to-histone interaction; the motif also contributes to the ability of the histone octamer to wrap DNA around itself. Areas of the histone folds are presented at the outer face of the histone octamer and are thus able to make contact with DNA. Recall that the segment of DNA on a nucleosome core particle consists of approximately 147 bp. Of the available 14 minor grooves in this segment facing toward the histone octamer, 12 are in contact with these exposed histone fold regions. All the interactions of the histone folds are with the inner face of the DNA supercoil and extend over 121 bp, as may be seen in **Figure 3.3**. However, there is still uncertainty about the absolute contribution made by these interactions to the overall stability of the DNA–histone octamer complex because other structural elements within the histone amino acid sequences also seem to be involved in DNA binding.

The effectiveness of DNA binding is not equal at all of the aforementioned sites, and the ability of individual parts of the protein molecule to bind DNA seems to depend largely on the electrostatic potential of the nucleosome core particle surface. Calculations of this potential reveal positively charged regions located in the N-terminal histone tails and also in the nucleosome core, where the DNA makes nonspecific contacts with

Box 3.1 The histone fold

The four core histone proteins all share the structural domain known as the histone fold (**Figure 1a**), which consists of a long central helix connected to two helix–strand–helix (HSH) motifs at opposite ends. In more detail, the histone fold consists of an 11-residue helix (helix I), followed by a short loop and strand (strand I), a longer (approximately 27-residue) helix (helix II), another short loop and strand (strand II), and a final 11-residue helix (helix III). The exact number of residues in each helix and loop or strand segment seems to vary by one or two from histone to histone. The sequences of the fold regions of the four mammalian core histones are shown in Figure 1b, but such proteins typically dimerize to form structures of the types shown in Figure 1c, in which the strand from the N-terminal HSH motif of one fold pairs with the C-terminal strand of its partner. The pairing of the two folds

generates a smoothly curving outer surface containing the three regions (black) that dock to the inner face of nucleosomal DNA (Figure 1d). Such dimers form initially between histones H3 and H4 followed by the association of two of the dimers to form an (H3–H4)$_2$ tetramer. Histones H2A and H2B have specific small amino acid chains known as docking sequences that allow them to bind to the (H3–H4)$_2$ tetramer to form the histone octamer needed for the nucleosome core particle.

Figure 1 Structure, sequence, and dimerization of the histone fold polypeptide. (a) The histone fold, (b) fold region sequences of the four mammalian core histones, (c) the histone dimer, (d) histone dimer docking nucleosomal DNA. [Adapted from Ramakrishnan V (1997) *Annu. Rev. Biophys. Biomol. Struct.* 26, 83. With permission from Annual Reviews, Inc.]

(a)

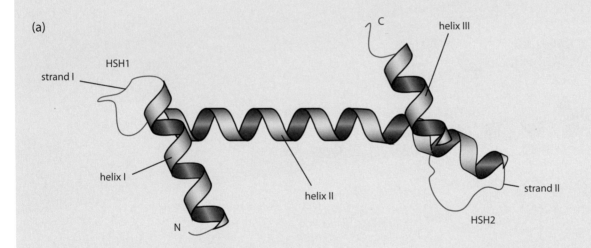

(b)

		loop/				loop/		
	helix I		strand I		helix II		strand II	helix III
H2A	(25)	PVGRVHRLLRKGNYAERVGAGAPVYLAAVLEYLTAEILELAGNPARDNKKTRIIPRHLQLAIRND	(38)
H2B	(36)	YSIYVYKVLKQVHPDTGISSKAMGIMNSFVNDIFERIAGEASRLAHYNKRSTITSREIQTAVRLL	(24)
H3	(66)	FQRLVREIAQDFKTDLRFQSSAVMALQWASEAYLVGLFEDTNLCAIHAKRVTIMPKDIQLARRIR	(4)
H4	(29)	TKPAIRRLARRGGV-KRISGLIYEETRGVLKVFLENVIRDAVTYTEHAKRKTVTAMDVVYALKRQ	(9)

(c)   pairing of N-terminus of one histone fold polypeptide (lilac) with the C-terminus of another (gray)

(d)

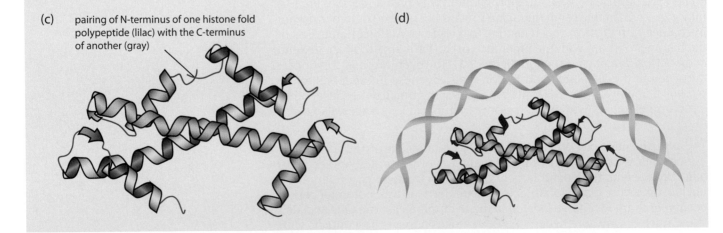

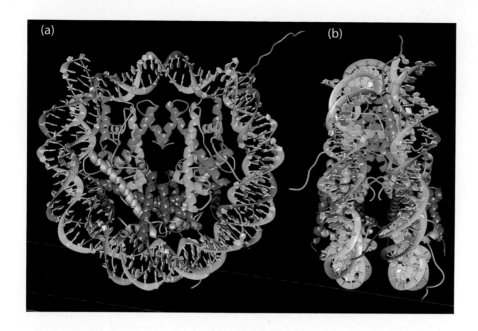

Figure 3.3 Interaction of DNA with the histone octamer. Part (a) represents the front view and part (b) represents the side view. This shows the binding of DNA to the polypeptide surfaces presented by the histone octamer to form the structure of the nucleosome. The N-terminal polypeptide "tails" of individual histones protrude from the globular nucleosome surface as shown. [From Luger K, Mäder AW, Richmond RK et al. (1997) *Nature* 389, 251. With permission from Macmillan Publishers Ltd.]

the protein surface (**Figure 3.4**). This interaction between DNA and the nucleosome can be explained by the helical nature of the DNA macromolecule, which ensures that its phosphate backbone with its inherent overall negative charge is presented to the amide groups (blue areas in Figure 3.4) of the histone protein main chain. Conversely, more negatively charged regions are found near the centre of the exposed face of the nucleosome core particle, and these do not normally make contact with the DNA.

The interaction of DNA with these particular sites must be very strong, because it requires substantial distortion of the DNA macromolecule. The wrapping of DNA around the nucleosome core particle therefore represents a considerable energetic investment, one that is probably only partly relieved by the neutralizing effect of the positively charged amino acids on the negative charges of the DNA phosphodiester backbone. The number and nature of these nucleic acid–histone interactions and the fact that they form primarily with the non-sequence-specific regions of DNA, namely the phosphodiester backbone, suggest that it should be possible to form nucleosomes with any DNA sequence. However, other factors may come into play.

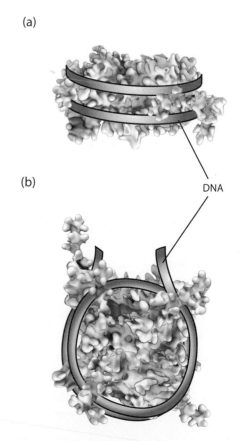

Figure 3.4 Electrostatic surface potential of the nucleosome core particle. The electrostatic potential of the nucleosome core particle ranges from +20 (blue) to −20 (red). (Protein Data Bank accession code (PDB) 1KX5.) (a) Side view of the nucleosome core particle. The amino acids lysine and arginine are positively charged by virtue of having side chains with amino groups that either can be easily ionized (the lysine γ-amino) or are already charged (the arginine guanido group), and the positively charged regions responsible for DNA binding are highly enriched in these amino acids. Note the distribution of positive charges (blue) around the DNA interface and the core histone tails (although the tails themselves have lower charge density). (b) Top view of the nucleosome core particle. This surface has several negative charges (red) in the exposed protein surface of the histone octamer. [Adapted from Mariño-Ramírez L, Kann MG, Shoemaker BA & Landsman D (2005) *Expert Rev. Proteom.* 2, 719. With permission from Expert Reviews Ltd.]

It is not clear whether the underlying DNA sequence is able to influence the exact position at which the nucleosome is formed. For example, do some sequences of DNA bases resist the required steric distortion more than others? If so, does this mean that the histone octamer is unable to form a nucleosome at those exact positions and is forced to 'move along' slightly to a region where it can more easily bend the DNA as required? Positioning of the histone octamer according to DNA sequence preferences would depend on the sum of the energetic costs arising from DNA structural adaptation at each of the 12 important binding sites. It is also possible that the other, more minor, histone–DNA interactions (for example, there are 146 possible hydrogen bond formation sites on the histone octamer) may fine tune nucleosome positioning. This could be one of the features controlling the variation in the lengths of linker DNA sequences.

The binding of histone proteins via their histone fold motifs to form the histone octamer and produce nucleosomes occurs spontaneously in experiments *in vitro* in which histones and DNA are simply mixed together. However, not long after the first description of the nucleosome as the basic chromatin unit, a series of *in vitro* experiments showed that adding *Xenopus laevis* egg extract to the histone–DNA mixture speeded up the process of nucleosome formation. This suggested the presence of additional factors in the *Xenopus* egg that enhance nucleosome formation; this ultimately led to the isolation of nucleoplasmin as the first of the **histone chaperones**. Histone chaperones are a group of acidic proteins that bind histones and participate in chromatin assembly and disassembly during transcription and DNA replication. A large number of histone chaperones have been discovered. Four of these molecules are shown in **Table 3.1**. Chromatin assembly consists of an initial stepwise core histone transfer by a core histone chaperone (nucleosome assembly protein 1; NAP1) and subsequent chromatin maturation by ACF (ATP-utilizing chromatin assembly and remodeling factor) (**Figure 3.5**). These ordered multiple steps can be separated or may lead directly to the formation of a nucleosomal array.

Most histone chaperones have a preference for binding either histones H3–H4 or H2A–H2B. Despite numerous attempts, there is currently no unified view of the mechanism by which histone chaperones promote nucleosome assembly, although current thinking seems to centre on their ability to inhibit nonspecific interactions between DNA and histone proteins so that the electrostatic potential of appropriate areas of the histone octamer surface can promote the binding of DNA.

TABLE 3.1 REPRESENTATIVE HISTONE CHAPERONES

Chaperone	Family and homologues	Histone	Related function
Nucleoplasmin	NPM1/numatrin/nucleophosmin NPM2 NPM3 NLP	H2A/H2B	H2A/H2B storage in oocytes Sperm chromatin decondensation Ribosome biogenesis Apoptotic chromatin condensation
NAP1	NAP1/NAP1L–NAP1L4	H2A/H2B	Histone H2A variant exchange Linker histone deposition Histone shuttling
N1/N2	NASP, N1/N2, Hif1	H3/H4	Storage of H3/H4 pools in oocytes Linker histone deposition Linker histone transport
CAF1	P150, p180, cac1, p60		Telomere silencing Cell cycle regulation

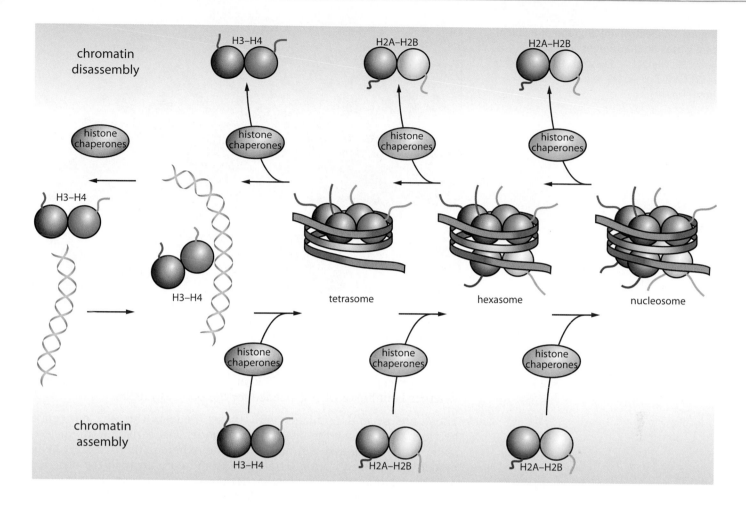

3.2 CHROMATIN ARCHITECTURE

Chains of nucleosomes organize into chromatin fibers

Although the structure of the first level of DNA folding, the nucleosome core, is known at atomic resolution, the structure of the second level of folding, whereby a string of nucleosomes folds into a fiber with an approximate diameter of 30 nm, remains undetermined. Early evidence for the presence of a 30 nm chromatin fiber *in vivo* came from electron microscopic analysis of the Balbiani ring genes in *Chironomus tentans* midges (**Figure 3.6**) and X-ray diffraction studies of cell nuclei that showed nucleosome spacings of 30–40 nm.

Recent structural analyses using *in vitro* reconstituted chromatin fibers have led to the proposal of two models for the 30 nm chromatin fiber that differ in topology, dimension, and nucleosome packing density (**Figure 3.7**). The first model is derived from measurements of the physical dimensions of long linear nucleosome arrays visualized by electron microscopy and is referred to as a one-start helix. This describes a 30 nm chromatin fiber in which successive nucleosomes are adjacent to one another. It has a diameter of 34 nm with a packing density of 11 nucleosomes per 11 nm. The second model is known as the two-start helix, which is based on a zigzag arrangement of nucleosomes that stack on top of one another, forming a 24–25 nm diameter fiber with a packing density of five or six nucleosomes per 11 nm. A consequence of these models is that straight sections of linker DNA (that is, the DNA sequences between the

Figure 3.5 Models for nucleosome assembly and disassembly. [Adapted from Das C, Tyler JK & Churchill ME (2010) *Trends Biochem. Sci.* 35, 476. With permission from Elsevier.]

Figure 3.6 Electron micrograph of Balbiani ring genes in *Chironomus tentans*. [From Skoglund U, Andersson K, Björkroth B et al. (1983) *Cell* 34, 847. With permission from Elsevier.]

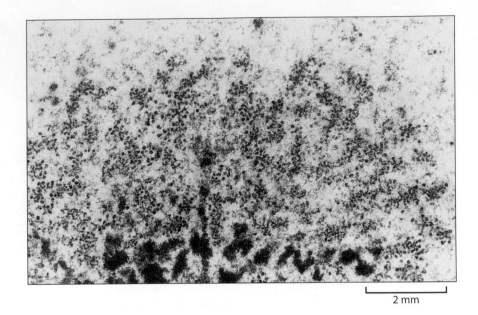

2 mm

nucleosomes) are required to link nucleosomes in the two-start helix, whereas bent linker DNA is needed for the one-start helix.

There are several problems with these models. The key difference between them is that the interdigitated model was derived from nucleosome arrays saturated with linker histone, whereas the two-start helix model was derived from a tetra-nucleosome core array in the absence of linker histone. The linker histone H1 does not form part of the nucleosome core but seems to act as a 'clamp' that helps to stabilize the association of DNA with the histone octamer (**Figure 3.8**).

Some studies have suggested that the linker histone is not essential for chromatin compaction; however, a gene knockout of three of the six murine H1 gene variants is lethal at the embryonic stage, whereas single knockouts modulate gene expression and affect chromatin compaction. Additionally, *in vivo* experiments show that linker histone variants control chromatin dynamics during early embryogenesis. Consequently, it seems likely that linker histones do something important.

Another potentially important problem with these compaction models is that they were created under the assumption that the length of the linker DNA between the nucleosomes was equal. Nature does not seem to follow this pattern: adjacent nucleosomes are joined by linker DNA that varies in length from 0 to 80 bp in a manner that depends on the tissue and the species, giving rise to different separations between nucleosomes. The addition of a linker histone into the compaction process not only constrains an additional 20 bp of DNA but also determines the trajectory of the DNA entering and leaving the nucleosome, which in turn directs the relative positioning of successive nucleosomes. From these data we can conclude that neither model explains the structure of the (approximately) 30 nm chromatin fiber in all cases. In all probability, the structure of the chromatin fiber differs depending on the type of cell in which compaction is occurring.

From the above information, it seems that compaction behavior depends both on the nucleosome repeat length and the linker histone. Only a nucleosome repeat length of 197 bp can form the 30 nm chromatin structure, and even then only with the cooperation of a linker histone.

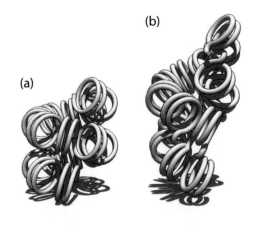

Figure 3.7 Two proposed models for the structure of the 30 nm chromatin filament. (a) The one-start helix. (b) The zig-zag arrangement of nucleosomes referred to as a two-start helix. Note that the histone octamers have not been included in this diagram for clarity, and only the loops of DNA are shown. (Courtesy of Richard Wheeler.)

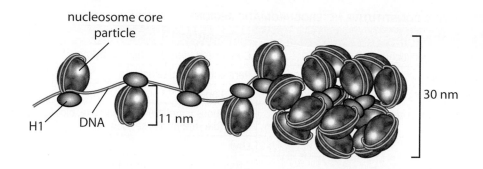

nucleosome core
particle

H1   DNA   11 nm   30 nm

Figure 3.8 Stabilization of nucleosome by the linker histone H1. The core nucleosome is highlighted in brown, histone H1 in orange, and DNA in blue. In this diagram, the nucleosomes are arranged according to the two-start helix model.

Other nucleosome repeat lengths show only limited linker-histone-dependent compaction, resulting in thinner and topologically different fibers. Nucleosome repeat lengths of all sizes are found in nature, but they do have a very distinct distribution of occurrence: those based on 188 and 196 bp are by far the most common, whereas shorter ones, such as the 165 bp repeats found in *Saccharomyces cerevisiae*, are comparatively rare. The reason for the differences in DNA length in nature is not understood. What happens to the chromatin fibers in the next stage of compaction is even less clear. It is clear only that there is a fundamental requirement for a mechanism to compact the chromatin further, because even the 30 nm fiber stage is not sufficient to package the entire DNA macromolecule into a cell's nucleus.

Chromatin fibers are further organized into euchromatin and heterochromatin

Whatever the mechanism that packages DNA into the nucleus, it is one that must be able to generate different types of chromatin compaction in distinct regions of the chromosome, because variations in chromosome packaging also differentiate regions of transcriptional activity from regions that are transcriptionally silent. As cells leave mitosis in the cell cycle, large regions of each chromosome become decondensed and disperse in the nuclei of each daughter cell. These regions can be seen by staining with Giemsa or quinacrine, dyes that are normally used for chromosome 'banding' or karyotyping procedures. Giemsa is specific for the phosphate groups of DNA. Quinacrine binds to the adenine/thymine-rich regions. Each chromosome has a characteristic banding pattern, but heavily stained regions tend to be late-replicating and AT-rich, whereas lightly stained regions tend to be early-replicating and GC-rich. The material in the lightly staining regions is referred to as euchromatin and contains most of the transcriptionally active genes. Some chromosomal regions, and whole chromosomes in certain cases, remain condensed throughout the cell cycle; this material, termed heterochromatin, commonly includes regions surrounding the centromeres and telomeres.

The locations and densities of heterochromatinized regions are not fixed. Some seem to be heterochromatic in all cell types and at all stages of differentiation; these regions are considered to be constitutive heterochromatin and include the heterochromatin that typically surrounds centromeres and the telomeres and repetitious sequences such as satellite DNA. Other chromosomal regions that may be heterochromatic in some cells and euchromatic in others are termed facultative heterochromatin; an example is the mammalian Barr body. **Table 3.2** gives an indication of the contrasting characteristics of euchromatin and heterochromatin that are typical of the genomes of eukaryotic cells.

The Barr body provides a particularly striking example of heterochromatin-based gene silencing. This nuclear structure (**Figure 3.9**) is found

TABLE 3.2 DISTINCTIONS BETWEEN EUCHROMATIC AND CONSTITUTIVE HETEROCHROMATIC REGIONS OF COMPLEX GENOMES

Feature	Euchromatin	Heterochromatin
Interphase appearance	Decondensed (lightly staining)	Condensed (densely staining; pyknotic)
Chromosomal location	Distal arms	Pericentromeric, telomeric
Replication timing	Throughout S phase	Late S phase
Sequence composition	Unique DNA, dispersed middle repetitious sequences	Repetitious (satellite) DNA, satellite, blocks of middle repetitious sequences
Gene density	Variable	Low
Nuclear location	Dispersed	Often clumped (nuclear periphery, around nucleoli)
Meiotic recombination	Significant	Undetectable
DNA methylation (vertebrates and plants)	CpG islands hypomethylated near transcribed genes	Extensively methylated
Histone acetylation	High	Low
Nucleosome spacing	Irregular	Regular
Nuclease accessibility	Variable	Low

exclusively in the cells of female mammals and represents the transcriptionally inactivated X chromosome; such wholesale inactivation is the means by which mammals achieve compensation for the unequal number of X chromosomes in males and females. Heterochromatinization (that is, the process of transforming transcriptionally active euchromatin into heterochromatin) of the Barr body is initiated at specific sites on the X chromosome, from which heterochromatin spreads outward rapidly to cover the entire chromosome. Once established, the silent state is stably inherited through subsequent rounds of cell division, thus achieving epigenetic regulation of one copy of each of the genes encoded on the X chromosome. Surprisingly, the heterochromatinization process that forms the Barr body does not seem to compress the single X chromosome into a space any smaller than it occupied previously.

The global term of chromatin 'compaction' has been applied to the process by which nucleosomes organize themselves into apparently more densely packed structures, but despite its dense appearance from G-banding (that is, staining with Giemsa), some evidence suggests that the Barr-body X chromosome occupies a nuclear volume similar to that of its active homologue. This implies that the cytological appearance of the heterochromatic Barr body is the result of a peculiar biochemical composition and/or a distinctive architecture, but not of a more compact or condensed folding. The apparent lack of compaction may be specific to X-chromosome silencing, because several lines of evidence suggest that the volumes occupied by other chromosomes (or chromosomal regions) do expand and contract depending on the levels of heterochromatinization. This variation in observed compaction could, of course, be related to the phase of the mitotic cell cycle at which the volume measurements were made, because the newly replicated DNA generated during S phase may not immediately form the higher-order compact structures typical of chromosomes during interphase.

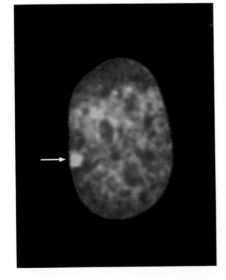

Figure 3.9 The Barr body. A dense object lying along the inside of the nuclear envelope in cells of female mammals, representing a highly condensed, inactivated X chromosome. [From Chadwick BP & Willard HF (2003) *Semin. Cell Dev. Biol.* 14, 359. With permission from Elsevier.]

A variety of mechanisms are involved in compacting chromatin beyond the 30 nm fiber stage

Having defined the type of compacted chromatin, we need to understand how the process of compaction works. The initial chromatin fiber created by the DNA–nucleosome interaction is roughly 40 times shorter than the

DNA chain from which it is made, but at the same time the presence of the nucleosomes reduces its flexibility, so that its coil size is still much larger than the diameter of the cell nucleus. Greater compaction requires nucleosome–nucleosome attraction, a mechanism that needs to be tunable so that some fractions of the fiber are dense and transcriptionally passive while others are more open and active.

The notion that attraction between nucleosomes might be responsible for at least some of the observed compaction is supported by recent experiments that point toward a mechanism called the **histone tail bridging model**. Figure 3.2 shows the protrusion of polypeptide sequences from the N-termini of the histone polypeptide sequences. These act as tails or flexible extensions of the eight core proteins and carry several positively charged residues. The tails extend considerably outside the globular part of the nucleosome and, according to the tail bridging model, they may be attracted to adjacent nucleosome core particles because of the overall negative charge conferred on them by the bound DNA. However, this simple model cannot explain why heterochromatin is more heavily compacted than euchromatin, because individual nucleosome core particles and the histone tails that they attract should in theory be structurally quite similar (at least in terms of charge distribution around the nucleosome). In view of this, it is highly probable that there are other mechanisms for controlling nucleosome interactions, and one of the most striking and perhaps the best characterized is that of the **polycomb-group (PcG) proteins**.

Polycomb proteins were identified first in *Drosophila melanogaster*, where PcG genes maintain repression of the homeobox (Hox) genes that establish and preserve the anterior–posterior axis of the insect's body plan during development. PcG proteins were proposed to alter chromatin structure to maintain gene repression, but it was surprisingly difficult to get direct evidence of this until studies using electron microscopy showed that **polycomb repressive complex 1 (PRC1)** was able to transform arrays of nucleosomes from their 'beads on a string' conformation into highly compact structures in which the individual nucleosomes could not be distinguished. There is some degree of uncertainty about the site of the interaction between the polycomb complex and the nucleosomes. It is possible that compaction could occur when the polycomb complex promotes interaction of the linker DNA between the nucleosomes or even the linker histone protein H1. However, the few studies of this mechanism available in the literature seem to suggest that the polycomb complex is more likely to bind the protruding N-terminal histone tails.

There are other proteins that mediate chromatin compaction that seem to make use of this method. For example, heterochromatin protein 1 (HP1) is recruited to the N-terminal tails of histone H3 by trimethylation of lysine 9. HP1 was first identified as a cytological component of *Drosophila* heterochromatin required for **position effect variegation (PEV)**, the positionally dependent silencing of transcription that occurs within heterochromatin. The function of HP1 is mediated via two protein domains. An N-terminal 'chromodomain' and a C-terminal shadow chromodomain allow HP1 to serve as an adaptor for the recruitment of other proteins that can physically bind the nucleosomes together. Important members of this group of proteins are the Su(var)3Y9-related histone methyltransferases (HMTs) that catalyze H3 lysine-9 methylation (H3 K9Me) and thus are responsible for the establishment of the trimethylation of lysine 9 in the first place. This is one of the ways in which heterochromatinization can spread along well-defined regions of chromosomes and in which normally euchromatic genes can be silenced when they are artificially positioned in a normally heterochromatic region (which, of course, is how position effect variegation works), but we will leave a more detailed description of this mechanism until later (see Chapter 4).

Chromatin compaction restricts access to the information content of DNA

The foregoing sections have described how DNA is packaged into the cell's nucleus through its association with histone proteins to form nucleosomes that can subsequently be compacted into heterochromatic or euchromatic structures, depending on the cell-specific requirement for gene expression. In principle, DNA that is incorporated into constitutive heterochromatin should not be required for any transcriptional processes within the cell because for the most part these regions contain only satellite repeats, telomeric DNA, and other features that, although important for the structural integrity of the chromosomes, do not contain genes. Repression of genes in regions of facultative heterochromatin might be reversible in certain circumstances. However, if the phenotype of the cells in which the facultative heterochromatin is located is stable, there should be little requirement to remodel such silenced regions. The vast majority of genes are incorporated into euchromatic regions of chromosomes, so our discussion of how the cell can extract information from compacted chromatin will focus initially on this more accessible chromatin architecture.

How does the cell access the genetic information of the DNA when it needs to make messenger RNA? Although nucleosome packaging is clearly essential, it does present something of a problem to the cell because DNA wrapped in nucleosomes is sterically occluded from many protein complexes that must act on it; how such complexes gain access to nucleosomal DNA is not known with absolute certainty. It could be imagined that the linker DNA between individual nucleosomes might be accessible to the gene translation machinery, but this may not be of significance because it is unlikely that a whole gene would be contained on such a relatively short length of DNA. There must be some means of temporarily releasing the DNA macromolecule from its interaction with the nucleosomes, and this mechanism will probably need to operate either over very long stretches of DNA (in the case of very large genes or gene clusters such as the Hox genes) or be capable of sequential release of shorter DNA segments that are then transcribed and subsequently re-incorporated into their nucleosome packaging. The next chapter details the current state of knowledge as to how this might work.

KEY CONCEPTS

- DNA is packaged into the nuclei of most eukaryotic cells by wrapping it around histone octamers to form structures called nucleosomes. Four histone proteins (H2A, H2B, H3, and H4) form the nucleosome structure, and part of the N-terminus of the polypeptide that forms each of these proteins protrudes out of the nucleosome. This is very important for controlling the function of individual nucleosomes, because the amino acids in this N-terminal tail can be subjected to many post-translational modifications.

- The DNA–protein conjugates known as nucleosomes can form higher-order folded structures that compact the DNA more effectively into the volume of the cell's nucleus. These conjugates are known as chromatin. Most genes are incorporated into chromatin, but the ease with which the cell's transcriptional machinery can access the information contained within DNA defines the type of chromatin present. Actively transcribed genes are generally found in areas of euchromatin, whereas those that are normally repressed are generally found in areas of heterochromatin.

- The mechanisms that control chromatin compaction into heterochromatin or allow it to remain as euchromatin depend largely on post-translational modifications of the histone N-terminal polypeptides.

FURTHER READING

Alberts B, Johnson A, Lewis J et al. (2007) Molecular biology of the cell, 5th ed. Garland Science.

Arents G, Burlingame RW, Wang BC et al. (1991) The nucleosomal core histone octamer at 3.1 A resolution: a tripartite protein assembly and a left-handed superhelix. *Proc Natl Acad Sci USA* **88**:10148–10152.

Bartolomé S, Bermúdez A & Daban JR (1994) Internal structure of the 30 nm chromatin fiber. *J Cell Sci* **107**:2983–2992.

Beh LY, Colwell LJ & Francis NJ (2012) A core subunit of Polycomb repressive complex 1 is broadly conserved in function but not primary sequence. *Proc Natl Acad Sci USA* **109**:E1063–E1071 (doi:10.1073/pnas.1118678109).

Chandra T, Kirschner K, Thuret JY et al. (2012) Independence of repressive histone marks and chromatin compaction during senescent heterochromatic layer formation. *Mol Cell* **47**:203–214 (doi:10.1016/j.molcel.2012.06.010).

Dorigo B, Schalch T, Bystricky K & Richmond TJ (2003) Chromatin fiber folding: requirement for the histone H4 N-terminal tail. *J Mol Biol* **327**:85–96 (doi:10.1016/S0022-2836(03)00025-1).

Dutnall RN & Ramakrishnan V (1997) Twists and turns of the nucleosome: tails without ends. *Structure* **5**:1255–1259.

Francis NJ (2009) Mechanisms of epigenetic inheritance: copying of polycomb repressed chromatin. *Cell Cycle* **8**:3513–3518.

Francis NJ, Kingston RE & Woodcock CL (2004) Chromatin compaction by a polycomb group protein complex. *Science* **306**:1574–1577 (doi:10.1126/science.1100576).

Grau DJ, Chapman BA, Garlick JD et al. (2011) Compaction of chromatin by diverse Polycomb group proteins requires localized regions of high charge. *Genes Dev* **25**:2210–2221 (doi:10.1101/gad.17288211).

Hamiche A, Schultz P, Ramakrishnan V et al. (1996) Linker histone-dependent DNA structure in linear mononucleosomes. *J Mol Biol* **257**:30–42 (doi:10.1006/jmbi.1996.0144).

Huynh VA, Robinson PJ & Rhodes D. A method for the *in vitro* reconstitution of a defined "30 nm" chromatin fibre containing stoichiometric amounts of the linker histone. *J Mol Biol* **345**:957–968 (doi:10.1016/j.jmb.2004.10.075).

Karymov MA, Tomschik M, Leuba SH et al. (2001) DNA methylation-dependent chromatin fiber compaction in vivo and in vitro: requirement for linker histone. *FASEB J* **15**:2631–2641 (doi:10.1096/fj.01-0345com).

Lowary PT & Widom J (1989) Higher-order structure of Saccharomyces cerevisiae chromatin. *Proc Natl Acad Sci USA* **86**:8266–8270.

Lowary PT & Widom J (1997) Nucleosome packaging and nucleosome positioning of genomic DNA. *Proc Natl Acad Sci USA* **94**:1183–1188.

Lowary PT & Widom J (1998) New DNA sequence rules for high affinity binding to histone octamer and sequence-directed nucleosome positioning. *J Mol Biol* **276**:19–42 (doi:10.1006/jmbi.1997.1494).

Morales V & Richard-Foy H (2000) Role of histone N-terminal tails and their acetylation in nucleosome dynamics. *Mol Cell Biol* **20**:7230–7237 (doi:10.1128/MCB.20.19.7230-7237.2000).

Morales V, Giamarchi C, Chailleux C et al. (2001) Chromatin structure and dynamics: functional implications. *Biochimie* **83**:1029–1039 (doi:10.1016/S0300-9084(01)01347-5).

Ramakrishnan V (1997) Histone H1 and chromatin higher-order structure. *Crit Rev Eukaryot Gene Expr* **7**:215–230.

Ramakrishnan V (1997) Histone structure and the organization of the nucleosome. *Annu Rev Biophys Biomol Struct* **26**:83–112 (doi:10.1146/annurev.biophys.26.1.83).

Robinson PJ & Rhodes D (2006) Structure of the "30 nm" chromatin fibre: a key role for the linker histone. *Curr Opin Struct Biol* **16**:336–343 (doi:10.1016/j.sbi.2006.05.007).

Robinson PJ, An W, Routh A et al. (2008) 30 nm chromatin fibre decompaction requires both H4-K16 acetylation and linker histone eviction. *J Mol Biol* **381**:816–825 (doi:10.1016/j.jmb.2008.04.050).

Simon JA (2010) Chromatin compaction at Hox loci: a polycomb tale beyond histone tails. *Mol Cell* **38**:321–322 (doi:10.1016/j.molcel.2010.04.018).

Wolffe AP (1999) Chromatin: structure and function, 3rd ed. Academic Press.

Yang Z, Zheng C, Thiriet C & Hayes JJ (2005) The core histone N-terminal tail domains negatively regulate binding of transcription factor IIIA to a nucleosome containing a 5S RNA gene via a novel mechanism. *Mol Cell Biol* **25**:241–249 (doi:10.1128/MCB.25.1.241-249.2005).

4

MODIFYING THE STRUCTURE OF CHROMATIN

Genomes encode an intrinsic nucleosome organization in which transcription-factor-binding sites that need to be accessible are less likely to be occluded inside a nucleosome (for example, when they are part of euchromatin) than are other portions of the DNA. But patterns of nucleosome positioning reflect probabilistic biases only; they do not keep critical target sites nucleosome-free at all times under all physiological conditions. Nucleosomal DNA needs to be unwrapped at times to function.

We know from *in vitro* studies that nucleosomes undergo spontaneous conformational transitions that result in the partial unwrapping of nucleosomal DNA, making it transiently accessible. This points to site exposure as a possible means for protein complexes involved in transcription to have access to nucleosomal DNA. Two broad mechanisms exist that may allow such exposure to occur—one involving chromatin remodeling activities and the other involving chromatin modifying activities. Chromatin remodeling activities increase the accessibility of the genome to DNA-binding proteins by altering the DNA–histone interaction noncovalently. Chromatin modifying activities do so by introducing covalent modifications on histone tails or histone core proteins. Multiprotein complexes perform these two activities by using the energy derived from ATP hydrolysis to induce changes in the histone octamer–DNA interaction. First we examine how nucleosome remodeling factors are able to remodel chromatin and how this process functions in controlling gene expression.

4.1 CHROMATIN REMODELING

4.2 CHROMATIN MODIFICATION

4.1 CHROMATIN REMODELING

Chromatin remodeling transiently exposes DNA to binding proteins

Chromatin remodeling involves **ATP-dependent nucleosome remodeling factors**, whose global function is to modulate the access of transcription factors to chromosomal DNA. They do so by hydrolyzing ATP to noncovalently restructure, mobilize, or eject nucleosomes, causing them to disassemble nucleosomes or redistribute their locations in response to the binding of proteins such as transcription factors that control gene activity. Chromatin remodeling factors are recruited to particular chromatin regions by site-specific DNA-binding proteins. However, this presents another problem: how do those proteins find their binding sites on the DNA if they are still occluded by nucleosome occupancy? It is possible that some ATP-dependent remodeling factors may act on nucleosomes ubiquitously, without a requirement for specific binding. This would mean that chromatin is inherently accessible to specific types of protein that mediate this effect. However, it is difficult to see how this idea might work. If such binding were truly ubiquitous, one would expect that areas of densely packed heterochromatin might be as subject to similar remodeling as the more accessible euchromatin, but this is not so. Hence the exact mechanism by which ubiquitous binding could induce exposure only at specific sites is not clear. Of the two possibilities, the one with site-specific binding proteins acting as intermediaries seems the more likely.

Chromatin remodeling is mediated by the SWI/SNF family of proteins in eukaryotes

The catalytic subunits of chromatin remodeling complexes are members of the Snf2 family of DNA-dependent ATPases. This large family of ATPases can be grouped into subfamilies that are structurally and functionally conserved in the animal and plant kingdoms. Three subfamilies are implicated in transcriptional regulation: the **SWI/SNF ATPases** (named after the founding member Swi2/Snf2), the **imitation switch (ISWI) ATPases**, and the **chromodomain and helicase-like domain (CHD) ATPases**.

The eukaryotic protein complex SWI/SNF is probably one of the main contributors to chromatin remodeling by ATP hydrolysis, so we will focus on this one in the following description of ATP-dependent chromatin remodeling. There are many variants of this complex represented in the SWI/SNF protein superfamily, and all are ATP-dependent chromatin remodelers. However, SWI/SNF proteins do not by any means include the entire range of ATP-dependent chromatin remodeling systems.

SWI/SNF is a relatively large protein complex comprising up to 12 subunits; it is highly conserved across eukaryotic species and goes by several names, depending on the species under consideration. Examples of these are ySWI/SMF in yeast, BAP in *Drosophila*, and BAF in humans (**Figure 4.1**). First discovered in yeast, the complex is found in low abundance in cells, suggesting that it is not a general component of chromatin. The complex disrupts nucleosome structure, increases the binding of transcription factors to nucleosomes, mobilizes histone octamers along the same stretch of DNA, transfers histone octamers to different DNA fragments, displaces histone H2A/H2B dimers, generates superhelical torsion in DNA, and binds preferentially to DNA and nucleosomes without

Figure 4.1 Comparison of the protein domains of the SWI/SNF superfamily of ATP-dependent chromatin remodeling complexes across three eukaryotic species. These complexes have been purified in yeast, flies, and mammals and consist of 9–15 subunits. SWI/SNF ATPases are the central catalytic subunits (brown), and the other subunits such as Snf5 are involved in binding the nucleosome and are required to modulate the DNA–histone interaction. [Adapted from Kwon CS & Wagner D (2007) *Trends Genet.* 23, 403. With permission from Elsevier.]

yeast

ySWI/SNF

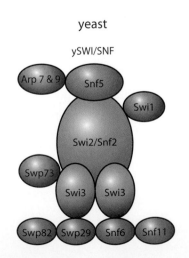

Drosophila

BAP

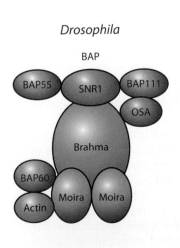

human

BAF

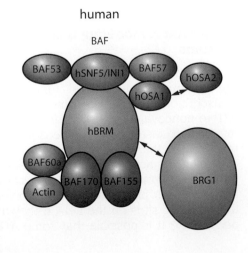

any DNA-sequence specificity. It seems to interact with transcription factor proteins that help it to localize to a specific region of chromatin or to the gene whose transcription level is to be altered by nucleosome repositioning.

The SWI/SNF ATPase is the central subunit of SWI/SNF chromatin remodeling complexes. Additional accessory subunits form tissue-specific holocomplexes together with the core complex. These accessory subunits can modulate the recruitment and activity of the complex. The ATPase activity of the Swi2/Snf2 subunit in yeast is stimulated by double-stranded DNA and is critical for the transcriptional activation and chromatin remodeling functions of the SWI/SNF complex.

The ATPase activity of SWI/SNF implies that whatever it does to the DNA requires a reasonable degree of energy. It seems that the complex is able to make the DNA loop away from the surface of the histone octamer. Although we know the positions at which SWI/SNF binds to nucleosomal DNA, that does not tell us how this protein complex is able to slacken the interaction of DNA with the histone octamer. It is probable that joint interactions occur among SWI/SNF, the nucleosomal DNA, and the histone octamer that allow the SWI/SNF protein complex to push apart the existing histone–DNA binding.

Chromatin remodeling by SWI/SNF works by repositioning nucleosomes

The remodeling process seems to begin with a small movement of the SWI/SNF complex over approximately 20 bp of DNA upstream from its original binding position. The DNA at this new position becomes anchored to the SWI/SNF complex, which then relocates, pulling the DNA with it and creating a small loop (**Figure 4.2**). The energy source for the translocation is most probably ATP hydrolysis, although quite how this allows the three-dimensional structure of the protein complex to move in this manner is far from clear. As more ATP hydrolysis occurs, a larger DNA loop is created that is primarily free of the histone core and is accessible to binding proteins.

Subsequently, the translocation and anchor domains of SWI/SNF release their DNA contact. The DNA loop migrates across the histone octamer, away from the translocation site and away from the location at which DNA was initially displaced, with eventual reassociation of DNA with the histone surface. The loop continues to propagate until it reaches the site at which the DNA would normally exit from the nucleosome, whereupon the movement of DNA around the nucleosome is completed. One of the consequences of this process is that the linker DNA between the nucleosomes becomes progressively shortened as it is pulled into the SWI/SNF complex by the formation of the loop. Ultimately, if this continues far enough, one might expect the nucleosome to collide with its immediate neighbor; however, if the SWI/SNF complex is operating in a similar manner on the adjacent nucleosomes, the overall effect will be one of nucleosome repositioning that will most probably result in the process recognized as **nucleosome sliding**.

This description of the basis of chromatin remodeling gives us a reasonable idea of how the DNA–histone interaction in the nucleosome can be relaxed or altered to expose the DNA to the transcriptional machinery of the cell. However, as was noted earlier, we are still left with the problem of how the transcription factors responsible for recruiting the chromatin remodeling complex find their binding sites in the first place, because it does so at a time when the DNA containing this site is still tied up within the structure of a nucleosome.

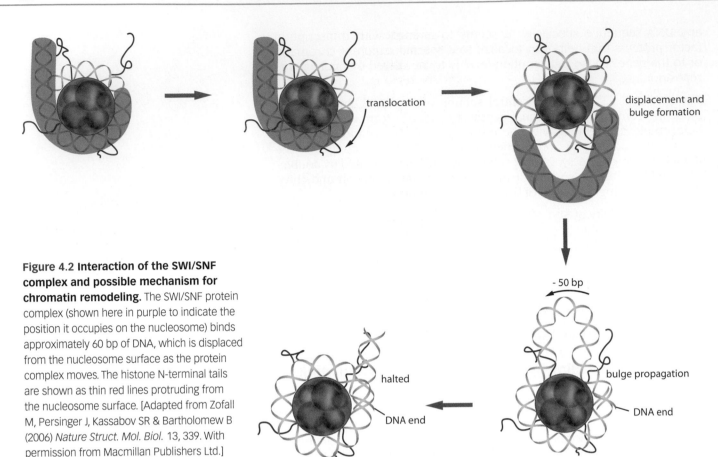

Figure 4.2 Interaction of the SWI/SNF complex and possible mechanism for chromatin remodeling. The SWI/SNF protein complex (shown here in purple to indicate the position it occupies on the nucleosome) binds approximately 60 bp of DNA, which is displaced from the nucleosome surface as the protein complex moves. The histone N-terminal tails are shown as thin red lines protruding from the nucleosome surface. [Adapted from Zofall M, Persinger J, Kassabov SR & Bartholomew B (2006) *Nature Struct. Mol. Biol.* 13, 339. With permission from Macmillan Publishers Ltd.]

Transcription factor binding sites are often located in regions of low nucleosome occupancy

Transcription factors (TFs) typically bind to specific DNA sequences in a gene's promoter regions, upstream of the sequences to be transcribed. The DNA sequences bound by transcription factors are therefore important components in the regulation of gene expression, because they determine which genes the different transcription factors will regulate. Binding sites for many transcription factors have been characterized. Importantly, genome-wide localization studies have shown that for highly transcribed genes, transcription factors localize to DNA elements closely upstream of transcription start sites in regions that are usually relatively depleted of nucleosomes in comparison with genes having lower levels of expression. Moreover, active transcription-factor-binding sites—that is, those bound by transcription factors—are usually depleted of nucleosomes in comparison with inactive (cryptic) sites.

But what determines the nucleosome occupancy of a specific DNA sequence? A wealth of literature evidence describing computational studies of the human genome suggests that it is the underlying DNA sequence that regulates the DNA–histone interactions needed to position the nucleosomes precisely. The prevailing **"statistical positioning" theory** of nucleosome organization was first proposed by Kornberg more than 25 years ago. This theory, for which considerable experimental evidence exists, posits that nucleosomes are stochastically positioned along the genome and are distributed between boundary events that comprise nucleosome-free regions, such as those known to be found at the promoters of transcriptionally active or poised genes. The repetitive nucleosomal structure is dynamically punctuated by short regions where regulatory factors can bind.

The present understanding of chromatin biology suggests that there are indeed genomic sequences that enhance nucleosomal occupancy, as well as sequences that nucleosomes seem to avoid. The latter include nucleosome-free regions enriched in gene regulatory regions, CpG islands, and transcription termination sites; those flanked by H2A.Z (an alternative form of histone H2A); those 200 bp upstream of the start codon of RNA polymerase II (Pol II)-transcribed genes in yeast; and also polycomb response elements and p53 binding sites.

Additionally, there is evidence that the primary DNA sequence can facilitate the bending of the helix around the histone octamer by presenting AA dinucleotides at those positions where the phosphodiester backbone of the helix faces toward the histone core. A sequence favoring nucleosome wrapping is therefore composed of AA dinucleotides spaced about 10 bp apart. TT dinucleotides are observed about 5 bp in either direction of the AA dinucleotides, because this is the position where the complementary helix faces the core complex. This particular arrangement and periodicity of dinucleotides along a stretch of DNA is referred to as a **nucleosome positioning sequence (NPS)**. The binding of nucleosomes is highly dependent on the individual DNA sequence (**Figure 4.3**).

The affinity of different DNA sequences for nucleosomes may vary 1000-fold or more. The genome thereby seems to encode, at least partly, its own packaging by positioning nucleosomes using these NPSs. Unfortunately, this "genomic code" is subtle and diffuse, and it is often difficult to distinguish it from random noise. This explains the considerable gap between recognition of the gene-encoding function of DNA and the discovery of nucleosome position encoding.

It is important to stress that the DNA–nucleosome interaction is dynamic, and the nucleosome occupancy is considered to be in a state of thermodynamic equilibrium. Although nucleosomes can form on almost any DNA sequence, the binding affinity is highly dependent on the specific sequence. Hence, the probability that a certain sequence is occupied by a nucleosome depends partly on the DNA sequence itself and partly on other factors. The epigenetic architecture of the genome is a major determinant of this binding affinity, because this can affect the competitive binding of transcription factors if the probability of nucleosome formation is lower in the regions of their binding sites. Once a transcription factor has located and bound to a specific site, it can recruit chromatin remodeling proteins such as those described in Section 4.2 to begin the process of exposing the underlying DNA sequence in readiness for transcription.

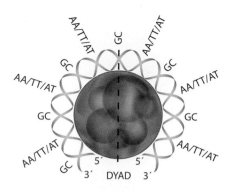

Figure 4.3 Appropriately spaced AA and TT dinucleotides facilitate the interaction of DNA with the histone octamer. The dashed vertical line represents the dyad axis, which is the twofold axis of symmetry common to most nucleosomes. [Adapted from Arya G, Maitra A & Grigoryev SA (2010) *J. Biomol. Struct. Dyn.* 27, 803. With permission from Taylor & Francis Ltd.]

4.2 CHROMATIN MODIFICATION

Spontaneous conformational changes and covalent modifications can also expose DNA to transcription factors

A mechanism that exposes DNA to binding proteins has been observed that is more intrinsic to the nucleosomes themselves. As was mentioned at the beginning of the chapter, nucleosomes spontaneously undergo large-scale conformational transitions in which stretches of their wrapped DNA partly unwrap from the histone protein surface. This can ultimately provide spontaneous access to the entire nucleosomal DNA length, with access to the outer stretches of the DNA being particularly rapid (these spontaneous conformational transitions can unwrap a nucleosome in as little as 250 ms).

An additional mechanism that has been suggested to have a role in site exposure involves covalent modification of the linker histone H1, although this is probably not a spontaneous process. It has long been believed that the primary function of linker histones (LHs) is to help create and/or maintain the compact higher-order structure of the chromatin fiber (see Chapter 3; Figure 3.2). Indeed, there is abundant evidence that the highest compaction can be attained only when such proteins are present. Because the compact fiber should be refractory to transcription, linker histones have been thought of as nonspecific repressors. It was suggested that linker histones may have a more specific role in transcription regulation by acting at the level of critically placed individual nucleosomes rather than by general compaction of the fiber. Interesting results on the selective effect of linker histones on transcription of individual genes were also obtained with an *in vivo* system for inducible overexpression of different histone H1 subtypes in cultured mouse cells. Overexpression of H1o, the differentiation-specific variant, led to reduced steady-state transcription levels for all Pol II genes studied.

A reasonable mechanism for providing site exposure is the "unpeeling" of a portion of the DNA from the histone core of the nucleosome, and this seems to occur spontaneously on chromatin that has been depleted of histone H1. Therefore the presence of linker histones bound to nucleosomes occupying gene regulatory regions would be expected to act as a block against such unpeeling.

To understand how linker histone binding could exert such a blocking action, we must consider the way in which these histones interact with the nucleosomal structure. Most linker histones possess a well-defined three-dimensional structure: a short N-terminal randomly coiled basic portion of the molecule is followed by a structured globular domain and a long C-terminal unstructured basic tail (**Figure 4.4**).

The globular domain is believed to be situated at or near the entry–exit point of the DNA into the nucleosome core particle, and the binding of the globular domain at this position allows the C-terminal tail to interact with both the incoming and outgoing linker DNA helices, bringing them close together, to enable the formation of a so-called "stem" structure 30 bp in length (**Figure 4.5**). Thus, the linker histone-binding site may be seen as a "gate," which may be either closed when the protein is bound or opened when the protein is released (**Figure 4.6**), to allow invasion of the nucleosome by protein factors. In such a scenario, the problem of gaining access to a DNA site covered by a nucleosome reduces to the problem of removing a particular linker histone. Mechanisms for controlling the timing and specificity of this removal are likely to involve epigenetic modification of the linker histone.

Figure 4.4 Structural model for linker histone (LH) globular domain binding to nucleosomal DNA. (a) Three DNA-binding domains are present on histone H1 and its variant form histone H5. The domains are shown in gray, green, and purple. (b) Model of the interaction of histone H1 with the nucleosome. N and C represent the N-terminus and C-terminus, respectively. [Adapted from Cui F & Zhurkin VB (2009) *Nucleic Acids Res.* 37, 2818. With permission from Oxford University Press.]

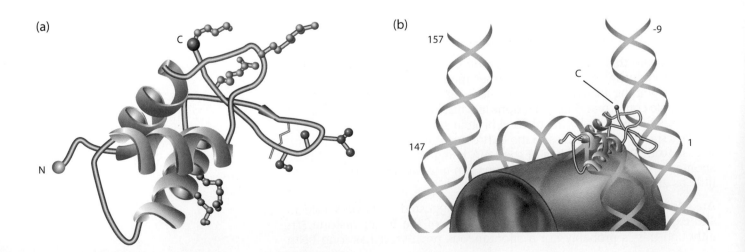

(a)

(b)

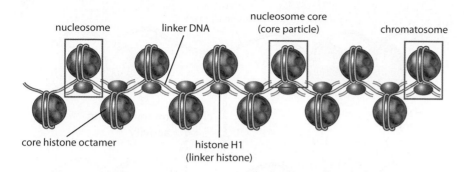

nucleosome   linker DNA   nucleosome core (core particle)   chromatosome

core histone octamer   histone H1 (linker histone)

Figure 4.5 Contribution of histone H1 to the 30 nm stem structure of chromatin, showing the relative positions of H1 binding and the linker DNA.

Epigenetic modification of DNA or histones regulates nucleosome occupancy and repositioning

With the exception of some apparently less than useful sections of the genome (such as endogenous retroviruses), pretty much all of the genes it encodes will need to be accessed at some point during the development or adult life of an organism. The reorganization of chromatin structure allows this to happen by periodically allowing the transcriptional machinery to access the DNA sequence; as we have seen, the underlying sequence of DNA bases is at least partly responsible for controlling such access. However, if a cell is to assert total control over gene expression, it needs more robust ways to force DNA to associate with histones and, more importantly, to force them apart again at very specific times and locations. This requirement for nucleosomal disruption provides an opportunity for the regulation of transcription initiation through the control of nucleosome stability or positioning at the 5' end of transcription units. Although the 5' chromatin must be "loose" enough to allow the access of proteins needed to initiate transcription, the stability of downstream nucleosomes must be sufficient to prevent "leaky" transcription from inappropriate genes without inhibiting the process of transcriptional elongation. During elongation, nucleosomes are disassembled in front of the polymerase to allow the latter's passage but are rapidly reassembled after the complex has passed.

There is substantial evidence that epigenetic modifications of DNA and/or chromatin proteins are the basic means for controlling this regulation of nucleosome occupancy and positioning, although this is by no means equally applicable to all regions of the genome. In the yeast *Saccharomyces cerevisiae*, transcriptional initiation seems to be under the control of histone modifications at the 5' and 3' regions of several genes that in turn regulate the nucleosomal composition of those regions. There is strong evidence that histone acetylation in particular promotes the disruption

Figure 4.6 The "gate" function of linker histone. (a) When a linker histone (LH) such as histone H1 is bound to the nucleosome, it holds the entry-point and exit-point strands of DNA in a conformation that prevents the access of transcription factor (TF). Once the LH is removed, binding of TF can take place. (b) Three-dimensional representation of the binding of histone H1 to the entry and exit DNA.

(a)

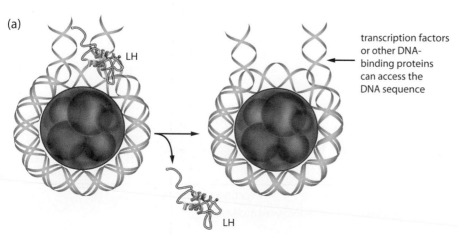

LH

transcription factors or other DNA-binding proteins can access the DNA sequence

LH

(b)

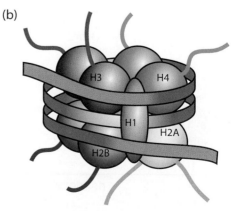

H3   H4

H1

H2A

H2B

Figure 4.7 Deposition of H2A.Z nucleosomes. A 22 bp sequence, here called the "nucleosome exclusion element," induces a nucleosome-free region (NFR). Transcription-factor binding may be required for NFR formation, but H2A.Z is not. Deposition of H2A.Z nucleosomes occurs preferentially at the edges of nucleosome-free regions and requires the activity of chromatin remodeling complexes (such as Swr1 protein). [Adapted from Lieb JD & Clarke ND (2005) *Cell* 123, 1187. With permission from Elsevier.]

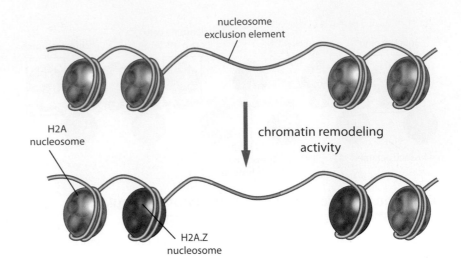

of nucleosomes at promoters in advance of initiation. Presumably the degree of acetylation is itself controlled by (1) the underlying DNA sequences and (2) the recruitment of histone modification complexes by transcription-factor-driven assembly of the pre-initiation complex.

Additionally, there is substantial evidence that nucleosome function can be modified by the incorporation of histone variants. H2A.Z is known to localize preferentially to the promoter regions of actively transcribed genes. It shows a particular preference for being incorporated into the two nucleosomes that flank a short nucleosome-free region upstream of the genes (**Figure 4.7**). This region is usually found about 200 bp upstream of the first codon of the coding sequence and generally includes the transcription start site.

As shown in Figure 4.7, the **nucleosome exclusion element** is immediately bookended by nucleosomes containing H2A.Z, but the nucleosomes adjacent to these are almost always constructed with normal H2A histones. Precisely what determines this demarcation is only now being understood, and it seems likely that combinatorial patterns of DNA methylations and histone post-translational modifications dictate the positions at which transcription factors can bind. The actions of chromatin remodeling complexes can also define the functional elements of the gene and keep them separate from one another. A particularly striking example of this is the apparently extensive methylation of lysine 36 on histone H3 across the coding sequences of many yeast genes, in complete contrast to their promoter elements, which are devoid of this modification (**Figure 4.8**).

Figure 4.8 Extensive H3K36 methylation marks the coding sequences of many genes. Note also the presence of nucleosomes with higher levels of histone acetylation immediately adjacent to the nucleosome-free region (NFR) and H2A.Z deposition site, which may contribute to the disruption of nucleosome structure ahead of the advancing DNA polymerase complex. [Adapted from Lieb JD & Clarke ND (2005) *Cell* 123, 1187. With permission from Elsevier.]

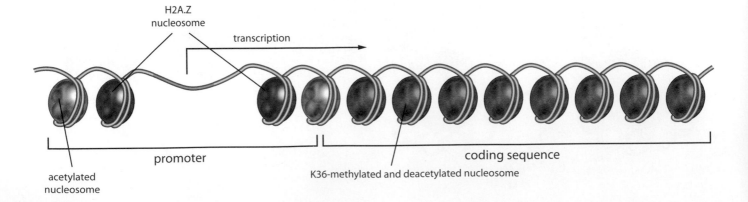

Up to this point we have described the ways in which nucleosomes are formed and how DNA is presented to the transcriptional machinery of the cell from this architectural format. In the next chapter we describe how both the DNA and histone proteins can undergo various covalent modifications and what implications these have for the control of gene transcription.

KEY CONCEPTS

- The transcriptional machinery needs to access the DNA sequence, but the structure of the nucleosome often prevents this from happening or at least makes it difficult. To get around this problem, cells need to perform chromatin remodeling to expose key regions of the DNA required for interaction with proteins controlling transcription and maintenance of the genome.

- Chromatin remodeling is an active, ATP-consuming process. In eukaryotes, remodeling is performed by the SWI/SNF superfamily of proteins; however, there are a few instances in which exposure of the DNA sequence may occur spontaneously without the influence of chromatin remodeling protein complexes.

- Sections of DNA sequence that are important for controlling gene function seem to be in regions of low nucleosome occupancy. Transcription factors bind to specific sites in the gene promoter, and these sections of DNA often have lower numbers of nucleosomes than adjacent coding sequences. It seems probable that the underlying sequence of DNA bases controls nucleosome numbers and locations.

- Covalent modification of histones through acetylation or methylation of key amino acids on the N-terminal tails is the most probable mechanism controlling chromatin remodeling and changes in nucleosome occupancy.

FURTHER READING

Andrews AJ & Luger K (2011) Nucleosome structure(s) and stability: variations on a theme. *Annu Rev Biophys* 40:99–117 (doi:10.1146/annurev-biophys-042910-155329).

Annunziato AT (2005) Split decision: what happens to nucleosomes during DNA replication? *J Biol Chem* 280:12065–12068 (doi:10.1074/jbc.R400039200).

Annunziato AT (2012) Assembling chromatin: the long and winding road. *Biochim Biophys Acta* 1819:196–210 (doi:10.1016/j.bbagrm.2011.07.005).

Böhm V, Hieb AR, Andrews AJ et al. (2011) Nucleosome accessibility governed by the dimer/tetramer interface. *Nucleic Acids Res* 39:3093–3102 (doi:10.1093/nar/gkq1279).

Chodaparambil JV, Edayathumangalam RS, Bao Y et al. (2006) Nucleosome structure and function. *Ernst Schering Res Found Workshop* (57):29–46.

Clapier CR & Cairns BR (2009) The biology of chromatin remodeling complexes. *Annu Rev Biochem* 78:273–304 (doi:10.1146/annurev.biochem.77.062706.153223).

Crosio C, Heitz E, Allis CD et al. (2003) Chromatin remodeling and neuronal response: multiple signaling pathways induce specific histone H3 modifications and early gene expression in hippocampal neurons. *J Cell Sci* 116:4905–4914 (doi:10.1242/jcs.00804).

Euskirchen GM, Auerbach RK, Davidov E et al. (2011) Diverse roles and interactions of the SWI/SNF chromatin remodeling complex revealed using global approaches. *PLoS Genet* 7:e1002008 (doi:10.1371/journal.pgen.1002008).

Geiman TM & Robertson KD (2002) Chromatin remodeling, histone modifications, and DNA methylation – how does it all fit together? *J Cell Biochem* 87:117–125 (doi:10.1002/jcb.10286).

Gupta S, Dennis J, Thurman RE et al. (2008) Predicting human nucleosome occupancy from primary sequence. *PLoS Comput Biol* 4:e1000134 (doi:10.1371/journal.pcbi.1000134).

Hogan GJ, Lee CK & Lieb JD (2006) Cell cycle-specified fluctuation of nucleosome occupancy at gene promoters. *PLoS Genet* 2:e158.

Kondo T & Raff M (2004) Chromatin remodeling and histone modification in the conversion of oligodendrocyte precursors to neural stem cells. *Genes Dev* **18**:2963–2972 (doi:10.1101/gad.309404).

Lubliner S & Segal E (2009) Modeling interactions between adjacent nucleosomes improves genome-wide predictions of nucleosome occupancy. *Bioinformatics* **25**:i348–i355 (doi:10.1093/bioinformatics/btp216).

Mariño-Ramírez L, Kann MG, Shoemaker BA & Landsman D (2005) Histone structure and nucleosome stability. *Expert Rev Proteomics* **2**:719–729 (doi:10.1586/14789450.2.5.719).

Mizuguchi G, Vassilev A, Tsukiyama T et al. (2001) ATP-dependent nucleosome remodeling and histone hyperacetylation synergistically facilitate transcription of chromatin. *J Biol Chem* **276**:14773–14783 (doi:10.1074/jbc.M100125200).

Mizuguchi G, Shen X, Landry J et al. (2004) ATP-driven exchange of histone H2AZ variant catalyzed by SWR1 chromatin remodeling complex. *Science* **303**:343–348 (doi:10.1126/science.1090701).

Morse RH (ed.) (2012) Chromatin remodeling: methods and protocols. Humana Press.

Peterson CL & Workman JL (2000) Promoter targeting and chromatin remodeling by the SWI/SNF complex. *Curr Opin Genet Dev* **10**:187–192 (doi:10.1016/S0959-437X(00)00068-X).

Petty E & Pillus L (2013) Balancing chromatin remodeling and histone modifications in transcription. *Trends Genet* **29**:621–629 (doi:10.1016/j.tig.2013.06.006).

Saha A, Wittmeyer J & Cairns BR (2006) Chromatin remodelling: the industrial revolution of DNA around histones. *Nat Rev Mol Cell Biol* **7**:437–447 (doi:10.1038/nrm1945).

Tolstorukov MY, Sansam CG, Lu P et al. (2013) Swi/Snf chromatin remodeling/tumor suppressor complex establishes nucleosome occupancy at target promoters. *Proc Natl Acad Sci USA* **110**:10165–10170 (doi:10.1073/pnas.1302209110).

Tsui K, Dubuis S, Gebbia M et al. (2011) Evolution of nucleosome occupancy: conservation of global properties and divergence of gene-specific patterns. *Mol Cell Biol* **31**:4348–4355 (doi:10.1128/MCB.05276-11).

Turner BM (2007) Chromatin remodelling machines. In Turner BM (ed.). Chromatin and gene regulation: mechanisms in epigenetics, pp 146–171. Blackwell Science Ltd.

van Bakel H, Tsui K, Gebbia M et al. (2013) A compendium of nucleosome and transcript profiles reveals determinants of chromatin architecture and transcription. *PLoS Genet* **9**:e1003479 (doi:10.1371/journal.pgen.1003479).

Vignali M, Hassan AH, Neely KE & Workman JL (2000) ATP-dependent chromatin-remodeling complexes. *Mol Cell Biol* **20**:1899–1910 (doi:10.1128/MCB.20.6.1899-1910.2000).

CHAPTER FIVE

DNA METHYLATION

Most DNA in eukaryotes is composed of the four bases adenosine (A), cytosine (C), guanine (G), and thymine (T). As early as 1948 the existence of a fifth base, 5-methylcytosine, was suggested, although it was also clear that cytosine and 5-methylcytosine were often interconvertible. DNA methylation in eukaryotes involves the addition of a methyl group to carbon 5 on the pyrimidine or the purine ring of a base in a reaction that is catalyzed by a class of enzymes known as the **DNA methyltransferases**. **Figure 5.1** shows the methylation of cytosine.

DNA methylation is a common epigenetic modification and has been implicated in gene expression control (usually repression) in several species, although it is by no means a mechanism universal to all eukaryotes. *Saccharomyces cerevisiae* and *Drosophila melanogaster* have little detectable DNA methylation and yet perform selective gene expression. This has led to doubts about the true significance of DNA methylation during development and tissue-specific gene expression. Stacked against these observations are the aberrant development and early embryonic lethality of mice in which DNA methyltransferase activity has been knocked out. The latter findings support the notion that DNA methylation has an important role in gene expression, at least in the context of mouse development. Some researchers have suggested that the primary function of DNA methylation in mammalian cells is to silence invading parasitic DNA sequences, such as those of viruses. However, as we will see later in the chapter, this does not correlate well with the effects of its absence on murine development and with the existence of other, more subtle DNA methylation-related consequences such as **genomic imprinting** (the selective expression of certain genes from only one of the parentally derived alleles, with the other allele being repressed by DNA methylation) and inactivation of the X chromosome in female mammals.

5.1 PATTERNS OF DNA METHYLATION

The distribution of DNA methylation in eukaryotic genomes is not uniform. Some regions, quite often those associated with DNA sequences involved in gene regulation rather than gene-coding sequences, are heavily methylated, whereas others are relatively sparsely methylated. The greater part of the DNA methylation (80%) is found on the cytosine of CpG dinucleotides, which are distributed throughout the genome but are also found concentrated in regions known as **CpG islands**. These regions fall into two classes depending on the concentration of CpG dinucleotides found in the promoter region of genes normally transcribed

Figure 5.1 Schematic representation of the methylation of cytosine. The forward reaction (such as methylation) is mediated by DNA methyltransferase enzymes using *S*-adenosylmethionine as the methyl group donor. Demethylation is performed by DNA demethylase enzymes.

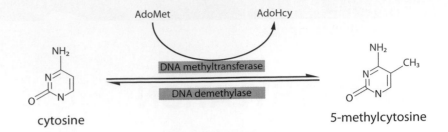

by RNA polymerase II. First, promoters with lower CpG concentrations (approximately one CpG per 100 nucleotides) are usually found in tissue-specific genes that are expressed only in certain cell types, and second, genes that have higher than normal concentrations of CpG dinucleotides (approximately a tenfold enrichment over the rest of the genome) in regions that are about 1000 bases (1 kilobase) in length. These enriched regions—the CpG islands—can easily be detected in sequence maps of gene promoters and are so conspicuous as to be one of the most reliable features for predicting the positions of hitherto unknown promoters in mammalian genomes.

CpG-rich islands are infrequently methylated

Methylation of the CpG islands in this second class of genes is generally found at low levels whether the gene is undergoing active transcription or not, hence the initial suggestion that CpG islands were not subject to methylation. However, small numbers of genes in the CpG island-rich class, such as the *MAGE* genes that become methylated during the course of normal mammalian development, contain short (350–600 bp) CpG islands. These are methylated in somatic cells but not in gametes. Normally, *MAGE* genes are not expressed in somatic tissues, but treatment of somatic cells in culture with DNA demethylating agents such as 5-azacytidine causes up-regulation of their expression. Of course this type of artificial induction is not observed *in vivo*, but the experiment does indicate that the methylation is functional (and not just decorative) in these genes. However, for the most part, in genes with CpG island-rich promoter sequences, the CpG islands remain consistently devoid of methylation throughout the development of the embryo and the life of the adult organism, regardless of the expression level of the gene. This implies that DNA methylation does not make a significant contribution to the transcription control mechanisms of such genes. There are exceptions to this rule for CpG island-rich genes, and the best examples are found in cells associated with many types of cancers, wherein transcription is strongly repressed upon unscheduled *de novo* methylation of CpG islands. This phenomenon is rarely observed in normal cells growing in physiological conditions, and the mechanisms that lead to such aberrant methylation remain largely unknown. The epigenetic phenomena related to cancer are discussed in Chapter 16.

CpG-poor islands are frequently methylated

In contrast, CpG-poor promoters are frequently methylated in the genomes of gametes and cells during early development. As previously stated, these genes are always tissue-specific and show precise spatio-temporal expression control during development. Excellent examples of this are provided by the early silencing of genes crucial for development of the pre-implantation embryo, such as *OCT4* and *NANOG* (**Figure 5.2**), which show limited methylation of their promoters until the epiblast stage of development and thereafter become increasingly methylated, leading to repression of gene activity.

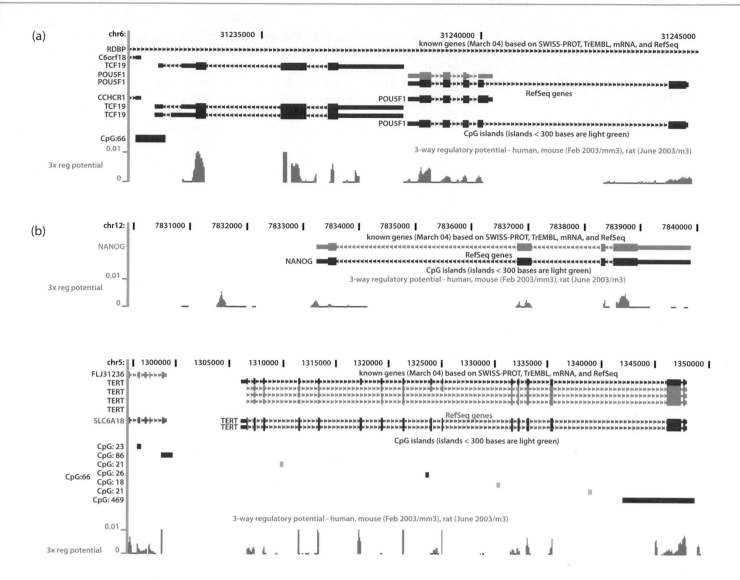

The *TERT* gene, which encodes the reverse transcriptase subunit of the telomerase holoenzyme complex, is also involved in early embryonic development but differs from such genes as *OCT4* in that it has CpG islands in both its promoter and coding sequences. *TERT* is also unusual in that some cell types, such as tissue-specific stem cells of adults, retain *TERT* expression (and therefore telomerase activity) at reduced levels compared with the pre-implantation embryo. It is probably not surprising that the *TERT* expression levels do not respond in a uniform manner to promoter methylation, because the absolute levels of 5-methylcytosine seem to differ among the different types of *TERT*-positive cells. For example, it is possible to have well-defined regions of the *TERT* promoter methylated in the genome of one cell type but unmethylated in another, despite a requirement for TERT expression in both of these cells. Telomerase activity is observed in numerous types of cancer cells because it is an important determinant of cellular immortality. Thus there is considerable interest in working out how differences in the methylation of *TERT* promoter regions contribute to expression control.

Disrupting the ability of the promoter sequence to drive gene expression is probably one of the more important functions of DNA methylation, at least in vertebrates. However, for this to occur, the sequence information contained in the coding DNA also must be held in a relatively open

Figure 5.2 Regions of CpG islands in gene promoters. (a) Genomic location of the coding and promoter sequences of the pluripotency-associated transcription factor *POU5F1* (*OCT4*) showing the positions of CpG islands and regions of high regulatory potential. The CpG islands are indicated by light or dark green bars, which represent sparse (less than 300 methylated bases) or dense (more than 300 methylated bases) levels of DNA methylation. Typically, the positions of the CpG islands correlate with regions of known regulatory potential. Abbreviation: chr6, chromosome 6. (b) Similar analysis of the human *hTERT* gene encoding the catalytic subunit of the telomerase holoenzyme complex. [From Kent WJ, Sugnet CW, Furey TS et al. (2002) *Genome Res.* 12, 996 (http://genome.ucsc.edu/)].

chromatin conformation. The requirement for this is perhaps not very exacting, because it is possible to remodel chromatin as the transcription initiation complex passes, but if the DNA sequence is heavily methylated and leads to chromatin condensation, it may not allow the transcription complex access to begin synthesis of messenger RNA (mRNA). Even in the smaller CpG-enriched regions found in the CpG-poor promoters of tissue-specific genes, this alteration of chromatin structure may be sufficient to repress transcription. However, this still leaves the problem of how DNA methylation may be able to control the activity of genes that have CpG-rich promoters. The simple answer is that it probably does not, but this in itself prompts us to ask why the CpG islands are there at all. Perhaps the lack of DNA methylation-based expression control is acceptable, because the bulk of genes falling into this class are housekeeping genes whose expression is required in all cell types regardless of their identity or function; thus they always need to maintain their chromatin in an open conformation.

5.2 EFFECTS OF DNA METHYLATION ON TRANSCRIPTION

Proteins controlling cellular function interact with methylated DNA

So how does DNA methylation affect transcription from CpG-rich and CpG-poor promoters? Three basic mechanisms have been proposed. One model is based on the fact that methylated and unmethylated DNA lead to quite different chromatin structures. A key feature of chromatin structure in regions of unmethylated DNA is that it adopts an open, accessible conformation that permits nonchromatin proteins such as transcription factors to make contact with the DNA. If transcription factors are able to reach their binding sites, it is likely that they will assist in the recruitment of other components of the cell's transcriptional machinery so that synthesis of mRNA can begin. In contrast, methylation of DNA results in a chromatin structure that may prevent transcription factors from accessing DNA.

Another possibility is that the binding of a transcription factor to its normal recognition site within the control sequences of a gene may be specifically inhibited by the presence of methylated DNA at that site. Several transcription factors, including AP-2, c-MYC, E2F, and NF-κB (these are all involved in multiple gene expression control functions within the cell), recognize sequences that contain CpG residues, and this recognition has been shown to be inhibited by methylation. In contrast, other transcription factors, such as Sp1 (another transcription factor with a broad range of binding sites across the genome), are not sensitive to the presence of 5-methylcytosine within their binding sites, and still others have binding sites that do not possess a CpG dinucleotide and cannot therefore be affected by methylation. That is not to say that the expression of such genes is not influenced in some general fashion by DNA methylation, but it would be more likely to be affected by the chromatin architecture of the promoter regions if they happened to contain CpG dinucleotides or CpG islands, albeit outside the transcription-factor-binding sites.

A third potential mechanism is the direct binding of specific transcriptional repressors to methylated DNA. Two protein complexes have been identified that are good candidates for this type of repression, namely **methylcytosine-binding proteins 1 and 2 (MECP1 and MECP2**; see **Box 5.1**). These proteins bind to the 5-methylcytosine residues of CpG dinucleotides regardless of their genomic position. However, because

MECP1 is known to bind only multiple, symmetrically methylated CpG sites, a threshold number of methylated CpGs are required to attract MECP1 molecules in sufficient number to have any effect on the transcriptional control of the gene. This ability to recruit the MECP1 protein categorizes genes on the basis of their response to MECP1-based repression, with genes containing large numbers of CpG dinucleotides showing the most obvious repression. Genes of this type typically exhibit greatly reduced repression in MECP1-null cells, even when the genes are already methylated. However, another class of genes in which the numbers of CpG dinculeotides are much lower also show MECP1-mediated

Box 5.1 Possession of a methyl-binding domain is common to a wide range of proteins

MECP1 and MECP2 are members of a large family of methylated-DNA-binding proteins. Comparison with other methyl-binding proteins defined the **methyl-binding domain (MBD)** as a motif about 75 amino acid residues long that seems to be conserved in a wide range of proteins involved in epigenetic modification of the genome. Sequence comparisons of four human MBD family proteins (MBD1–MBD4) that function independently of MECP1 and MECP2 show the presence of 16 strictly conserved residues within the MBD domain. MBD proteins interact with methylated DNA in the major groove, where the two methyl groups from the methyl-CpG (mCpG) point towards the exterior of the double helix (**Figure 1**). Several residues from the L1 loop, connecting the β2 and β3 strands of MBD, and the α helix make several contacts with the sugar-phosphate backbone on each strand of the DNA molecule.

Four conserved residues (Arg 22, Tyr 34, Arg 44, and Ser 45) in MBD1 are involved in recognizing the methyl-CpGs via a complex set of interactions. It seems that each side chain interacts with DNA in a bivalent manner, in which the polar moiety of each of these residues contacts a C or G base and their hydrophobic regions stack around the methyl groups. Such bivalent contacts from each important amino acid side chain may explain why both the CpG dinucleotide and the two methyl groups are strictly required for efficient recognition by the MBD. Subtle variations in this configuration might abolish binding. X-ray crystallographic and nuclear magnetic resonance studies suggest that one MBD is only able to bind one symmetrically methylated CpG at any one time.

Despite our knowledge of the structure of this protein motif, we are not certain how specific targeting of MBD-containing proteins is achieved. One possible model would be that each MBD protein randomly occupies any available methylated CpG. In this scenario, the relative abundance of each MBD protein within a cell together with the methylation density of a DNA region will dictate the MBD occupancy of individual methylated CpG sites. A different model envisions that other factors may influence the distribution of MBD proteins within a cell nucleus, making it nonuniform and nonrandom, with each MBD protein occupying unique sites in the genome. This model posits that a subset of genes would be affected by the loss of one MBD protein but not another and is supported by the recent observation that MBD1, MBD2, and MECP2 do not share binding sites in primary human fibroblasts. Such specific interactions probably require the involvement of other DNA-binding partner proteins, but we currently lack experimental evidence for this occurrence.

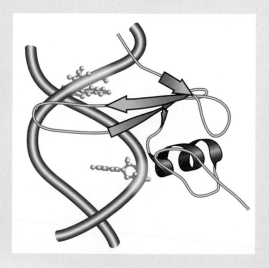

Figure 1 The methyl-binding domain is common to all proteins that bind methylated DNA. Proteins capable of binding methylated DNA contain a specific domain of approximately 70 residues, the methyl-CpG-binding domain (MBD), which is linked to additional domains associated with chromatin. The MBD folds into an α/β sandwich structure comprising a layer of twisted β-sheet backed by another layer formed by the α1 helix and a hairpin loop at the C-terminus. The methylated CpG is shown in orange and the other regions of the DNA surrounding it are shown in blue. [Adapted from Khorasanizadeh S (2004) *Cell* 116, 259. With permission from Elsevier.]

repression, but of weak promoters only. A much stronger promoter (such as one containing an SV40 enhancer) that is subject to relatively low CpG methylation is less likely to undergo repression via MECP1. This suggests that the binding of the MECP1 complex with sparsely methylated DNA is unstable and can easily be overcome by stronger promoters.

MECP1 consists of several subunits with distinct functions related to its DNA binding and its repressive activities. One of its subunits, the pericentriolar material protein 1 (PCM1), now known as methyl-CpG-binding domain protein 1 (MECP1 or MBD1)—which is needed for assembly of the centrosome—has a specific methyl-CpG-binding domain but also two cysteine-rich domains that are also found in DNA methyltransferases, such as DNA (cytosine-5-)methyltransferase (DNMT1), and in other proteins. Although we understand why MECP1 should bind to its target CpGs, we still have no clear idea of the exact way in which these domains can mediate repression.

MECP2 is slightly different. To begin with, it is much more abundant in the cell, and it is able to bind single methylated CpGs. We also have a clearer understanding of its mode of action, because it has fairly distinct methyl-CpG-binding and transcriptional repressor domains. The latter is able to inhibit transcription even when the target CpG is some distance from a known promoter element, which suggests that MECP2 may interact with the transcriptional machinery in some way or be able to disrupt the stability of the pre-initiation complex.

Transcription factors and methylated-DNA-binding proteins can repress transcription

So how does the binding of transcription factors and methylated-DNA-binding proteins function as a transcriptional repression system? There is some evidence to suggest that DNA methylation has little impact on gene expression until the affected DNA has been assembled into some form of condensed chromatin structure. Several proteins containing methyl-binding domains (MBDs) are known to cooperate with enzymes such as histone deacetylases and methylases that can modify the structure of chromatin and prevent access of the transcriptional machinery to the DNA. For example, MECP2 interacts with DNMT1, co-repressor of RE1 silencing transcription factor (CoREST), nuclear co-repressor (NCoR/SMRT) proteins that repress the transcription of several genes, histone H3 lysine 9 methylase, RNA splicing factors, and chromatin-remodeling proteins such as α-thalassemia/mental retardation syndrome X-linked (ATRX) and Brahma (Brm1)-related switch/sucrose nonfermentable (SWI/SNF) complex. However, we do not know whether this is a direct binding of MECP2 to these partners or whether it requires other co-repressor complexes that are present in all cell types. In any case, the reports describing these bindings to MECP2 are somewhat contradictory, because purification of the MECP2 protein from nuclear extracts of mammalian cells has produced results that range from complete lack of binding to proteins such as the paired amphipathic helix protein Sin3A, to the isolation of high-molecular-weight complexes that contain MECP2 in addition to lots of other proteins, only some of which are characterized.

The situation is slightly clearer with MBD2 and MBD3, which co-purify with the large nucleosome remodeling and histone deacetylation complex (NURD). This complex does lots of epigenetic restructuring. It has ATPase-dependent chromatin-remodeling activity to help DNA loop away from the nucleosomes, and it can also remove acetyl groups from the N-terminal tails of individual histones. Although MECP2 and MBD2 are likely to be responsible for the initial recruitment of NURD complexes to chromatin assembled on methylated DNA, studies *in vitro* and *in vivo* suggest that the chromatin-remodeling activities of NURD further facilitate the binding of MBD proteins to methylated sites that are not readily accessible on nucleosomal templates and that by doing so they stimulate MBD-mediated gene repression.

It is possible that MBD-containing proteins may have functions that are independent of such complexes as NURD and do not require the presence of co-repressors. For example, we know that MECP2 forms discrete complexes with nucleosomes and is able to cause compaction of the beads-on-a-string type of array in an *in vitro* model in which no co-repressor proteins are present. We also know that MBDs can localize to pericentric regions of heterochromatinization referred to as a **chromocenter** (regions that stain heavily with Wright–Giemsa stain in the nucleus). These tend to be part of the noncoding regions of the genome, and so any methylation events taking place here may not be directly relevant to the mechanisms by which gene coding chromatin is controlled; however, it does demonstrate the redundancy of functions that many of these epigenetic modifier proteins seem to possess. Localization to regions such as the pericentric microsatellite (regions of repeated DNA sequence near the centrome—the midsection of a chromosome) implies some degree of control over the organization of heterochromatin architecture that may affect the way in which whole chromosomes are constructed for fitting into the nucleus. If DNA methylation is the marker that designates such regions for incorporation into heterochromatin, it would seem to be a most powerful yet simple means of controlling the genome.

5.3   THE MOLECULES THAT METHYLATE DNA

Global cytosine methylation in mammals is established by a complex interplay of three independently encoded DNA methyltransferases (DNMTs): DNMT1, DNMT3A, and DNMT3B. Another potential DNA methyltransferase, DNMT2, has been cloned and partly characterized; however, at the time of writing, its activity seems to be the transfer of methyl groups to RNA not DNA—hence the suggestion that it should be renamed tRNA aspartic acid methyltransferase 1 (TRDMT1). In view of this, our discussion will be restricted to the three characterized enzymes. All three of these synthesize the same product, 5-methylcytosine, but they do so in different molecular contexts and are commonly classified as being either "*de novo*" (DNMT3A, DNMT3B) or maintenance (DNMT1) methylases. We introduced the basic form of the chemical reaction that methylates cytosine in Figure 5.1; **Figure 5.3** gives a more expanded form showing the areas of the molecules that interact with the DNA methyltransferases, and the general organization of domains for the DNA methyltransferases is shown in **Figure 5.4**.

De novo methylation of cytosine establishes the methylation pattern

De novo methylation is the establishment of 5-methylcytosine at genomic loci where none has previously existed. It is responsible for the increase in global DNA methylation that occurs early in embryonic development. This type of methylation is generated by the DNMT3 family of enzymes including two active DNMTs (DNMT3A and DNMT3B) and one regulatory factor called the DNMT3-like protein (DNMT3L), which, although not a DNA methyltransferase in its own right, is essential for the generation of 5-methylcytosine. Evidence for this is obtained from mice from which DNMT3L is absent: their phenotypes are indistinguishable from those resulting from inactivation of the DNMT3A enzyme. The cooperation

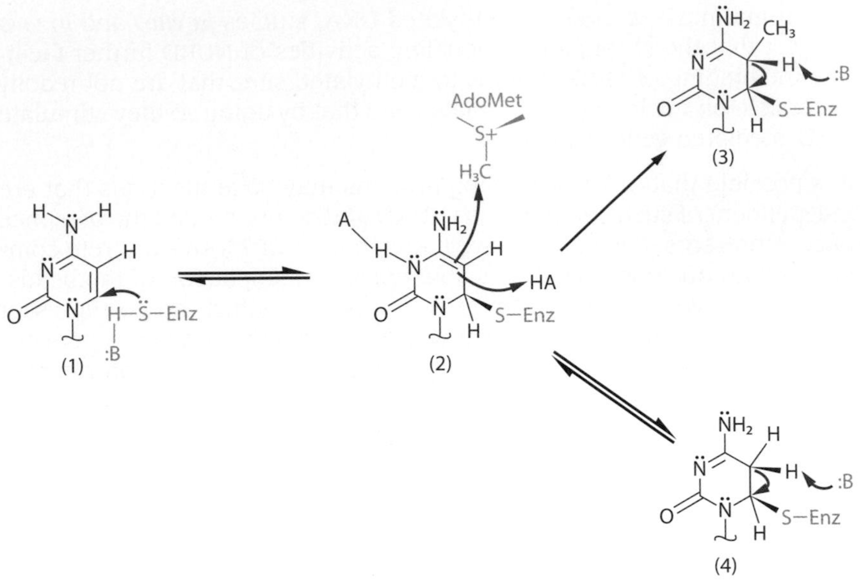

Figure 5.3 The chemistry of DNA methylation. The methyl group comes from the donor molecule *S*-adenosyl methionine (AdoMet). (1) An electron pair from the sulfur atom in a cysteine thiol group is transferred to C6 of cytosine. (2) This transfer allows the conjugated double bond to shift onto C5. Protonation of N3 helps out, and together they allow the formation of a covalent bond to the methyl carbon of AdoMet. (3) Base-catalyzed dehydration removes the extra hydrogen from C5, which results in 5-methylcytosine (the base is indicated by "B:") (4) This structure shows the position of the conjugated double bond when N3 is not protonated. The DMNT1 catalytic subunit can catalyze DNA methylation because it has amino acids positioned in its active site that can bind to hemi-methylated DNA and react with cytosine C6 as well as binding the AdoMet donor in just the right place for it to react with cytosine C5.

between these two proteins is further underlined by co-immunoprecipitation experiments that demonstrate the co-localization of DNMT3A and DNMT3L in the cytoplasm and nucleus.

Dnmt3a from mice and humans are highly homologous proteins comprising a 908-residue polypeptide chain with a variable region at the N-terminus, a PWWP domain (a ubiquitous eukaryotic protein module of about 80 amino acids with the strongly conserved sequence Pro-Trp-Trp-Pro) that may be involved in nonspecific DNA binding, a cysteine-rich zinc-binding domain comprising six CXXC motifs (where the CXXC domain is a protein motif capable of selective binding to unmethylated DNA) and a C-terminal catalytic domain. The amino acid sequence of DNMT3L is very similar to that of DNMT3A and DNMT3B in the Cys-rich three-zinc-binding domain, but it lacks the conserved residues in the C-terminal domain that are required for DNA methyltransferase activity.

DNMT3A and DNMT3L are both required for the *de novo* methylation of most DNA sequences. The minimal regions required for interaction between DNMT3L and DNMT3A (or DNMT3B), and for stimulated activity, are in the C-terminal domains of both proteins. The "logic" behind this interaction becomes a little clearer if we look at the dimensions of

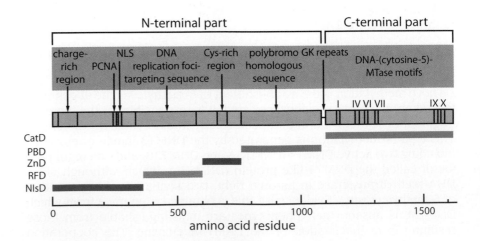

Figure 5.4 The relative positions of protein domains in mammalian DNA methyltransferases. [Adapted from Fatemi M, Herman A, Pradhan S & Jeltsch A (2001) *J. Mol. Biol.* 309, 1189. With permission from Elsevier.]

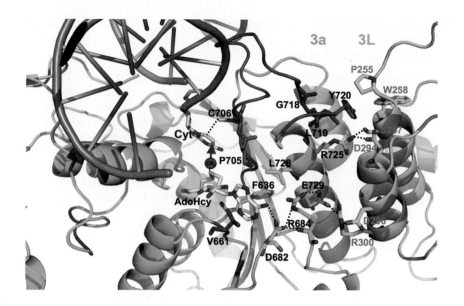

Figure 5.5 The intermolecular interactions between Dnmt3a and Dnmt3L may also explain the observation that Dnmt3L increases the binding of AdoMet by Dnmt3a. Bound AdoHcy in the structure interacts with Dnmt3a. Dnmt3a is shown in green, and Dnmt3L is in blue. AdoHcy is shown in yellow with its heterocyclic portion in yellow and blue. The catalytically active site of Dnmt3a is shown in pink. These interactions involve the adenine ring of AdoHcy making flanking van der Waals contacts with the phenyl ring of Phe 636 and with Val 661, and its exocyclic amino group (N6) making a hydrogen bond with Asp 682 (indicated by the black dashed line). The importance of these interactions is indicated by the observation that Asp 682 and Phe 636 of Dnmt3a are critical for stabilizing a network of interactions involving both Dnmt3a and Dnmt3L. These include Asp 682 (3a) interacting with Arg 684 (3a), Arg 684 (3a) interacting with Glu 729 (3a), Glu 729 (3a) interacting with Arg 300 (3L), Arg 300 (3L) interacting with Asp 296 (3L), Phe 636 (3a) interacting with Leu 726 (3a), and Arg 725 (3a; the residue next to Leu 726) interacting with Asp 294 (3L). The interface of Dnmt3a–3L therefore provides an extensive network of stabilizing polar interactions that enhances the binding of AdoMet by Dnmt3a. [From Cheng X & Blumenthal RM (2008) *Structure* 16, 341. With permission from Elsevier.]

the DNMT3A–C/DNMT3L complex, which has been shown to be 16 nm long. This is slightly more than the diameter of a single core nucleosome (11 nm), suggesting that the dimensions of the complex have evolved very specifically to enable it to interact with and methylate DNA that is combined with histone proteins into nucleosomes. The complex contains two monomers of DNMT3A-C and two of DNMT3L-C, which form a tetramer with two 3L–3A interfaces and one 3A–3A interface (3L–3A–3A–3L). The integrity of this structure is essential to its function, and substitution of key amino acids, even in the apparently noncatalytic regions, eliminates the DNA methyltransferase activity completely. DNMT3L seems to stabilize the conformation of the active-site loop of DNMT3A, which contains the catalytic nucleophile (Cys 706), via interactions with the C-terminal portion of that loop (Gly 718-Leu 719-Tyr 720). Cys 706 contains the thiol group that performs the nucleophilic attack on C6 of unmethylated cytosine, as shown in Figure 5.3. **Figure 5.5** shows the interactions between the active-site regions of DNMT3A and DNMT3L in relation to the binding of the complex to DNA.

Existing patterns of DNA methylation are maintained

Because DNA methylation of certain gene control elements is essential to cellular identity and function, it is important to maintain those genomic regions modified by 5-methylcytosine. This function is performed by the alternative DNA methyltransferase DNMT1, which methylates cytosine that are adjacent to 5-methylcytosine in the newly synthesized DNA strands of dividing cells. It is able to do this because its preferred substrate is hemi-methylated double-stranded DNA. Mice deficient in Dnmt1 lose about 90% of their DNA methylation and die early in embryogenesis, and mouse embryonic stem (ES) cells deficient in Dnmt1 die when induced to differentiate. Dnmt1 conditional knockout mouse fibroblasts die within a few cell divisions after gene deletion, but the reduction in Dnmt1 activity must be almost complete for death to occur: even minimal expression of this enzyme seems able to maintain methylation, and the cells retain viability. The mammalian (mouse, human) DNMT1 complex contains 1620 amino acid residues distributed among several domains credited with having distinct functions that, when combined, permit the complex to methylate cytosine in hemi-methylated double-stranded

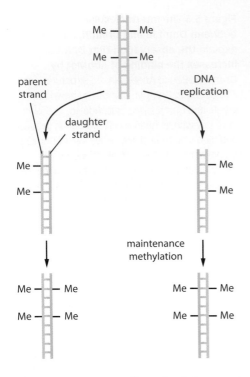

Figure 5.6 Representation of maintenance methylation. Each of the daughter strands synthesized by DNA polymerase I will only be a faithful copy of the sequence of bases found on the parent strands. The DNA polymerase cannot distinguish 5-methylcytosine from cytosine; DNMT1 is therefore required to bind to the sites when 5-methylcytosine is located opposite its unmethylated cytosine partner (hemi-methylated DNA) and add the missing methyl group.

DNA. The chemical reaction catalyzed by DNMT1 is identical to that of *de novo* DNA methylation detailed in Figure 5.3. A simplified diagram of the maintenance methylation process is shown in **Figure 5.6**.

The catalytic subunit of DNMT1 has binding sites for *S*-adenosylmethionine (AdoMet) and DNA and an active site capable of performing the reactions, but the other subunits of the DNMT1 enzyme complex are needed for allosteric activation of the catalytic subunit. Basically, this means that when the other subunits bind to hemi-methylated DNA, they alter their three-dimensional structure in such a way that they "twist" the DNMT1 complex, making the active site work more efficiently. Together, these attributes—specificity of binding and allosteric activation—help DNMT1 to methylate cytosines on hemi-methylated DNA 20 times faster than on unmethylated DNA. This works via the binding of 5-methylcytosine-containing DNA to the N-terminal parts of the DNMT1 complex, causing a direct interaction of the zinc-binding domain with the catalytic domain (**Figure 5.7**).

5.4 DNA METHYLTRANSFERASE ACTIVITY

There are several methods by which DNA methyltransferases may be subjected to control within the cell, but not all of these are able to exert a site-specific influence on the positions at which DNA methylation is established. This section describes the latter control methods.

Enzyme activity can be controlled by small molecules *in vivo*

As mentioned earlier, the catalytic activity of DNMT1 is greatly increased by the presence of its hemi-methylated DNA substrate. That activation mechanism does not apply to the DNMT3 enzymes, but there are other means of controlling the DNA methyltransferases. One of these is based on small molecules (by this we mean molecules that are smaller than proteins), particularly the polyamines putrescine, spermidine, and spermine. Cells require optimal levels of these polyamines for their growth and differentiation and are therefore equipped with many intricate mechanisms for controlling the amounts of these molecules in the cell.

The polyamine biosynthetic pathway consists of two highly regulated enzymes, ornithine decarboxylase (ODC) and *S*-adenosylmethionine decarboxylase (AdoMetDC), and two constitutively expressed enzymes, spermidine synthase and spermine synthase. Any cell that becomes polyamine-deficient is severely limited in its ability to grow and proliferate. There is also evidence suggesting that polyamines have a role in apoptotic cell death. So how do they work?

It is interesting that when ODC is overexpressed in many cell types, the cells undergo a form of malignant transformation and show characteristic dysregulation of their growth control. Conversely, inhibition of ODC counteracts this effect. Less attention has been paid to the fact that ODC inhibition alters the differentiation profile of teratocarcinoma stem cells. Because cell differentiation is heavily dependent on DNA methylation, this observation implies that the activity of polyamines may influence the activity of DNA methyltransferases. **Figure 5.8** shows that the biochemical interactions between polyamine synthesis and DNA methylation are, in fact, linked through the feedback inhibition of AdoMetDC by polyamines.

Decarboxylated AdoMet (dcAdoMet) normally acts as an aminopropyl group donor in polyamine biosynthesis, but it can also act as a competitive inhibitor of the DNA methyltransferases by displacing the real methyl donor, AdoMet, and when present in fivefold excess it can inhibit DNA methyltransferase activity completely. Under normal physiological conditions the steady-state level of dcAdoMet is very low and is maintained by a balance between the AdoMetDC activity (which is regulated by polyamine concentrations) and the rate of use sufficient to make polyamines. Such low concentrations are unlikely to affect DNA methylation to any great extent under normal conditions. However, this system could have an impact in some disease states, such as cancer, when the activities of enzymes such as ODC may be disrupted.

Another competitive inhibitor of virtually all methyltransferase enzymes, including the DNA methyltransferases, is S-adenosylhomocysteine (AdoHcy), which is the normal by-product of methyl donation (see Figure 5.1). This molecule can have a significant impact on human health because increased levels of homocysteine in the blood plasma are associated with increased risk of cardiovascular disease. Increased intracellular concentrations of AdoHcy in humans can result in a lower AdoMet/AdoHcy ratio, diminished methylation capacity, and global DNA hypomethylation, presumably through the ability of AdoMet to inhibit the enzymes needed for both maintenance and *de novo* DNA methylation. Conversely, the AdoMet methyl donor is an inhibitor of DNA demethylase enzymes.

DNA methyltransferase activity can be controlled transcriptionally

A relatively easy way to control global activities of enzymes is to alter how much of the enzyme is present in the cell, and the DNA methyltransferases are no exception. For example, the 5' region of the DNMT1 gene has several possible binding sites for transcription factors that can alter the ability of its promoter to initiate transcription (**Figure 5.9**). Several lines of evidence point toward the conclusion that DNMT1 transcription is initiated from at least four independent transcription initiation clusters and that DNMT1 has the capacity to respond to two nodal cellular regulatory programs signaling through c-Jun and the retinoblastoma gene product (Rb).

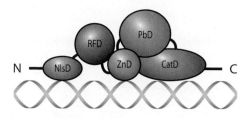

Figure 5.7 Interactions of the protein domains of DNMT1 and its binding with DNA during allosteric activation of the catalytic subunit (CatD). Interaction between the zinc-binding domain (ZnD) and CatD are essential for allosteric activation. Other domains present in DNMT1 are the replication-foci-directing domain (RFD), which helps to localize DNMT1 to sites of DNA replication. The proliferating cell nuclear antigen-binding domain (PbD) ensures that DNMT1 can interact with other components of the DNA replication machinery. The N-terminus contains the nuclear localization signal domain needed to ensure correct subcellular localization of the enzyme after assembly in the cell's cytoplasm. [Adapted from Fatemi M, Hermann A, Pradhan S & Jeltsch A (2001) *J. Mol. Biol.* 309, 1189. With permission from Elsevier.]

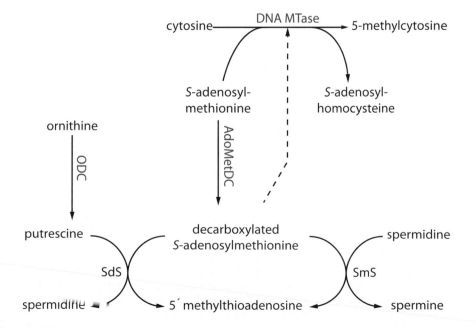

Figure 5.8 Metabolic pathway showing the coupling between polyamine synthesis and DNA methylation. Adverse effects on enzymes (pink text) by AdoMetDC are indicated by dashed lines. DNA methyltransferase (DNA MTase) catalyzes the methylation of cytosine to 5-methylcytosine exclusively at CpG dinucleotides. The syntheses of spermidine and spermine are catalyzed by spermidine synthase (SdS) and spermine synthase (SmS), respectively.

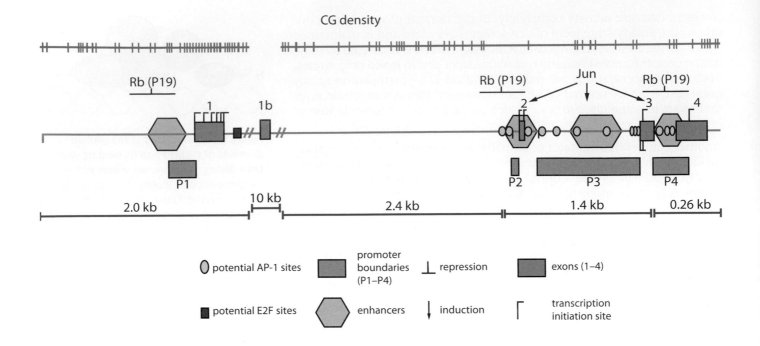

Figure 5.9 Transcription-factor-binding sites on the 5′ region of DNMT1. The line at the top of the diagram shows the density of CG bases across the 5′-control sequences of the gene relative to the positions of control elements in the physical map of the region. The position of putative binding sites for the transcription factors AP-1 and E2F, areas in which the retinoblastoma protein (Rb) represses gene activity, and regions that enhance gene activity through binding of the protein c-Jun are indicated. [Adapted from Bigey P, Ramchandani S, Theberge J et al. (2000) *Gene* 242, 407. With permission from Elsevier.]

Transcription factors are of immense importance in biology. Many of these factors are under a degree of control from signal transduction systems that transfer instructions from the cell's exterior to regulate parameters such as growth and chemotaxis. A particular example of this is the direct interaction between the signaling effector protein signal transducer and activator of transcription 3 (STAT3) and the DNMT1 promoter. STAT3 promotes malignant transformation in various cells by impacting a wide range of functions. Depending on the cell type, the normal function of STAT3 is to mediate cell proliferation, inhibit programmed cell death, and promote angiogenesis. It accomplishes these diverse tasks primarily by inducing the expression of several genes that code for c-MYC, cyclin D1, Bcl-xL, survivin, vascular endothelial growth factor inhibitor (VEGF-I), and other proteins, including DNMT1. The DNMT1 promoter/enhancer region contains at least two binding sites for the STAT3 protein.

Yet another way to control the activity of an enzyme is to build it slightly differently, so that although it still does essentially the same job, it either does it in a different way or slightly better or worse. One of the easiest ways in which the cell can do this is by assembling the RNAs derived from various exons that encode the subunits of a protein in a different order—that is, by so-called alternative splicing. A particularly good example of this is the control of DNMT1 in germ cells, where alternative splicing of sex-specific exons controls the production and localization of the enzyme during specific stages of gametogenesis. In the mouse at least, an oocyte-specific 5′ exon is associated with the production of very large amounts of active DNMT1 protein that is truncated at the N-terminus, a conformation that seems to regulate the protein's sequestration to the cytoplasm during the later stages of oocyte development. In males, a spermatocyte-specific 5′ exon is used to produce nonfunctional DNMT1 during the crossing-over stage of meiosis.

Three alternative 5′ exons have been identified in the DNMT1 gene: one is specific to the growing oocyte, one to the pachytene spermatocyte, and one to all somatic cells and other germ-cell types of both sexes. DNMT1 behaves in somatic cells as a replication factor: it is nucleoplasmic through most of the cell cycle and associates with replication foci during S phase, as do other replication factors. Cytoplasmic DNMT1 has not been observed in any somatic cell during interphase. The situation in oocytes and early embryos is quite different. DNMT1 shows a nuclear distribution in early growing oocytes, a lack of nuclear staining and a uniform cytoplasmic distribution in later growing oocytes, and a subcortical distribution in ovulated oocytes. Reserves of DNMT1 protein in mature oocytes are very large and seem to function as a maternal store of DNA methyltransferase during early development. The importance of this ability to restrict or specify DNMT1 activity is that meiotic chromosomes have several features that make them vulnerable to DNA methylation during the crossing-over process because their very structure makes them easy targets for the enzyme. Prevention of such promiscuous methylation is vital for germ-cell development, and it has been suggested that germ cells protect their meiotic DNA by excluding DNMT1 from the vicinity of the chromosomes at all stages necessary for germ-cell development.

5.5 METHYLATION REGULATION AT SPECIFIC GENE LOCI

Histone interaction with DNA methyltransferases affects where DNA is methylated

Even the previous example does not explain how one gene regulatory sequence can undergo methylation in preference to another. In Chapter 3 we saw that DNA is packed into the cell's nucleus partly through its interaction with complex arrangements of histone proteins called nucleosomes. We will be looking at the structure and biochemistry of these rather useful proteins in great detail later (see Chapter 6), but an important point—actually, a *very* important point—is that DNA methylation and histone biochemistry seem to be intimately related and in some cases inseparable. Most of a histone protein's structure is tied up in the core of the nucleosome, where it concerns itself with binding to the other histones in that complex and providing the molecular "hooks" to allow DNA to wrap itself around the nucleosome in precise but readily variable patterns. There are also the N-terminal polypeptide sequences of histone H3, which protrude from the nucleosome core; their positions suggest a possible interaction with DNMT3L.

A really great feature of histone N-terminal tails is that their individual amino acids can be subjected to a wide range of post-translational modifications that can provide the cell with information for controlling gene transcription that goes beyond the information contained in the underlying sequence of the DNA itself. We discuss the specific mechanism of this so-called "histone code hypothesis" later, in Section 6.7.

One of the principal histone post-translational modifications is methylation of the side-chain amino group of lysine residues located at specific positions along the length of the N-terminal tail. Because the side chain of lysine is a primary amino group, it is possible to add three methyl groups sequentially, which can alter the ability of the group to interact with other segments of the lysine residue itself and with the other amino acids in the N-terminal tail. Methylation of the fourth lysine residue of the N-terminal tail of histone H3 is frequently present on the nucleosomes at the promoters of genes that are either undergoing active transcription or at least may be expected to up-regulate their transcription very rapidly when required. Methylation of histone H3 lysine 4 (H3K4) has been suggested to protect gene promoters from *de novo* DNA methylation in somatic cells. Because of the interaction between DNMT3L and the histone H3 N-terminal tail, the mammalian *de novo* DNA methylation machinery may be able to translate patterns of H3K4 methylation, which are not themselves preserved during chromosome replication, into heritable patterns of DNA methylation that mediate transcriptional silencing of the affected sequences. Precisely how this might work in terms of steric interaction of the modified N-terminal tails and the DNA methyltransferase complex is not absolutely clear at the time of writing. Peptide interaction assays have shown that DNMT3L specifically interacts with the extreme N-terminus of histone H3; this interaction was strongly inhibited by H3K4 methylation but was insensitive to modifications at other positions. Co-crystallization of DNMT3L with the amino tail of H3 showed this tail to be bound to the cysteine-rich zinc-binding domain of DNMT3L, and substitution of key residues in the binding site eliminated the H3–DNMT3L interaction. These data suggest that DNMT3L is a probe of H3K4 methylation, and if the methylation is absent, then DNMT3L induces *de novo* DNA methylation.

The possibility that histone modification patterns are a major factor controlling where and when DNA methylation occurs is interesting, but it still does not explain what is special about a genetic locus that specifies that DNA methylation should occur at this position as opposed to other regions of the genome. Even assuming that histone modification patterns are the only explanation, what dictates this pattern in the first place? One possibility is that such placement of methylated cytosine is due to the underlying DNA sequence. In the next section we discuss the factors that may direct DNA methyltransferases to specific DNA sequences.

Transcription factors may control DNA methyltransferases

The promoter regions of virtually all genes have numerous binding sites for transcription factor proteins, just as they do for DNMT1. Many such proteins are available within the cell, but it would be a mistake to assume that each gene has its own transcription factor. In some cases this may well be true, but more often it is the combinatorial pattern of generic transcription-factor-binding sites that determines the cell-specific context of gene expression. Transcription factor proteins can either activate or repress gene activity. In the case of DNA methylation, we are specifically interested in how defined proteins may cause gene down-regulation through the relatively simple act of binding to a precise sequence of bases in the control regions, and the repressive activities of the MYC transcription factor provide some good examples of this type of control.

MYC is an essential mediator of cell growth and proliferation through its ability to regulate transcription, both positively and negatively. MYC is thought to repress transcription through functional interference with other transcriptional activators. Genes repressed by MYC do not seem to require its direct association with DNA, but rather MYC is recruited to core promoters through protein–protein interactions with positively acting transcription factors, such as TFII-I, NF-Y, and Miz-1. The current model for MYC-mediated gene silencing is that MYC associates with these activators and passively interferes with their transactivation function. An example of this is the gene *p21Cip1*. It is silenced by MYC through the association of MYC with MIZ1, which binds specifically to this gene. However, MYC also silences transcription by active recruitment of the DNA methyltransferase co-repressor DNMT3A.

Interactions with proteins other than transcription factors do not seem to disturb the activity of the DNA methyltransferases. This is reflected in the emerging technology behind artificial gene silencing using hybrid proteins in which the catalytic domains of mouse Dnmt3a and Dnmt3b are fused to the DNA-binding domain of the yeast GAL4 protein and synthetically engineered zinc-finger domains. This has been used to achieve targeted DNA methylation and the selective silencing of gene expression and could have applications in basic research. Eventually it could be used to directly correct some of the epigenetic defects observed in many cancers.

In the case of gene activation, it is necessary to keep the DNA methyltransferases away from the promoter regions. Transcription factors seem to be able to do this. For example, STAT4 is required for the differentiation of interferon-γ-secreting helper T type 1 (Th1) cells from naive T cells during the response of the immune system to infection. This process is induced by the cytokine IL-12 (interleukin 12), but STAT4 is the effector molecule that ends up being directed to the nucleus as a result of the activation of IL-12 signaling. Once in the nucleus, STAT4 binds to the IL-18 receptor 1 (IL-18R1) gene promoter (among others), thus promoting increased acetylation of the local histones, which reduces the ability of DNMT3A to interact with this locus. The ultimate result is decreased DNA methylation of the *IL18R1* gene, which allows T-cell differentiation to proceed.

Noncoding RNA may control DNA methyltransferases

Large chunks of the DNA macromolecule are devoted to the encoding of genes, but possibly even larger segments are not. Although the functions of these regions are still the subject of investigation, many noncoding regions encode RNA molecules that have been implicated in transcriptional control of the coding areas. In this regard, **microRNAs (miRNAs)** are of particular interest. miRNAs are single-stranded RNA molecules about 21–23 nucleotides in length; they are encoded by genes from whose DNA they are transcribed, but they are not translated into polypeptides. Each primary transcript (a *pri-miRNA*) is processed into a short stem-loop structure called a *pre-miRNA* and finally into a functional miRNA. Mature miRNA molecules are partly complementary to one or more mRNA molecules, and their main function is to down-regulate gene expression. However, they also have binding sites at diverse genomic loci, which implies that they have other functions than this.

miRNAs are initially synthesized (**Figure 5.10**) in the nucleus by RNA polymerase II as long, capped, and polyadenylated transcripts called pri-miRNAs. They may be expressed from their own transcriptional units but usually are embedded within other host genes, very often as part of their introns or, more rarely, included within exonic regions. It is not uncommon to find them clustered as a group of several miRNAs that are transcribed together and later processed into independent, mature molecules. The initial step of this maturation process is cleavage of the primary miRNA into shorter (70–90 nucleotides) pre-miRNAs through the action of the enzyme Drosha. Subsequent events include export from the nucleus, further trimming by the type III ribonuclease Dicer, and incorporation into the argonaute-containing RNA-induced silencing complex (RISC). RISC delivers mature miRNAs to their mRNA targets through imperfect Watson–Crick base pairing along the 3′ untranslated regions (3′ UTRs), resulting in either translational arrest or degradation of the transcripts.

Our knowledge of the mechanisms controlling the synthesis and activity of miRNAs is still in its infancy. Most data suggest that some degree of control is exercised by DNA methylation; specifically, there is evidence suggesting that expression of various miRNAs is regulated by methylation of CpG islands situated close to their coding sequences. Comparison of

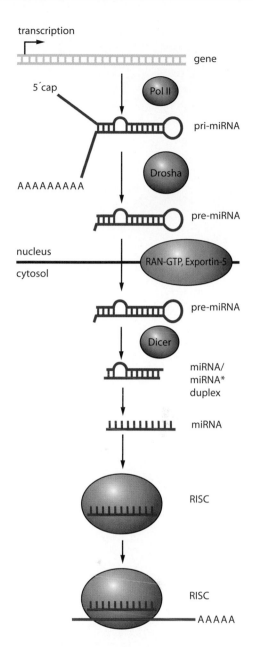

Figure 5.10 Biogenesis of microRNA.
MicroRNAs are transcribed by RNA polymerase II (Pol II) followed by polyadenylation addition of a cap sequence to the 5′ end of the RNA molecule and splicing to form a large primary miRNA (pri-miRNA). This undergoes cleavage to a pre-miRNA approximately 60 bp in length. The pre-miRNA is exported to the cytosol, where it is cleaved into miRNA sequences about 22 bp in length. These associate with the RNA-induced silencing complex (RISC), which controls the binding of miRNA to the target mRNA and subsequent target degradation. Drosha catalyzes the cleavage of large pre-miRNAs into the 22 bp fragments. RAN-GTP and Exportin mediate the nuclear export of pre-miRNAs into the cytoplasm.

the miRNA expression profiles between DNMT1- and DNMT3B-knockout cell lines and their wild-type counterparts reveals that around 10% of the known miRNAs were overexpressed as a consequence of the knockout, suggesting that they are normally controlled by DNA methylation. The proximity of the methylated CpG islands to the miRNA coding sequence does support the idea that they are the repressive elements. However, it is also possible that DNA methylation could regulate miRNA expression in an indirect manner if one or more transcription factors that are normally involved in their up-regulation were themselves down-regulated by methylation of their promoters. As happens with protein-coding genes, an aberrant pattern of methylation of CpG islands near or within miRNA genes could result in a misregulated expression of key miRNAs and ultimately in pathogenic alterations, including tumorigenesis.

It is still early days in the investigation of how noncoding RNAs can affect cell function, but a good example is the ES-cell-specific cluster miR-290 through miR-295 (the notation system for miRNAs simply involves the prefix miR- followed by a number indicating the order of naming or discovery) found in mice. This region encodes six miRNAs (miR-290 through miR-295) that share a 5′-proximal AAGUGC motif. The cluster (for brevity referred to as the miR-290 cluster) increases its expression during pre-implantation development and remains high in undifferentiated ES cells, but decreases after ES cell differentiation. In undifferentiated ES cells, the miR-290 cluster miRNAs suppress a transcriptional repressor that targets genes encoding *de novo* DNA methyltransferases, with the result that DNMT3A is strongly expressed in these cells.

Although findings such as the one above demonstrate the involvement of miRNAs in controlling the expression of single genes, current data suggest that there are enormous numbers of potential miRNAs active within the nucleus. This perhaps implies that they share a common regulatory logic with transcription factors. If this were true, we would expect cells to express sets of miRNAs combinatorially in much the same way that they express a wide range of transcription factors. In fact, cooperative action of miRNAs has been identified through reporter gene assays. Reporter genes are artificial constructs in which the promoter of a gene of interest drives the expression of a protein that can be easily detected in the cell, for example green fluorescent protein or β-galactosidase (whose presence can be shown by its reaction with the enzyme substrate 5-bromo-4-chloro-3-indolyl-β-D-galactoside (X-gal) to produce a colored dye.

As might be evident from the above discussion, miRNAs are an important new and rapidly developing area of research in biology, and work using ES cells provides us with a good insight into how miRNA might control DNA methylation. miRNAs bearing the AAGUGC seed (miRNA seed sequences must be perfectly complementary to a target sequence in the mRNA that they target for degradation), largely represented by the miR-290 cluster, are the functionally dominant miRNAs in mouse ES cells (mESCs) and have therefore been implicated in the maintenance of mESC pluripotency. *De novo* DNA methylation in differentiating ES cells also seems to be controlled by this cluster. This is evident from analyses of the RNA molecules present in the mESCs (in other words, transcriptome analyses), the results of which show that mESCs lacking the Dicer enzyme (see Figure 5.10 for its function) are defective in differentiation. Essentially, if the cells cannot differentiate they are no longer pluripotent; they just stay as ES cells.

The Dicer-knockout cells are unable to methylate the promoter regions of their *Oct4* gene, despite the accumulation of repressive histone modifications (trimethylation and dimethylation of lysines 27 and 9 on histone H3, respectively) and even the addition of a strong differentiation-inducing agent (all-*trans* retinoic acid, or ATRA) to the culture medium. If the knockout cells are placed in medium containing no retinoic acid, they begin to remove these repressive methylation marks and apparently function normally as ES cells once again.

The loss in the ability of Dicer-knockout cells to differentiate can be explained as follows. The AAGUGC seed sequence found in the miRNAs encoded by the miR-290 cluster works by recognizing the complementary sequences in the mRNAs of target genes (this is true for all miRNA seed sequences, not just AAGUGC). Bioinformatic analysis of genome sequences predicts 250 primary targets of the miR-290 cluster, and one of these is the gene *Rbl2* (retinoblastoma-like protein 2), a factor known to contribute to the repression of *Dnmt3* genes. However, it has also been reported that the miR-290 family of miRNAs (also expressed in mESCs) can directly target both *Dnmt3a* and *Dnmt3b*. DNMT1 may also be regulated by Rbl2, because Dicer-knockout ES cells down-regulate this enzyme and show decreased global DNA methylation even when not forced into a differentiation pathway that relies heavily on DNMT3 A/B.

It seems that miRNAs are able to exert some control over the extent of DNA methylation through post-translational regulation of the DNA methyltransferase enzymes. At our current state of knowledge this seems to be a relatively unspecific mechanism, in that altering expression levels of DNA methyltransferases should influence the levels of 5-methylcytosine at all genomic loci that are normally subjected to this modification. It seems more likely that the recruitment of DNA methyltransferases to specific loci is mediated by the actions of transcription factors that recognize particular sequences of DNA. However, it is also possible that the expression of these transcription factors is itself controlled by miRNA, and therefore the latter may still be able to influence the methylation of specific genes, albeit indirectly.

DNA methylation is not the only epigenetic modification to be controlled by miRNAs. They may have a decisive role in the control of chromatin structure by targeting the post-transcriptional control of several chromatin-modifying enzymes, as suggested by bioinformatic predictions of miRNA target genes. These data indicate that several proteins such as the methyl-CpG-binding proteins (Section 5.2), chromatin accessibility complexes, and enzymes responsible for the post-translational modification of histone proteins are all potential targets of miRNA regulation. The types of epigenetic modification that these proteins control are quite different from those controlled by DNA methylation, but they are just as important for regulating the structure of the nucleosomes that controls access to the information contained in the underlying DNA sequence.

Noncoding RNA can influence chromatin regulation directly

The mechanism of miRNA-based regulation described in the preceding section relies on the suppression of an already active gene transcription through binding of the miRNA to the corresponding mRNA via the activity of RISC. This is not the only mechanism by which regulation may be effected. There is evidence to suggest that many of the proteins involved in chromatin modifications (as well as transcription factors) have the capacity to bind RNA or complexes containing RNA. These include DNA methyltransferases and DNA methyl-binding domain proteins, heterochromatin protein 1 (HP1), the multi-KH domain protein DPP1 (which suppresses heterochromatin-mediated silencing in *Drosophila*), domains commonly found in chromatin-remodeling enzymes, and effector proteins such as SET domains, tudor domains, and chromodomains.

Moreover, certain noncoding RNAs are known to interact with specific sites on the 3′ untranslated regions of the *DNMT1*, *DNMT3A*, and *DNMT3B* genes.

One of the better examples of miRNA function is the regulation of gene clusters such as those of *HOX* genes, which contain several miRNAs needed for the regulation of *HOX* gene expression. Activation of *HOX* gene clusters is an early event in embryonic development because individual members have important roles in patterning of the body axis. Their functions require precise control of spatio-temporal expression to provide positional information for the cells of the developing embryo. *HOX* genes show a high degree of evolutionary conservation across eukaryotic species, which highlights their primitive origin and their importance as master regulators of development.

In humans, the *HOX* genes are organized into four clusters (*HOXA* to *HOXD*) located on different chromosomes (7p15, 17q21.2, 12q13, and 2q31). Each cluster contains nine to eleven member genes that encode relatively small gene products, but these are not normally expressed at the very early embryonic stages because there is little need for a body plan when all the cells are largely the same and not yet ready to contribute to the formation of structures such as organs. The really clever thing about the *HOX* clusters is that they are able not only to restrict the expression of certain cluster members to certain types of cells but also to do this depending on where those cells happen to be in the body. For example, fibroblasts from the human scalp express a rather different profile of *HOX* genes than those from lower body extremities, despite being identical cells as far as we can tell. This pattern seems to persist into adulthood and results from a remarkable feature of *HOX* cluster function known as **co-linear activation**. In essence, *HOX* genes are expressed in a precise spatio-temporal pattern along the anterior–posterior axis that corresponds to their position within the cluster. Expression of genes at the 3′ end begins at an early stage of development in the anterior segments of the embryo, whereas those at the 5′ end are up-regulated later in the posterior segment.

5.6   GENOME FUNCTION CONTROL ACROSS SPECIES

The greater part of the studies leading to the data presented in this chapter have been performed on mammalian cells or the yeast *S. cerevisiae*. However, the use of DNA methylation as a genome-control mechanism seems to occur in a much wider range of animal and plant species. The precise mechanism of gene control often differs among individual species, but there are few organisms for which there is no evidence of DNA methylation. Even then, there may be certain circumstances in which they do methylate a portion of their genome. So far, the presence of 5-methylcytosine has been reported in several invertebrates belonging to various orders, and varying levels of methylation affecting different functions have been reported. In particular, whereas vertebrate genomes (including fishes, amphibians, reptiles, birds, and mammals) are globally methylated, with 60–90% of CpG dinucleotides being methylated at cytosine residues, invertebrate genomes (including those of insects, molluscs, echinoderms, annelids, priapulides, bryozoans, and cnidarians) generally consist in large part of long segments of nonmethylated DNA and short stretches of heavily methylated DNA.

In this chapter we have shown that DNA methylation is an epigenetic modification of DNA that is important for regulating development and maintaining post-developmental cellular identity. In addition, we have described the mechanisms that may control the activity of the DNA methylation system, so our next step is to understand how the association of DNA with histone proteins can modulate (or undergo modulation by) DNA methylation. The next chapter details these post-translational modifications of histones.

KEY CONCEPTS

- DNA methylation is not uniform across the human genome and tends to be enriched in CpG islands usually found in gene promoter regions. Those CpG islands with lower CpG concentration are more likely to be methylated.

- The principal methylation event is the formation of 5-methylcytosine.

- DNA methylation is performed by DNA methyltransferases. These can establish *de novo* DNA methylation (DNMT3A/DNMT3B) or maintain existing methylation patterns during DNA replication (DNMT1).

- DNA methyltransferases function as parts of protein complexes that modify chromatin structure.

- Histone modification patterns, transcription factors, and miRNAs direct the DNA methyltransferases and their associated chromatin-structure-modifying complexes to specific locations in the genome.

FURTHER READING

Antequera F (2003) Structure, function and evolution of CpG island promoters. *Cell Mol Life Sci* **60**:1647–1658 (doi:10.1007/s00018-003-3088-6).

Bestor TH (2000) The DNA methyltransferases of mammals. *Hum Mol Genet* **9**:2395–2402 (doi:10.1093/hmg/9.16.2395).

Chen CC, Wang KY & Shen CK (2012) The mammalian de novo DNA methyltransferases DNMT3A and DNMT3B are also DNA 5-hydroxymethylcytosine dehydroxymethylases. *J Biol Chem* **287**:33116–33121 (doi:10.1074/jbc.C112.406975).

Cheng X (1995) Structure and function of DNA methyltransferases. *Annu Rev Biophys Biomol Struct* **24**:293–318 (doi:10.1146/annurev.bb.24.060195.001453).

Cheng X & Blumenthal RM (2008) Mammalian DNA methyltransferases: a structural perspective. *Structure* **16**:341–350 (doi:10.1016/j.str.2008.01.004).

Deaton AM & Bird A (2011) CpG islands and the regulation of transcription. *Genes Dev* **25**:1010–1022 (doi:10.1101/gad.2037511).

Denis H, Ndlovu MN & Fuks F (2011) Regulation of mammalian DNA methyltransferases: a route to new mechanisms. *EMBO Rep* **12**:647–656 (doi:10.1038/embor.2011.110).

Edwards JR, O'Donnell AH, Rollins RA et al. (2010) Chromatin and sequence features that define the fine and gross structure of genomic methylation patterns. *Genome Res* **20**:972–980 (doi:10.1101/gr.101535.109).

Ehrlich M (2003) Expression of various genes is controlled by DNA methylation during mammalian development. *J Cell Biochem* **88**:899–910 (doi:10.1002/jcb.10464).

Flanagan JM & Wild L (2007) An epigenetic role for noncoding RNAs and intragenic DNA methylation. *Genome Biol* **8**:307 (doi:10.1186/gb-2007-8-6-307).

Gardiner-Garden M & Frommer M (1987) CpG islands in vertebrate genomes. *J Mol Biol* **196**:261–282 (doi:10.1016/0022-2836(87)90689-9).

Holz-Schietinger C & Reich NO (2012) RNA modulation of the human DNA methyltransferase 3A. *Nucleic Acids Res* **40**:8550–8557 (doi:10.1093/nar/gks537).

Meng CF, Zhu XJ, Peng G & Dai DQ (2010) Role of histone modifications and DNA methylation in the regulation of O6-methylguanine-DNA methyltransferase gene expression in human stomach cancer cells. *Cancer Invest* **28**:331–339 (doi:10.3109/07357900903179633).

Qin W, Leonhardt H & Pichler G (2011) Regulation of DNA methyltransferase 1 by interactions and modifications. *Nucleus* **2**:392–402 (doi:10.4161/nucl.2.5.17928).

Rinn JL & Chang HY (2012) Genome regulation by long noncoding RNAs. *Annu Rev Biochem* **81**:145–166 (doi:10.1146/annurev-biochem-051410-092902).

Rottach A, Frauer C, Pichler G et al. (2010) The multi-domain protein Np95 connects DNA methylation and histone modification. *Nucleic Acids Res* **38**:1796–1804 (doi:10.1093/nar/gkp1152).

Schneider K, Fuchs C, Dobay A et al. (2013) Dissection of cell cycle-dependent dynamics of Dnmt1 by FRAP and diffusion-coupled modeling. *Nucleic Acids Res* **41**:4860–4876 (doi:10.1093/nar/gkt191).

Song J, Rechkoblit O, Bestor TH & Patel DJ (2011) Structure of DNMT1-DNA complex reveals a role for autoinhibition in maintenance DNA methylation. *Science* **331**:1036–1040 (doi:10.1126/science.1195380).

Tost J (ed.) (2009) DNA methylation: methods and protocols, 2nd ed. Humana Press.

Turker MS & Bestor TH (1997) Formation of methylation patterns in the mammalian genome. *Mutat Res* **386**:119–130 (doi:10.1016/S1383-5742(96)00048-8).

van Wolfswinkel JC & Ketting RF (2010) The role of small non-coding RNAs in genome stability and chromatin organization. *J Cell Sci* **123**:1825–1839 (doi:10.1242/jcs.061713).

Vavouri T & Lehner B (2012) Human genes with CpG island promoters have a distinct transcription-associated chromatin organization. *Genome Biol* **13**:R110 (doi:10.1186/gb-2012-13-11-r110).

Walsh CP & Bestor TH (1999) Cytosine methylation and mammalian development. *Genes Dev* **13**:26–34 (doi:10.1101/gad.13.1.26).

Xu GL & Bestor TH (1997) Cytosine methylation targetted to pre-determined sequences. *Nat Genet* **17**:376–378 (doi:10.1038/ng1297-376).

Zhou H, Hu H & Lai M (2010) Non-coding RNAs and their epigenetic regulatory mechanisms. *Biol Cell* **102**:645–655 (doi:10.1042/BC20100029).

POST-TRANSLATIONAL MODIFICATION OF HISTONES

As we have seen in earlier chapters, the structures of individual histone proteins are well suited to formation of the histone octamer and its ability to present appropriate amino acids at its exterior for stabilizing DNA binding. We also saw (Section 3.1 and Box 3.1) that apart from the histone fold motif common to all proteins of this class, each histone has several features that are uniquely its own. Probably the most relevant of these to the epigenetic control of the genome is the labile terminus, more commonly referred to as the histone's N-terminal polypeptide tail. These tails consist of chains of amino acids that protrude from the core of the nucleosome (**Figure 6.1**) and are subjected to post-translational modification by a large number of cellular systems.

Figure 6.1 gives some appreciation of the large number of different modifications that are possible. Whereas about 75% of the histone octamer mass forms the globular structure onto which the DNA is wrapped, the remaining 25% forms long N-terminal tails rich in lysine and arginine residues. In all, eight N-terminal tails protrude from the surface of the nucleosome core, where they are free to make numerous intermolecular contacts that depend on the pattern of modifications to which they are subjected.

Although many of the functions of the genome are influenced by the methylation state of DNA, the importance of histone modification for gene expression control cannot be understated. Understanding the structures of the N-terminal tails is central to the whole business of genome control by histone modification, and Figure 6.1 shows the amino acid sequences of these crucial structures and the types of covalent modification to which they are subject.

Even at a most fundamental level, a plethora of different post-translational modifications correlate with regions of silenced or transcriptionally active chromatin. There is substantial evidence that the primary mechanism by which many specific post-translational modifications of the histone tail function is the recruitment of regulatory proteins that probably affect chromatin structure indirectly. However, numerous *in vitro* studies have provided evidence that the histone octamer N-terminal tails may also have a direct role in the modulation of chromatin compaction (Section 3.2). At present we are not certain whether all histone tail modifications directly regulate chromatin compaction, but some of them, such as acetylation and phosphorylation, can alter the charge of the tails and therefore have the potential to directly influence chromatin through electrostatic mechanisms.

6.1 ACETYLATION AND METHYLATION OF LYSINE

The simplest modifications of the N-terminal tails of histones are the acetylation and methylation of lysine residues. Acetylation of lysine is almost always associated with active gene transcription. Methylation quite often also is present on active gene chromatin, but there are circumstances in which it can do the exact opposite—often very effectively. The reason for putting these two apparently contrary modifications in the same section is that unlike most of the others, such as the phosphorylation

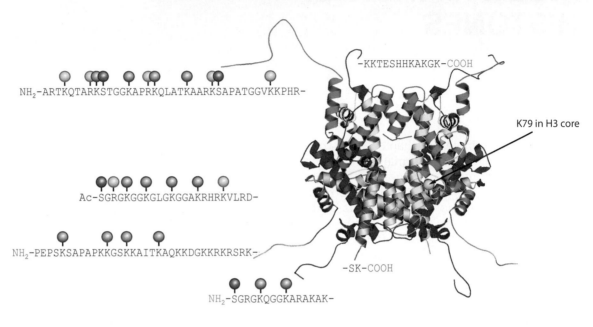

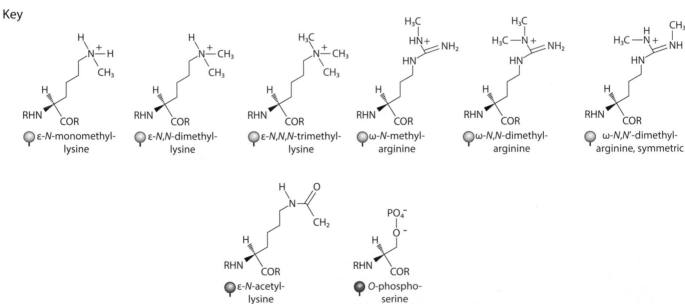

Figure 6.1 Generalized structure of histones in the nucleosome core and protrusion of their N-terminal tails from the globular section of the nucleosome. This figure shows the amino acid sequences of the histone N-terminal "tails" and some of the possible covalent modifications that can take place on histones such as acetylation and methylation of lysines, methylation of arginines and the phosphorylation of serines and threonines. The bulk of the modifiable amino acids are present in the N-terminal tails but there are some exceptions to this such as H3 lysine 79 (K79) which can be methylated despite being in the core nucleosome domain of H3. The key shows the types of modification possible and the amino acids to which they can be applied, for example, purple hexagons indicate lysine acetylation, blue circles indicate methylation whilst purple squares represent phosphorylation. [Adapted from Khorasanizadeh S (2004) *Cell* 116, 259. With permission from Elsevier.]

of serine residues, these two are frequently interchangeable and represent opposite effects on genome regulation. That is not to say that such "cross-talk" does not exist between other histone modifications because, for example, histone lysine methylation has also been shown to be cross-regulated by histone ubiquitination, but the substitution of acetyl for methyl groups and vice versa is far more common.

Lysine is often acetylated in histone tails

Acetylation of the histone octamer N-terminal histone tails is a prevalent and reversible histone modification whose levels, as already noted, correlate with transcriptionally active chromatin. The effect of acetylating lysine residues by acetyltransferases is to decrease the net positive charge of the very basic histone tails and is therefore expected to modulate electrostatic histone tail interactions. It is believed that acetylation causes unfolding of the chromatin fiber and thereby permits transcription, a notion that is consistent with the results of *in vitro* experiments using randomly hyperacetylated histones. As is evident from **Figure 6.2**, the structure of the side chain of lysine is very different when it has been amended by the addition of either acetyl or methyl groups to the ε-amino group. Acetylation in particular removes the net positive charge of the ε-amino group (although it has to be said that the presence of that charge is very dependent on physiological pH).

Acetylated lysines can be found at several positions on the terminal tails of all four histones. Histone H3, for example, has lysines at positions 4, 9, 14, 18, 23, 45, 36, and 79 that are known to undergo acetylation in response to changes in gene transcription. Some of these can also be methylated, which induces a rather different set of responses; H3 lysine 9 (H3K9), for example, is generally a part of actively transcribed chromatin when it is acetylated, but when this acetylation is replaced by methylation, gene repression occurs. The effect of widespread acetylation on chromatin structure is illustrated in **Figure 6.3**, which shows that histone N-terminal tails have a propensity to bind to other tails in the same or other nucleosomes only when they lack acetyl groups. The presence of acetyl groups disrupts this interaction, allowing the tails to spring apart

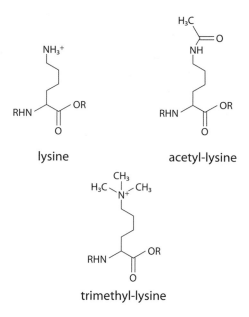

Figure 6.2 Acetylation and methylation of lysine in the histone N-terminal tails. These structures show the addition of acetyl or methyl groups to the ε-amino group of lysine. The bottom structure shows the trimethylation of lysine, but monomethylated and dimethylated states also exist.

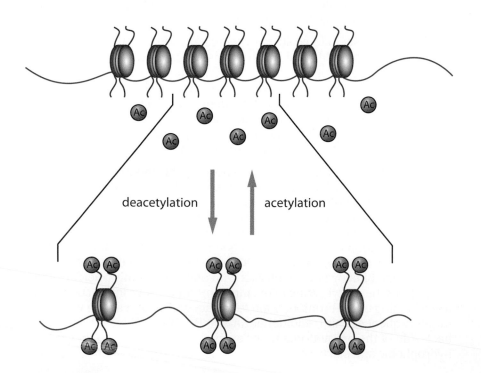

Figure 6.3 Acetylation forces the N-terminal tails apart to open up the chromatin. The acetyl groups are represented by Ac. Acetylation of the 11 nm chromatin fiber results in a more "open" chromatin conformation.

and open up the chromatin so that nucleosome remodeling complexes, transcription factors, and other epigenetic modification enzymes can gain access.

Such steric interactions might be enough to maintain an open conformation if all of the adjacent chromatin is organized in a similar manner (in other words, if it all has an acetylated, euchromatin-type structure). But what happens if the gene is situated next to another region, one that does not have a strong acetylation pattern? Given the semi-flexible nature of euchromatin, it would be surprising if the possibility of an interaction between the two areas did not exist. This could mean that precise expression control of the supposedly active gene could be modulated by the adjacent chromatin region. So it makes sense for the cell to have a system that reinforces the open conformation of acetylated chromatin and can reduce the possibility of interaction with adjacent less acetylated regions. Fortunately just such a system has evolved in the form of the **bromodomain**.

Proteins with bromodomains recognize and bind to acetylated histones

The bromodomain is an evolutionarily conserved protein module that functions as an acetyl-lysine (AcK)-binding domain. By recognizing regions of acetylated histones and by recruiting other proteins capable of remodeling chromatin conformation, bromodomain-containing proteins are able to reinforce the open conformation of chromatin where higher levels of histone acetylation are present. Many proteins involved in genome regulation have bromodomains as part of their structures.

Most, if not all, bromodomains adopt a conserved structural fold consisting of a left-handed bundle of four helices (α_Z, α_A, α_B, α_C) plus interhelical ZA and BC loops that are of variable length and sequence (**Figure 6.4**). The latter loops constitute a hydrophobic pocket that both stabilizes the structure and interacts with the acetyl-lysine. Despite the conserved structural fold, the overall sequence similarity of the bromodomain family is not high, with members of this large protein family showing major sequence variations in the ZA and BC loops. Nevertheless, the amino acid residues engaged in acetyl-lysine recognition are among the most highly conserved residues, as one might expect.

Bromodomains seem to be strongly represented in the structures of transcription factor proteins. This makes sense because transcription factors probably need to recognize the open conformation of chromatin at the genes they are meant to induce. Sequence-specific lysine acetylation recognition by a bromodomain is dependent on how the bromodomain recognizes target protein residues flanking the acetyl-lysine, and that, in turn, will most probably depend on other structural elements in the transcription factor or other bromodomain protein.

Salient features of histone recognition by bromodomains emerged from structural and biochemical analyses and can be defined as three major contact points in the conserved bromodomain structural fold. First, acetyl-lysine recognition by most, if not all, bromodomains occurs within a hydrophobic pocket embedded between the ZA and BC loops when the BC loop forms a hydrogen bond with the amide nitrogen of the acetyl-lysine. Second, ZA and/or BC loop residues at the entrance of the acetyl-lysine binding pocket interact with one or two amino acid residues adjacent to the acetyl-lysine in the target protein. These interactions reinforce binding of the bromodomain to the acetyl-lysine residing within the target sequence. Third, additional ZA and BC loop residues that face the back side of the bromodomain (for example, opposite the α_Z' helix) form hydrophobic and electrostatic interactions with target sequence residues.

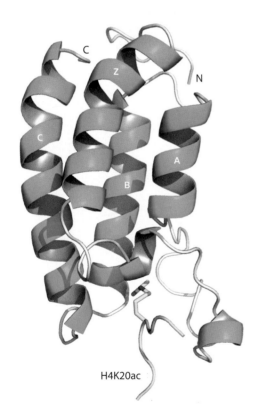

H4K20ac

Figure 6.4 Basic structure of the bromodomain binding to acetylated lysine (shown in yellow). The interhelical α_Z–α_A (ZA) and α_B–α_C (BC) loops constitute a hydrophobic pocket that recognizes the acetyl-lysine. [From Sanchez R & Zhou M-M (2009) *Curr. Opin. Drug Discov. Devel.* 12, 659. With permission from Roberto Sanchez, Mount Sinai School of Medicine.]

The interaction of individual bromodomains with other structural elements within a protein can achieve an impressive degree of specificity for an acetylated target. The bromodomain protein Brd2 selectively interacts with acetyl-lysine 12 (AcK12) on histone H4 to initiate transcription, whereas other bromodomain proteins—such as TAF(II)250 and P/CAF, which preferentially recognize AcK14 on histone H3—bind to other potential acetylation sites. Human polybromo protein 1 (PB1) is possibly one of the most impressive examples of this selectivity: it has five different bromodomains that can distinguish acetylation on lysines 4, 9, 14, and 23 of histone H3, with each domain able to bind its acetyl histone target site with high affinity and specificity (note that bromodomains 2 and 3 both bind acetylated lysine 9, which explains the discrepancy between the number of domains and the number of lysine residues).

So, having established that bromodomains recognize acetylated DNA, which helps a transcription factor enormously in finding its binding site, do they do anything else? Some of the ATPase-dependent chromatin remodeling complexes described in Section 4.1 (SWI/SNF-type proteins) contain bromodomains that help anchor the chromatin-modifying proteins on gene promoters. They have the interesting feature that the more acetyl groups there are on the chromatin, the stronger the chromatin remodelers bind.

The multiple methylation states of lysine can alter transcriptional response

Methylation of lysine can come in three forms; monomethyl, dimethyl, and trimethyl adducts are known. Of these, only the trimethyl form, being a quaternary amine, preserves the positive charge on the nitrogen atom of the ε-amino group. The other two are secondary and tertiary amines, respectively, and are uncharged at physiological pH (**Figure 6.5**).

Because of the multiple methylation states that are possible for a single lysine, the methylation of histone N-terminal tails can include a great deal more information than is possible with acetylation. An example of this is the difference between dimethylation and trimethylation of lysine 4 on histone H3, in which the dimethylated state is often associated with readiness for gene transcription (a so-called **permissive chromatin state**), whereas active transcription of the gene almost always requires the presence of trimethylated lysine at position 4 (H3K4me3). The effect is also determined by the position of the methylated lysines. H3K4 may be an excellent way to promote gene activity, but if lysine 9 on H3 becomes methylated (often by replacing acetyl groups), gene expression is decreased.

Because of its recruitment of heterochromatin protein 1 (HP1), methylation of lysine 9 on histone H3 is perhaps the best characterized example demonstrating the different effects of histone methylation. HP1 was first identified as a cytological component of *Drosophila* heterochromatin that

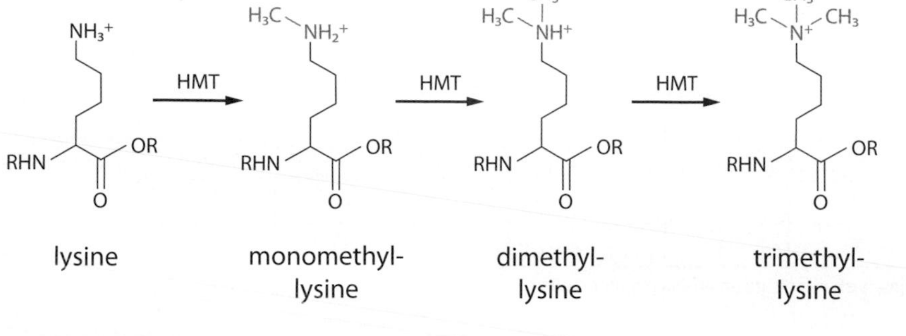

lysine   monomethyl-lysine   dimethyl-lysine   trimethyl-lysine

Figure 6.5 Three methylation states of lysine are generated through the activity of histone methyltransferases (HMT).

Figure 6.6 Mammalian chromodomain binding to H3 trimethyl-lysine 9.

is required for position effect variegation (PEV), the position-dependent silencing of transcription that occurs within heterochromatin. This protein is required for the large-scale compaction of chromatin that accompanies permanent gene repression, but one of the crucial features of HP1 and a range of other proteins that recognize and bind to methylated histone lysines is the possession of a **chromodomain**. The example in **Figure 6.6** shows the binding of a trimethyl-lysine 9 residue, but chromodomains are equally capable of recognizing trimethylated lysines in other positions on histone N-terminal tails. Their precise interaction is governed by other structural components of the proteins in which the chromodomains are incorporated. As in the bromodomain-containing proteins, chromodomain proteins serve as docking sites for effectors that aid in the recruitment of other enzymes that are necessary for regulating the locus in question.

As noted earlier, the complexity of epigenetic information conveyed by lysine methylation is enhanced by the fact that there are three methylated forms, and it has been suggested that distinct methyl-lysine marks recruit different proteins that, in turn, specify different transcriptional responses. There are several other domain types that are also found predominantly in chromatin-associated proteins, including Tudors, PHDs, SANTs, SWIRMs, MBTs, and PWWPs. Indeed, some of these domain types (Tudors, MBTs, and PWWPs) are structurally related to chromodomains and are known to recognize methylated lysines. However, they demonstrate considerable differences in specificity, depending on the nature of the methylation-recognizing domain, the type of protein into which the domain is incorporated, and the nature of other proteins that may bind to or be associated with the domain-containing protein. **Table 6.1** depicts much of our current knowledge concerning the binding of methylated states of lysine in histone N-terminal tails. The structures of several methyl-lysine-binding domains have been determined by X-ray crystallography, and three of these are depicted in **Figure 6.7**.

Lysine is not the only amino acid capable of being methylated. Arginine also has an amino functional group on its side chain, and its methylation is

TABLE 6.1 SUMMARY OF INTERACTIONS OF METHYL-LYSINE-BINDING DOMAINS

	H3K9me1	H3K9me2	H3K9me3	H3K4me1	H3K4me2	H3K4me3	H4K20me1	H4K20me2	H4K20me3	H3K79me2
Chromodomains										
HP1a	O	O	O	—	—	—	—	—	—	—
HP1	O	O	O	—	—	—	—	—	—	—
HP1g	O	O	O	—	—	—	—	—	—	—
CDY1	—	O	O	—	—	—	—	—	—	—
Tudor domains										
SMN	—	—	—	—	—	—	—	—	—	—
TDRD3	—	—	—	—	—	—	—	—	—	—
Pombe-1	—	—	—	—	—	—	—	—	—	—
53BP1	—	O	—	—	O	—	O	O	—	—
C20orf104	—	O	—	—	O	—	—	O	—	O
JMJD2A	—	—	O	—	—	O	—	O	O	—
MBT domains										
CGI-72	—	—	—	O	—	—	O	—	—	—
L(3)MBTL	—	—	—	O	—	—	—	O	—	—

Circles indicate interaction of the proteins with methylated lysines, dashes indicate an absence of interaction

(a)

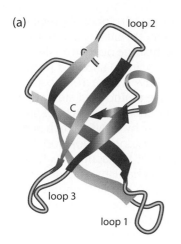

(b)

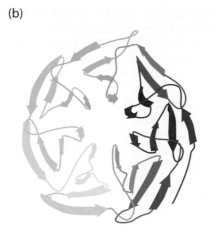

(c)

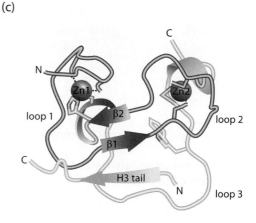

a common post-translational modification for several proteins. However, the functions of arginine methylation when applied to the histone tails are quite often antagonistic because of a specific interaction between the chromodomain proteins that regulate H3K4 and the arginine at position 2 of histone H3 (H3R2). When Arg 2 is methylated, this interferes with the ability of H3K4 methyl-binding proteins to attach to their target, which means that the H3K4 methylation signal, namely gene activation, is ignored regardless of the methylation state of the lysine at this position. The locus could have an apparently strong activation mark, such as H3K4me3, but this would be overcome by the methylation of H3R2.

6.2 PHOSPHORYLATION OF SERINE AND THREONINE

Whereas histone acetylation has essentially been coupled to active gene transcription, histone phosphorylation seems to be involved in a wider range of cellular processes. All core histones contain phospho-acceptor sites in their N-terminal domains. H2A can be phosphorylated on serine 1, H2B on serines 14 and 32, H3 on serines 10 and 28, and H4 on serine 1.

One of the better-studied examples of histone phosphorylation is that of H3S10, which in mammals has been shown to be involved in the initiation of the chromosome condensation process that is essential to mitosis. Histone H3 phosphorylation at serine 10 begins during prophase, with peak levels detected during metaphase and ultimately followed by a general decrease in the amount of phosphorylation during the progression through the cell cycle to telophase. Because the increase in H3S10 phosphorylation seems to be spread across large areas of the genome, this implies that phosphorylated H3S10 may not be controlling specific genes but, rather, is responsible for more general changes in chromosome structures that are needed to promote the proper segregation of chromosomes during cell division. Exactly how this might work is not clear, but H3S10 phosphorylation may have similarities to the effects of lysine acetylation (which stops the histone tails from binding to each other), except that in this case the influence operates not just on small areas of the genome but on whole chromosomes.

It is also not clear whether H3S10 phosphorylation has a direct functional role or simply paves the way for other modifications that can activate a gene. The possibility that H3S10 phosphorylation facilitates rather than triggers transcription arises because of the known interplay between H3S10 phosphorylation and other modifications of the H3 terminal tail, such as acetylation of lysines 9 and 14 and the prevention of lysine 9

Figure 6.7 Structures of some protein motifs that recognize methylated lysines in histone N-terminal polypeptides. (a) Tudor domain, (b) WD40 domain, and (c) PHD finger domain. [a, adapted from Selenko P, Sprangers R, Stier G et al. (2001) *Nat. Struct. Biol.* 8, 27. With permission from Macmillan Publishers Ltd. b, adapted from Stirnimann C, Petsalaki E, Russell RB & Müller CW (2010) *Trends Biochem. Sci.* 35, 565. With permission from Elsevier. c, adapted from Lallous N & Ramón-Maiques S (2011) Chromatin Recognition Protein Modules: The PHD Finger. eLS. With permission from John Wiley and Sons.]

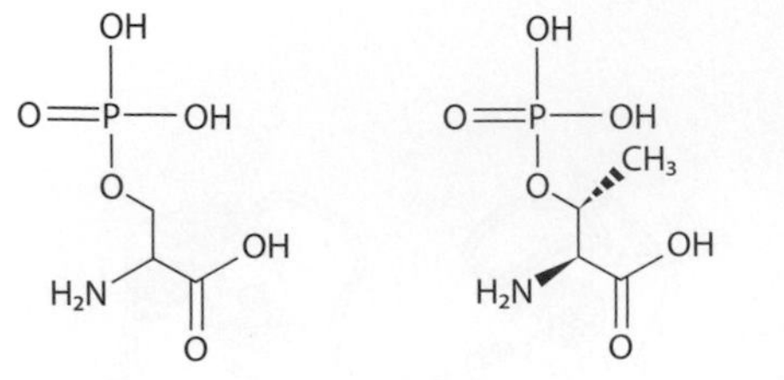

phosphoserine   phosphothreonine

Figure 6.8 Phosphorylation states of serine and threonine.

methylation. This latter effect is possible because H3S10 phosphorylation seems to prevent the SUV39H1 enzyme, which methylates lysine 9, from reaching its target site. Furthermore, one of the histone acetyltransferase enzymes (GCN5) that would normally acetylate lysine 9 shows a preferential binding to H3 terminal tails that already have serine 10 phosphorylated; this preference occurs because of a single amino acid side-chain link between the enzyme and the phosphorylated histone. This phosphorylated N-terminal arm is then bound preferentially by GCN5, which acetylates the same histone H3 N-terminal tail at the lysine 14 position. The acetylation of histone H3 leads to the induction of transcription. Taken together, these results suggest a synergistic mechanism that depends on the addition of each of these modifications to the N-terminal tails of histone H3 in the nucleosome particles of these loci and imply that H3S10 phosphorylation may only be a modification that predisposes a gene toward rapid induction when it is needed rather than being a direct induction system in its own right. Analyses of other genes (such as c-*jun*) that show evidence of H3S10 phosphorylation similarly suggest that the presence of a phosphorylated H3 terminal tail does not necessarily predispose that tail to acetylation at lysine 14, meaning that other factors are necessary to induce activation of the locus.

Serine and threonine modifications (**Figure 6.8**) provide additional examples of the versatility of the effects stemming from histone tail phosphorylations. Serine 139 in H2AX, a histone H2A variant, is phosphorylated by the ATR kinase in response to DNA damage, and this may cause an alteration of chromatin structure that facilitates DNA repair. Phosphorylation of serine 14 in histone H2B—a modification that is regulated by the Mst1 kinase—was found to be induced in apoptotic cells.

Phosphorylation of threonine, in contrast, is a much less common event but still may contribute a great deal to epigenetic control of the genome. As an example, threonine 119 in the C-terminal section of histone H2A influences nucleosome structure in a manner that seems to affect the regulation of transcription, DNA replication, and cell cycle progression. This is not, strictly speaking, a form of control that uses the histone N-terminal tails, but it demonstrates the enormous complexity of histone modifications as a means for regulating chromatin architecture.

6.3 ADDITION OF UBIQUITIN TO SPECIFIC LYSINES

Ubiquitin is a 76 amino acid polypeptide that can be enzymatically coupled to all sorts of proteins. Its major function in most cells is undoubtedly the "tagging" of worn-out or damaged proteins that are destined for degradation by the proteasome, but it also has uses in epigenetic genome control. The principal targets of ubiquitin are histones H2A (lysine 119) and H2B (lysine 120), although the exact positions of the target amino acids vary between species. The two lysines indicated above are for mammals, but they will suffice for the time being to demonstrate the importance of adding ubiquitin to these two histones. Many other applications of ubiquitination in the cell rely on the addition of multiple ubiquitin units (poly-ubiquitination), but histones only ever seem to be conjugated to a single ubiquitin moiety (monoubiquitination); however, even this was sufficient for ubiquitinated H2A to be originally considered a unique histone-like chromosomal protein, named A24 because the extra amino acids greatly alter the protein's molecular weight.

The functions of H2A and H2B ubiquitination are quite variable. The latter seems to be a prerequisite for the dimethylation and trimethylation of H3K4 and H3K79, although it does not seem to influence mono-methylation. This would suggest that gene activation will occur only if the ubiquitin group is present on lysine 120 of H2B, and it may thus be thought of as an activating histone mark. It is likely that this mark functions through an interaction between the ubiquitin and some of the enzymes that add methyl groups to H3K4 monomethyl lysines, such as COMPASS.

Some data indicate that, in addition to its function in regulating these initiation and early elongation steps in transcription, H2B ubiquitination might be directly required for transcription by RNA polymerase II. The process of transcriptional elongation of the nascent RNA molecule is hindered by the presence of nucleosomes, because RNA polymerase II needs to translocate along the DNA. H2B ubiquitination might assist the histone chaperones (FACT) in stimulating the passage of RNA polymerase II through a nucleosomal template. FACT can displace an H2A–H2B dimer from a nucleosome core, enhancing transcription elongation on the chromatin template.

Monoubiquitinated H2A (ubH2A) shows a quite different spectrum of activities. Whereas H2B ubiquitination is mostly an activating mark, the opposite is mostly true for H2A. In humans, ubiquitination of H2A is mediated by at least two different E3 ubiquitin ligases, Ring1B and 2A-HUB, both of which are associated with transcriptional silencing. Ring1B (in addition to Ring1A and Bmi1) is part of the polycomb repressive complex 1 (PRC1) and specifically adds ubiquitin to lysine 119 of H2A (**Figure 6.9**). This modification seems to work in conjunction with other repressive modifications, such as H3K27 trimethylation, by interfering with the loading of RNA polymerase II onto the DNA. Loss of H2A ubiquitination seems to have no effect on other modifications involved in silencing, such as lysine 27 H3 or lysine 9 H3 methylation; rather, it seems that PRC1-mediated H2A ubiquitination occurs downstream of Lys 27 H3 methylation, because knockdown of the enzyme that introduces the

Figure 6.9 Subunits of the polycomb repressive complexes PRC1 and PRC2. PRC2 complex contains EZH2, which trimethylates H3K27. This epigenetic mark can recruit PRC2, which carries the proteins RING1A and RING1B, which in turn can add ubiquitin (Ub) to lysine 119 on histone H2. [Adapted from Spivakov M & Fisher AG (2007) *Nat. Rev. Genet.* 8, 263. With permission from Macmillan Publishers Ltd.]

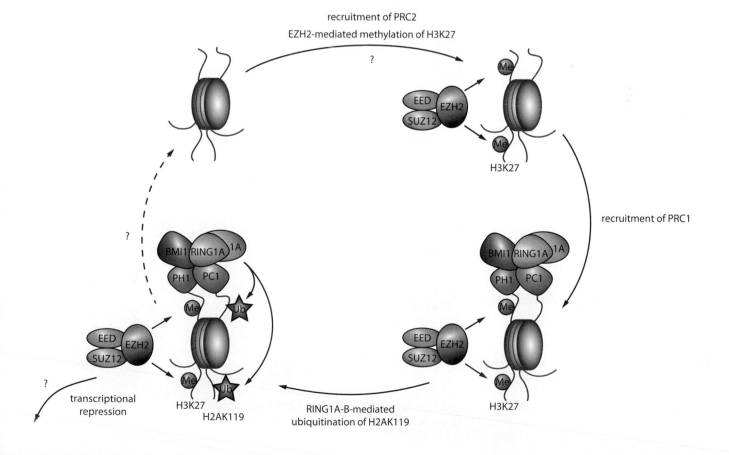

lysine 27 trimethylation decreases the amount of H2A ubiquitination. ubH2A does, however, seem to interact with some enzymes that dimethylate and trimethylate H3 lysine 4, which helps it to prevent the initiation of transcription.

Some of the roles of ubH2A in repression of transcription might relate to the finding that ubH2A enhances the binding of the linker histone H1 to reconstituted nucleosomes *in vitro*. Furthermore, mononucleosomes purified from a Lys 119→Arg mutant of H2A lack histone H1, indicating that H2A de-ubiquitination might cause the dissociation of linker histones from core nucleosomes. This idea is consistent with the structure of the nucleosome, in which the C-terminus of H2A seems to interact with linker histones. Thus, it is possible that ubH2A contributes directly to transcriptional repression by regulating higher-order chromatin structure, in addition to inhibiting H3 methylation on lysine 4.

6.4 SUMOYLATION OF LYSINES

The small ubiquitin-related modifier protein (SUMO) shares 18% sequence identity with ubiquitin and adopts a similar three-dimensional structure, but unlike ubiquitin it has no apparent role in targeting proteins for degradation by the proteasome. Addition of this group to lysines in the histone N-terminal polypepetide is facilitated by E3-SUMO ligases in a manner somewhat analogous to ubiquitination (**Figure 6.10**), although the interaction is reversible and specific isopeptidases release the SUMO moiety. SUMOylation can compete with other lysine modifications such as acetylation. For the most part, it seems to lead to the inhibition of transcription, often through the addition of one or more SUMO peptides to all four core histones; however, the extent of SUMOylation does not seem to be equal for all histones. The primary targets seem to be histones H2A and H2B (lysine 126 in H2A, and lysines 6, 7, and possibly 16 in H2B), but there is also evidence to suggest the involvement of all five lysines on the N-terminal tail of H4.

Analyzing the functions of this modification by using chromatin immunoprecipitation has been difficult because of the lack of appropriate antibodies. Most investigations have therefore tended to concentrate on the expression of mutant forms of histone H2B in which the target lysines

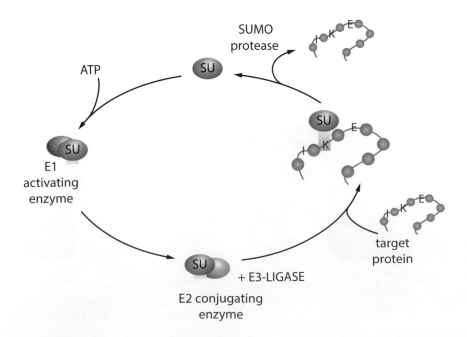

Figure 6.10 The SUMOylation pathway.
E1 activates SUMO (SU) in an ATP-dependent manner. E2 attaches SUMO to the lysine in the target protein, supported by E3-LIGASE. SUMO protease removes SUMO from the protein, which is now free to be reused in another cycle. [Adapted from Bauer DC, Bushe FA, Bailey TL & Bodén M (2010) *Neurocomputing* 73, 2300. With permission from Elsevier.]

have been replaced by alanine. Strains of the yeast *Saccharomyces cerevisiae* in which lysines 6, 7, 16, and 17A have been replaced showed enhanced basal transcription of multiple genes. This supports the view that SUMOylation is a repressive histone modification and that the sites occupied by this modification are frequently those that are occupied by acetyl groups during active transcription of the target gene.

Exactly how the SUMO moiety is able to repress gene activity is not clear. It is easy to imagine that the structural similarities between SUMO and ubiquitin might imply some degree of similarity in their modes of action, but the number of SUMO groups present at the promoter regions of several repressed genes in *S. cerevisiae* is low (less than a few percent). This suggests that the disruption of chromatin architecture or the prevention of interaction with the transcriptional machinery induced by this modification must be very effective indeed. SUMO has an effector region (the circled area in **Figure 6.11**) that seems to be essential for its repressive function. Replacement of basic amino acids in this region by acidic ones eliminates the repressive ability of a SUMO–H2B fusion protein; this underlines the importance of this protein segment, whose structure is very probably a major determinant in the strong repressive ability of SUMO even at very low stoichiometries. At present there are insufficient data to provide insights into the physical mechanism of repression of SUMOylated genes. It is possible that the effector region is able to distort nucleosomes sufficiently to affect their interaction with the transcriptional machinery; however, if the mark is present in low amounts, such a general approach is less likely to be the method of choice for preventing gene expression because it is difficult to imagine how a small number of SUMOylated histones could achieve sufficient disruption of transcriptional machinery binding. Alternatively, the presence of SUMO may disrupt transcription factor binding at a much more local level, but until hard evidence of such an interaction is available, this possibility remains speculative.

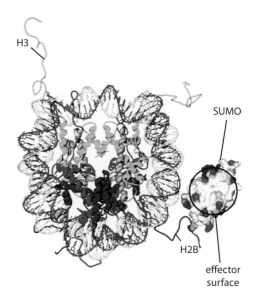

Figure 6.11 Structure of the SUMO–H2B interaction. This diagram shows a single SUMO molecule covalently linked to the N-terminal tail of histone H2B via lysine 6. [Adapted from Iñiguez-Lluhi JA (2006) *ACS Chem. Biol.* 1, 204. With permission from American Chemical Society.]

6.5 BIOTINYLATION OF HISTONES

Histones may also be modified by covalent attachment of the vitamin biotin, with potential involvement of five major histone classes. The classical role of biotin in metabolism is to serve as a covalently bound coenzyme for a bevy of carboxylases (cytoplasmic acetyl-CoA carboxylase α and mitochondrial acetyl-CoA carboxylase β, 3-methylcrotonyl-CoA carboxylase, propionyl-CoA carboxylase, and pyruvate carboxylase). There is evidence that biotin also has a role in cell signaling, and it is known to be involved in controlling the expression levels of at least 2000 genes. The covalent modification of proteins by biotin is normally mediated by the enzymes biotinidase and holocarboxylase synthetase (**Figure 6.12**). Specific lysine residues within the histone N-terminal regions seem to be targeted by this system, because biotinylated histones may be detected in human cells. This is accomplished by using streptavidin-linked fluorescent probes. (Streptavidin has a very high affinity for binding to the biotin moiety.)

A reaction mechanism by which biotinidase mediates the covalent biotinylation of histones has been proposed. It was suggested that cleavage of biocytin (biotin-ε-lysine), generated in the breakdown of biotin-dependent carboxylases, by biotinidase leads to the formation of a biotinyl-thioester intermediate at or near the active site of biotinidase. In a further step, the biotinyl moiety is transferred from the thioester to the ε-amino group of lysine residues in histones. Biotinidase and holocarboxylase synthetase are ubiquitous in mammalian cells, but the mechanism(s) directing them

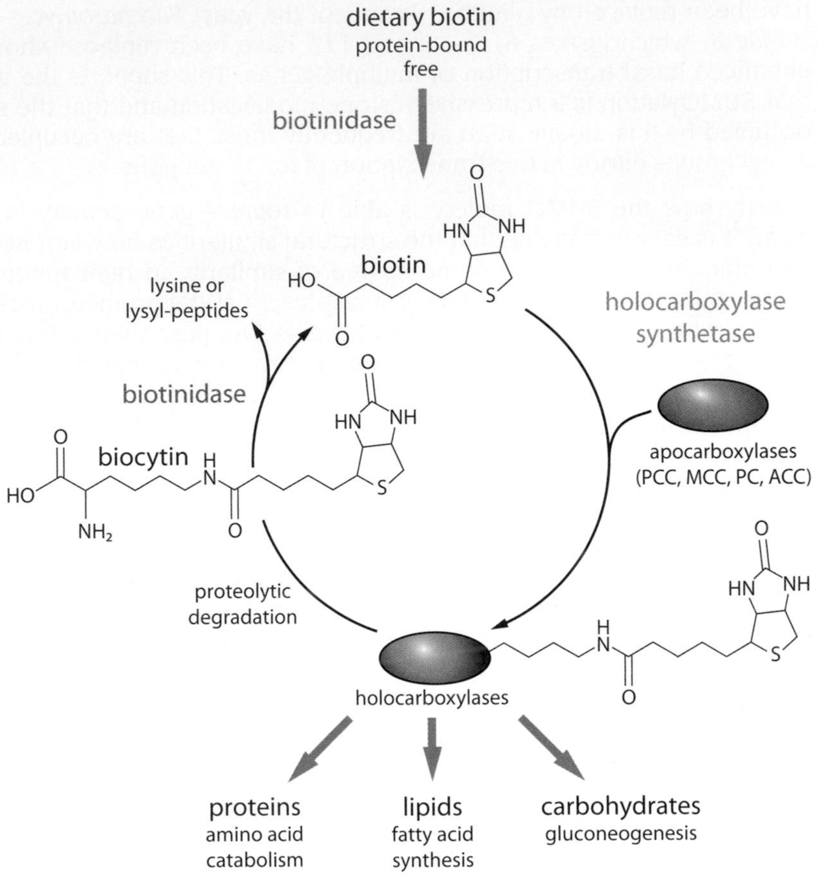

Figure 6.12 The biotin cycle.
Holocarboxylase synthetase catalyzes the biotinylation of specific lysine residue of apocarboxylase to make holocarboxylase. Biotinidase catalyzes the release of biotin from biocytin or its transfer to lysine on various proteins, including histones.

to add the biotin moiety to specific N-terminal tail lysines is currently unclear. The existence of such a mechanism is supported by the observation of 10 target lysines distributed across histones H2A, H3, and H4, as shown in **Figure 6.13**.

There are relatively few data that suggest a role for histone biotinylation in the epigenetic control of gene function. Studies using chicken erythrocytes have provided circumstantial evidence that biotinylated histones are enriched in transcriptionally silent chromatin, with particular involvement of lysines 8 and 12 on the N-terminus of histone H4. Importantly, stimulation of interleukin-2 (IL-2) expression by using mitogens was associated with a rapid depletion of biotinylated histones at the IL-2 locus in human lymphoid cells, which suggests that biotin is readily removed. However, the molecular systems to effect this change are thus far unknown.

There is some evidence that biotinidase itself may catalyze both the biotinylation and de-biotinylation of histones and that the choice of which process occurs may be determined partly by the presence of other histone post-translational modifications—a case of the epigenetic "crosstalk" also seen with other histone N-terminal modifications. It has also been suggested that alternative splicing of messenger RNAs could give rise to differentially active forms of biotinidase, but what regulates the choice of exons that become incorporated is unknown. Finally, it is possible that the biotin moieties may not be removed at all and that the depletion of biotinylated histones may simply occur via replacement of the entire histone molecule with a non-biotinylated copy.

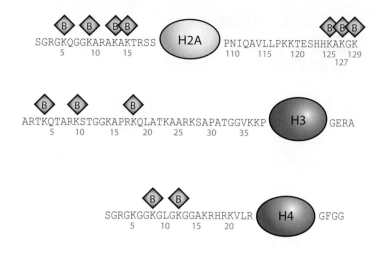

6.6 ADP-RIBOSYLATION OF HISTONES

Poly(ADP-ribosyl)ation is a post-translational modification capable of regulating gene expression either by modulating chromatin structure or by directly influencing the expression of specific genes. Histone H1—the linker histone molecule— is one of the best substrates for poly(ADP-ribose) polymerase, which builds or transfers ADP-ribose (ADPR) polymers onto H1 and other proteins in both a covalent and noncovalent manner. Covalent binding leads to the presence of short ADP-ribose chains on the modified protein, for example 8–10 units of ADPR on H1, whereas noncovalent binding leads to the attachment of long and branched polymers (such as 100–200 ADPR units) to specific domains on the protein.

It is clear that the presence of ADPR polymers on H1 changes its charge markedly and probably affects both H1–DNA and H1–H1 interactions, both of which are important in chromatin organization. Electron micrographs have shown that *in vitro* poly(ADP-ribosyl)ation of polynucleosomes leads to a significant relaxation of chromatin structure. However, this does not cause H1 to detach from the internucleosomal regions (**Figure 6.14**). If poly(ADP-ribosyl)ation does not cause or help linker histone displacement, what is the possible mechanism whereby this modification modulates the expression of specific genes? Recent evidence suggests that, *in vivo*, this modification has a regulatory role in protecting genomic DNA methylation patterns, particularly in maintaining the unmethylated state of CpG islands in the promoters of constitutively expressed housekeeping genes.

6.7 THE HISTONE CODE HYPOTHESIS

We have seen in the preceding sections that a range of potential post-translational modifications may be incorporated onto the N-terminal segments of the histone proteins and that the consequences of these modifications are diverse. Some of these contribute to gene activation when present on relatively well-defined genomic regions (usually but not exclusively in the gene control regions or promoters), and others are more typically associated with gene repression. As we have discussed in some detail (Chapter 3), the structures of individual histone proteins may affect the larger-scale structure of chromatin in such a way that genes are either exposed or hidden, depending on their genomic locations in a cell

Figure 6.14 Histone ADP-ribosylation controls chromatin organization. Upon ADP-ribosylation of histones by ADP-ribosyltransferases (ARTDs), the chromatin relaxes. Modification of histones in this manner is NAD⁺ dependent. ARH3 represents ADP-ribosylhydrolase 3 and PARG represents Poly (ADP-ribose) glycohydrolase. [Adapted from Messner S & Hottiger MO (2011) *Trends Cell Biol.* 21, 534. With permission from Elsevier.]

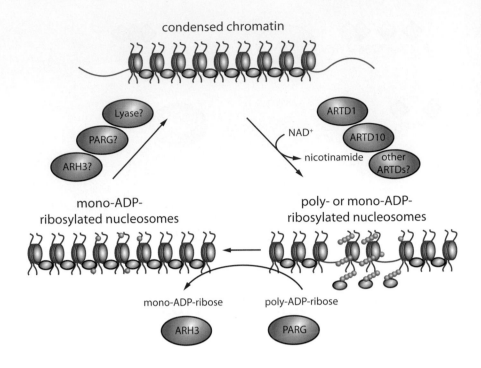

context-based manner. However, there is also substantial evidence suggesting that specific classes of proteins serve the gene control mechanism of the cell by recognizing and interrogating the patterns of histone modifications present at a genomic locus.

A central tenet of epigenetics seems to be that for histone modifications to have any function at all, they need to be present in substantial numbers for the cellular machinery to notice them. This is certainly true of the better-characterized modifications such as acetylation and methylation, but even SUMOylated lysine, with its strong repressive powers, need to be present in many copies if they are to have an effect. If you add to this the observation that such modifications rarely seem to operate alone, the situation becomes truly complex. For example, monomethylation and dimethylation of lysine 4 on histone H3 usually indicate that the gene possessing these marks has the potential for transcription, but these two marks alone are not sufficient to ensure gene up-regulation. Trimethylation of lysine 4 coupled to widespread acetylation around the transcription start site is needed to prepare the promoter and the start codon for the assembly of the transcription complex. Another example is the phosphorylation of serine 10 on histone H3, which seems to work in concert with the phosphorylation of serine 28 to ensure the proper segregation and condensation of chromosomes during mitosis and meiosis. However, because one of the other functions of H3 phosphorylation is to produce an open chromatin conformation (as discussed earlier), how can an identical modification occurring at the same amino acids of the H3 terminal tail result in an apparently opposite effect on chromatin structure? The simple answer is that it probably cannot do this—at least not alone. The serine 10 plus serine 28 phosphorylations on histone H3 must work in concert with other modifications to create these disparate effects.

It is apparent from studies performed over the past decade that a single histone modification does not function alone, and these results have led to the proposal of one the most important concepts in this discussion: the **histone code hypothesis**. This hypothesis states that multiple histone modifcations, acting in a combinatorial or sequential fashion on one or multiple histone tails, specify unique downstream functions. The existence of multiple modifications within a short stretch of the same histone tail still leaves us with the question of how a complex, multimark code is established in the first place. It is possible that certain modifications are able to influence the rate or efficiency with which later modifications are added, because the early modifications alter the structural characteristics of the N-terminal tail substrate for the enzymes that add later modifications. Many modifications are certainly close enough to one another to exert this type of control over the ability of enzymes to incorporate further modifications.

We have discussed some of the protein domains that might be able to recognize the different histone modifications, such as the bromodomain examined in Section 6.1. In the next chapter we will look at the enzymes that bring about N-terminal tail modifications in more detail.

KEY CONCEPTS

- Post-translational modification of histones mostly occurs on the amino acids that comprise the histone N-termini.

- These are frequently referred to as the N-terminal tails, and a wide variety of specific chemical modifications of several amino acids is possible.

- The precise combination of these modifications on different amino acids is able to convey a great deal of information to the cellular transcriptional machinery about the intended activity of the DNA in the chromatin region that is subjected to such patterns of histone modification.

- The most important modifications for epigenetic control of genome activity are probably the acetylation and methylation of specific lysine residues on the N-termini of histones H3 and H4.

- Several other modifications such as the phosphorylation of certain serine, tyrosine, and threonine residues in the N-termini also have important roles, as do the addition of ubiquitin, biotin, and the small ubiquitin-related modifier protein (SUMO).

- These modifications can either activate or repress gene expression. **Table 6.2** gives an overview of the respective effects of the histone modifications described in this chapter.

FURTHER READING

Black JC, Van Rechem C & Whetstine JR (2012) Histone lysine methylation dynamics: establishment, regulation, and biological impact. *Mol Cell* 48:491–507 (doi:10.1016/j.molcel.2012.11.006).

Cao J & Yan Q (2012) Histone ubiquitination and deubiquitination in transcription, DNA damage response, and cancer. *Front Oncol* 2:26 (doi:10.3389/fonc.2012.00026).

Deckert J & Struhl K (2001) Histone acetylation at promoters is differentially affected by specific activators and repressors. *Mol Cell Biol* 21:2726–2735 (doi:10.1128/MCB.21.8.2726-2735.2001).

Emre NC & Berger SL (2006) Histone post-translational modifications regulate transcription and silent chromatin in

Saccharomyces cerevisiae. Ernst Schering Res Found Workshop (57):127–153.

Fischle W (2008) Talk is cheap—cross-talk in establishment, maintenance, and readout of chromatin modifications. *Genes Dev* 22:3375–3382 (doi:10.1101/gad.1759708).

Fischle W, Wang Y, Jacobs SA et al. (2003) Molecular basis for the discrimination of repressive methyl-lysine marks in histone H3 by Polycomb and HP1 chromodomains. *Genes Dev* 17:1870–1881 (doi:10.1101/gad.1110503).

Fischle W, Tseng BS, Dormann HL et al. (2005) Regulation of HP1-chromatin binding by histone H3 methylation and

TABLE 6.2 SUMMARY OF THE ACTIVATING OR REPRESSIVE ACTIVITIES OF HISTONE MODIFICATIONS

Modification	Histone	Amino acid site	Function
Acetylation	H2A	Lysine 4	Transcriptional activation
	H2B	Lysine 5	Transcriptional activation
		Lysine 11 (*S. cerevisiae*)	Transcriptional activation
		Lysine 12 (mammals)	Transcriptional activation
		Lysine 15 (mammals)	Transcriptional activation
		Lysine 20	Transcriptional activation
	H3	Lysine 4 (*S. cerevisiae*)	Transcriptional activation
		Lysine 9	Transcriptional activation
		Lysine 14	Transcriptional activation
		Lysine 18	Transcriptional activation and DNA repair
		Lysine 23	Transcriptional activation and DNA repair
		Lysine 27	Transcriptional activation
		Lysine 56 (*S. cerevisiae*)	DNA repair
	H4	Lysine 5	Histone deposition, transcriptional activation, and DNA repair
		Lysine 8	Transcriptional activation and DNA repair
		Lysine 12	Histone deposition, transcriptional activation
		Lysine 16	Transcriptional activation and DNA repair
		Lysine 91 (*S. cerevisiae*)	Chromatin assembly
Methylation	H1	Lysine 26	Transcriptional silencing
	H3	Lysine 4	Permissive euchromatin (dimethylation), Transcriptional activation (trimethylation)
		Arginine 8	Transcriptional repression
		Lysine 9	Transcriptional repression (trimethylation)
		Arginine 17	Transcriptional activation
		Lysine 27	Transcriptional silencing, X-inactivation (trimethylation)
		Lysine 36	Transcriptional activation
		Lysine 79	Marker of euchromatin and transcriptional activation
	H4	Arginine 3	Transcriptional activation
		Lysine 20	Transcriptional silencing (monomethylation), Heterochromatin formation (trimethylation)
		Lysine 59	Transcriptional silencing
Phosphorylation	H1	Serine 27	Transcriptional activation, chromatin decondensation
	H2A	Serine 1	Mitosis, chromatin assembly, transcriptional repression and DNA repair
		Serine 139	DNA repair
	H2B	Serine 10	Apoptosis
		Serine 14	Apoptosis, DNA repair
		Serine 33 (*D. melanogaster*)	Transcriptional activation
	H3	Threonine 3	Mitosis
		Serine 10	Transcriptional activation
		Threonine 11	Mitosis
		Serine 28	Mitosis
	H4	Serine 1	Mitosis, chromatin assembly, DNA repair
Ubiquitination	H2A	Lysine 119	Spermatogenesis
	H2B	Lysine 120	Meiosis
Sumoylation	H2A	Lysine 126 (*S. cerevisiae*)	Transcriptional repression
	H2B	Lysine 6/lysine 7 (*S. cerevisiae*)	Transcriptional repression
Biotinylation	H3	Lysine 4	Role in gene expression unclear
		Lysine 9	Role in gene expression unclear
		Lysine 18	Role in gene expression unclear
	H4	Lysine 12	DNA damage response

phosphorylation. *Nature* **438**:1116–1122 (doi:10.1038/nature04219).

Gelato KA & Fischle W (2008) Role of histone modifications in defining chromatin structure and function. *Biol Chem* **389**:353–363 (doi:10.1515/BC.2008.048).

Jenuwein T & Allis CD (2001) Translating the histone code. *Science* **293**:1074–1080 (doi:10.1126/science.1063127).

Johansen KM & Johansen J (2006) Regulation of chromatin structure by histone H3S10 phosphorylation. *Chromosome Res* **14**:393–404 (doi:10.1007/s10577-006-1063-4).

Katan-Khaykovich Y & Struhl K (2005) Heterochromatin formation involves changes in histone modifications over multiple cell generations. *EMBO J* **24**:2138–2149 (doi:10.1038/sj.emboj.7600692).

Martin C & Zhang Y (2005) The diverse functions of histone lysine methylation. *Nat Rev Mol Cell Biol* **6**:838–849 (doi:10.1038/nrm1761).

Nathan D, Sterner DE & Berger SL (2003) Histone modifications: now summoning sumoylation. *Proc Natl Acad Sci USA* **100**:13118–13120 (doi:10.1073/pnas.2436173100).

Pérez-Cadahía B, Drobic B, Khan P et al. (2010) Current understanding and importance of histone phosphorylation in regulating chromatin biology. *Curr Opin Drug Discov Devel* **13**:613–622.

Resch MG (2011) The influence of histone orthologues, histone variants and post-translational modifications on the structure and function of chromatin. ProQuest Dissertation Publishing.

Robin P, Fritsch L, Philipot O et al. (2007) Post-translational modifications of histones H3 and H4 associated with the histone methyltransferases Suv39h1 and G9a. *Genome Biol* **8**:R270 (doi:10.1186/gb-2007-8-12-r270).

Rossetto D, Avvakumov N & Côté J (2012) Histone phosphorylation: a chromatin modification involved in diverse nuclear events. *Epigenetics* **7**:1098–1108 (doi:10.4161/epi.21975).

Sims RJ 3rd, Nishioka K & Reinberg D (2003) Histone lysine methylation: a signature for chromatin function. *Trends Genet* **19**:629–639 (doi:10.1016/j.tig.2003.09.007).

Struhl K (2001) Gene regulation. A paradigm for precision. *Science* **293**:1054–1055 (doi:10.1126/science.1064050).

Tian Z, Tolić N, Zhao R et al. (2012) Enhanced top-down characterization of histone post-translational modifications. *Genome Biol* **13**:R86 (doi:10.1186/gb-2012-13-10-r86).

Tweedie-Cullen RY, Brunner AM, Grossmann J et al. (2012) Identification of combinatorial patterns of post-translational modifications on individual histones in the mouse brain. *PLoS One* **7**:e36980 (doi:10.1371/journal.pone.0036980).

Wang Y, Fischle W, Cheung W et al. (2004) Beyond the double helix: writing and reading the histone code. *Novartis Found Symp* **259**:3–17.

Weake VM & Workman JL (2008) Histone ubiquitination: triggering gene activity. *Mol Cell* **29**:653–663 (doi:10.1016/j.molcel.2008.02.014).

Weake VM & Workman JL (2012) SAGA function in tissue-specific gene expression. *Trends Cell Biol* **22**:177–184 (doi:10.1016/j.tcb.2011.11.005).

Winter S & Fischle W (2010) Epigenetic markers and their crosstalk. *Essays Biochem* **48**:45–61 (doi:10.1042/bse0480045).

7

HISTONE MODIFICATION MACHINERY

Having discussed the post-translational modifications of histones in Chapter 6, we need to examine the molecular machinery that adds and removes the modifying groups. As we noted previously, the most important modifications for epigenetic control of genome activity are probably the acetylation and methylation of specific lysine residues on the N-termini of histones H3 and H4, so we begin this chapter by describing histone acetylation.

7.1 ENZYMES THAT ACETYLATE OR DEACETYLATE HISTONES

Extensive regions of chromatin characterized by large numbers of acetylated lysine residues on histone H3 are almost always associated with actively transcribed regions of the genome. Such acetylated regions tend to adopt chromatin conformations that are more readily accessible to the transcription machinery than areas in which acetylated histones are fewer or absent.

Acetyl groups are added by a class of enzymes known as histone acetyltransferases

Using acetyl-CoA as the acetyl donor, histone acetyltransferases (HATs) acetylate the ε-amino group of lysine residues. Acetylation removes the positive charge of lysine and alters interactions of histones with DNA and other chromatin-associated proteins. The acetyl group (usually in the form of acetylated lysine) is by far the best documented acyl group known to be able to modulate histone function in this manner and is, as was noted in Chapter 6, almost always associated with gene transcription. HATs and the more general group of enzymes that add acetyl groups to proteins have also been shown to use longer-chain acyl-CoAs to add the larger propionyl or butyryl to proteins such as histone H4 and p53, but the functions of these alternative modifications are as yet unknown. We will therefore restrict our discussion to what is known about histone acetylation.

Histone acetyltransferases add acetyl groups to specific lysine residues

HATs can be categorized into three major families: the MYST family (so called because a group of HATs, MOZ (human monocytic leukemia zinc-finger HAT), yeast YBF2, SAS2, and TIP60, all contain a conserved amino acid domain that became known as a MYST domain), the Gcn5-related N-acetyltransferase (GNAT) family, and the more recently characterized p300 family (**Box 7.1**). In addition, there are several putative acetyltransferases (such as Spt10) that contain motifs similar to those found in HATs but whose acetyltransferase activity remains to be confirmed. Although the target proteins of many of these latter enzymes are typically histones, it is important to note that many of them are also capable of acetylating non-histone proteins, and thus these enzymes could be generally termed **protein lysine acetyltransferases**. In terms of primary polypeptide sequences, many HATs seem unrelated, although most HATs have an acetyl-CoA-binding domain, or Motif A, and structurally similar active sites, as shown in Box 7.1.

As with most enzymes, the overall structure of a HAT has a great influence on its catalytic activity, which varies between individual examples of this enzyme class. The active site of a HAT is an efficient domain in its own right. It has fairly minimal requirements for enzyme activity and has the structural integrity required for performing such activity. For example, the catalytic domains of Gcn5, p/CAF, and p300 require only three to five residues on either side of the substrate lysine for efficient binding and catalysis. However, for some HATs, the presence of other protein domains results in a marked enhancement of activity. For example, studies comparing the kinetics of yeast HATs Esa1 and Rtt109 as individual subunits with their respective physiological complexes indicate that other subunits enhance the catalytic efficiency more than 50-fold, with only minor effects on the Michaelis constant (K_m). Esa1 minimally requires the presence of two other subunits, Yng2 and Epl1, to allow it to recognize nucleosomes and have full catalytic activity. However, these "helper" proteins do not seem to alter overall affinity for substrate; rather, they alter the conformation of the HAT domain, making it more active and thereby leading to enhanced rates of catalysis.

Recent studies on Esa1 and MOZ, both members of the MYST family of HATs, indicate that efficient acetylation is influenced by interactions with more distal regions of the substrate. The normal targets of Esa1 are a group of closely positioned lysines on the N-terminal tail of histone H4, but, strikingly, full-length histone H4 is acetylated 2000-fold faster than short peptides with the same sequence of amino acids (histone tail peptides) that would normally be found in close proximity to lysine. This indicates that interactions with the histone fold domain of H4 are crucial for high catalytic efficiency and the ability of the enzyme to acetylate four lysine residues within the same H4 tail. This is by no means a universal mechanism for promoting HAT activity: p300, Gcn5, and p/CAF seem to achieve similar catalytic constant (K_{cat}) values to those of the other enzymes, despite not possessing subunits similar to Yng2 and Epl1, and may therefore be thought of as being constitutive HATs.

The HAT enzyme p300 seems to have one of the broadest protein specificities of all the HAT class in that it prefers to acetylate lysines adjacent to positively charged amino acids either three positions downstream or four positions upstream of the target. This is perhaps not surprising, because it has been suggested that such a relaxation of specificity is necessary to allow the Theorell–Chance mechanism of p300 to work (see Box 7.1). Too close an association with the polypeptide might inhibit the formation of the more sterically complex ternary intermediate produced by this mechanism. Despite widespread occurrence of protein acetylation in eukaryotic cells, this modification is performed by relatively few HATs. A critical determinant of histone acetylation enzymes is that they must be restricted to the nucleus because this is their most likely site of action; this distinguishes them from the cytoplasmic acetyltransferases.

Box 7.1 The family of histone acetyltransferase enzymes and their mechanism of action

Most histone acetyltransferases (HATs) have similar domain structures in their active-site regions (**Figure 1a**). Table 1 shows the principal HAT families and their histone specificities. How HATs achieve acetyl transfer has been the subject of much debate and investigation. For the GNAT and MYST families of HATs, initial biochemical and structural work revealed a sequential mechanism of acetyl transfer. In this mechanism, acetyl-CoA and the protein substrate bind to form a ternary complex before any chemical catalytic step. An active-site glutamate (base) assists in deprotonating the lysine, allowing nucleophilic attack on the carbonyl carbon of acetyl-CoA. A putative tetrahedral intermediate forms, which then collapses to the reaction products, namely CoA and the acetylated protein (Figure 1b). This direct attack mechanism is similar to that described for the glycoside acetyltransferases, serotonin acetyltransferases, and amino acid acetyltransferases. Other HAT classes make use of slightly different catalytic mechanisms and there is less certainty as to their exact modes of action. The p300 enzyme for example makes use of a special type of sequential mechanism with a short-lived ternary complex, known as a Theorell–Chance mechanism, in which the backbone amide of Trp 1436 aids in deprotonating the substrate lysine residue for attack on the carbonyl carbon of bound acetyl-CoA. An active-site tyrosine residue is proposed to serve as a general acid and protonate the sulfhydryl of CoA after acetyl transfer.

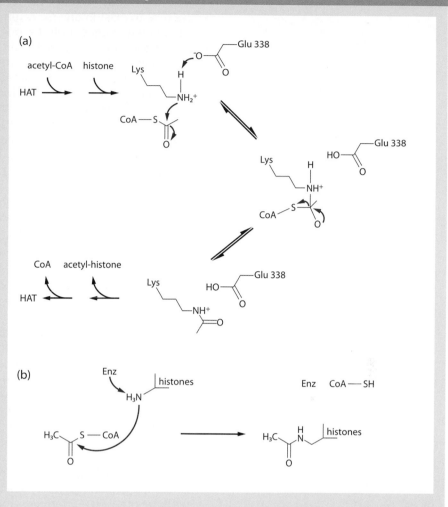

Figure 1 Mechanisms of acetyl transfer in (a) GNAT and MYST families and (b) p300 HATs.

TABLE 1 THE MAJOR FAMILIES OF HISTONE ACETYLTRANSFERASES

Common/gene name[a]	KAT designation[b]	Histone specificity	Known complexes
GNAT family of histone acetyltransferases[c]			
Gcn5	2	H3K9, H3K14, H3K36	SAGA, Gcn5/Ada2/Ada3, ATAC, TFTC
p/CAF	2B	H3K14	STAGA
MYST family of histone acetyltransferases			
Esa1 (Tip60 in *H. sapiens*)	5	H4K5, H4K8, H4K12, H4K16; Htz1K14	NuA4, Piccolo NuA4
Sas2 (MOF in *H. sapiens*)	8	H4K16	SAS-I
Sas3 (MOZ in *H. sapiens*)	6	H3K14, H3K23	NuA3
p300 family of histone acetyltransferases			
CBP	3A	H2AK5; H2B	Numerous
p300	3B	H2AK5; H2B	Numerous
Rtt109	11	H3K56, H3K9, H3K23	Rtt109–Vps75, Rtt109–Asf1

[a] Designations used in this review. [b] K (lysine) acetyltransferase nomenclature proposed by Allis et al. (2007) to simplify the identification and classification of new histone acetyltransferases. [c] GNAT, Gcn5-related N-acetyltransferase; MYST, MOZ, Ybf2/Sas3, Sas2, Tip60.

Although more protein acetyltransferases that are HATs may yet be discovered, it may be that only a limited number of HAT catalytic subunits are able to exist in diverse multi-subunit complexes, because their interactions with other proteins might regulate the catalytic activity of the acetyltransferase subunit and affect the multiple (enzymatic and binding) functions within the protein complex. Such interactions may help explain how HAT activity can be restricted to specific genomic loci despite very few results describing an obvious DNA binding function of most of the HATs known so far. There is some evidence suggesting that the presence of DNA is able to enhance the acetyltransferase activity of the MOZ enzyme. The possibility arises because this HAT has a zinc-finger domain that may be able to bind DNA, but how this might relate to sequence-specific DNA binding is not clear.

Histone deacetylase enzymes remove acetyl groups from histone lysine residues

Histone acetylation is a highly dynamic modification, and successful application of this in relation to other post-translational modifications relies on the cell's being able to remove the acetyl groups as rapidly as they were added. This activity is the preserve of the histone deacetylase (HDAC) group of enzymes, of which many members have been described. Phylogenetic analyses have led to the subdivision of the HDACs into four families (Box 7.2). Class I HDACs are homologous to yeast Rpd3 and include human HDACs 1–3 and 8; class II HDACs are homologous to yeast Hda1 and are further subdivided into classes IIa (including human HDACs 4, 5, 7, and 9) and IIb (HDACs 6 and 10) on the basis of domain organization and sequence homology. Class III HDACs are named after yeast Sir2 and are also called sirtuins; included among them are yeast Hst proteins 1–4 and human sirtuins 1–7. Class IV HDACs are related to human HDAC11. Class I, II, and IV HDACs share sequence and structural homology within their catalytic domains, and they share a related catalytic mechanism that does not require a cofactor but does require a zinc ion. In contrast, the sirtuins do not share sequence or structural homology with the other HDAC families and use a distinct catalytic mechanism that is dependent on the oxidized form of nicotinamide adenine dinucleotide (NAD^+) as a cofactor.

7.2 ENZYMES THAT METHYLATE OR DEMETHYLATE HISTONES

Methylation of specific lysines establishes a signal for gene activation or repression depending on the precise position of the lysine in the histone protein. We discussed these factors in Chapter 6, but histone methylation is thought to represent a more stable imposition of epigenetic information on the genome than acetylation does.

The histone methyltransferases add methyl groups to histone residues

There are many enzymes in eukaryotes that are capable of methylating a diverse range of amino acid positions on an equally large number of proteins. In this section we concentrate on the histone lysine methyltransferases and histone arginyl methyltransferases because they have the greatest impact on epigenetic control of gene transcription. However, it is important to acknowledge that these enzymes are really examples of a much broader group dedicated to protein methylation in general. The type of enzyme required depends largely on the position and identity of the amino acid to be methylated. This substrate specificity is determined

Box 7.2 Histone deacetylases (HDACs): classes, structures, and mechanisms of action

Class I/II HDACs

Structure determination for the isolated HDAC1 enzyme shows that it has a conserved deacetylase core domain consisting of an open α/β class of folds and a tubular active-site pocket containing the inhibitor and a bound Zn^{2+} ion at the base as shown in **Figure 1**.

The structural features present in the active site of HDLP suggest a mode of catalysis that has features of both metalloproteases and serine proteases in which the bound Zn atom mediates the nucleophilic attack of a water molecule on the acetylated lysine substrate, resulting in a tetrahedral oxyanion intermediate. The carbon–nitrogen bond of the intermediate is then broken, with the nitrogen of the scissile bond primed to accept a proton from an Asp-His charge relay, resulting in the formation of the acetate and lysine products as seen above. The presence of the Zn^{2+} ion is an essential structural feature for class I/II HDACs, although this is common to many enzyme systems. Interestingly, some members of the class have second or third monovalent metal-binding sites away from the active site. The high degree of conservation of the monovalent ligands among the

HDAC proteins suggest that, in addition to Zn^{2+} association, association with monovalent ions may be a conserved feature of class I HDAC proteins.

Class III HDACs

A key feature of the preceding two HDAC classes is their requirement for a Zn^{2+} ion, and this feature can be used to distinguish them from class III HDACs, which although they may possess Zn^{2+} ion-binding sites, this does not seem to be an absolute requirement for their activity. Class III HDACs, also known as sirtuins, do not share sequence or structural homology with the other HDAC families. The catalytic mechanism of class III is rather different because it depends on the presence of the oxidized form of nicotinamide adenine dinucleotide (NAD^+), and acetyl-lysine histone substrates are converted to nicotinamide, deacetylated histone lysine, and a metabolite 2′-O-acetyladenosine diphosphate ribose. It is established that NAD^+ and acetylated histones go into one side of the reaction and the listed products come out of the other, but the exact mechanism by which this is achieved is still a matter of debate. The following aspects of the sirtuin

(a)

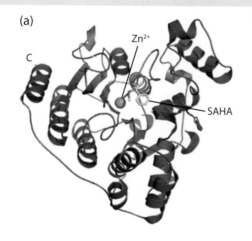

(b)

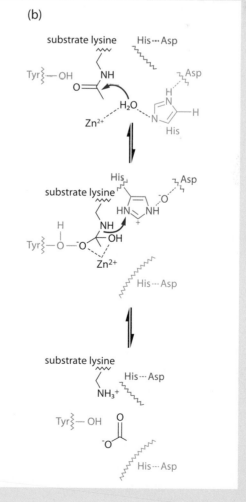

Figure 1 Reaction mechanism for class I and II HDACs. (a) Structure and (b) reaction mechanism of HDAC1. C, C-terminus. SAHA represents Suberoylanilide hydroxamic acid (an HDAC inhibitor). [a, from Hodawadekar SC & Marmorstein R (2007) *Oncogene* 26, 5528. With permission from Macmillan Publishers Ltd.]

Continued

Box 7.2 *(continued)*

reaction are generally accepted (see **Figure 2** for a diagram of the reaction mechanism). First, for the deacetylation reaction, the binding of acetyl-lysine is required for nicotinamide hydrolysis, and the reaction proceeds through the formation of an *O*-alkylamidate intermediate, involving attack of the acetyl-lysine acetyl group on carbon 1 of the ribose ring. It is also well accepted that the nicotinamide inhibition reaction occurs by base exchange of nicotinamide with an intermediate of the deacetylation reaction to re-form β-NAD$^+$ through a transglycosidation reaction, but it is not clear which face of the nicotinamide ribose ring is approached by the acetyl group. Approach from different faces requires very different nucleophilic substitution mechanisms. One option is bimolecular nucleophilic substitution (S$_N$2-like reaction) in which nicotinamide hydrolysis from the β-face of the C1 position of the nicotinamide ribose ring is mechanistically linked to nucleophilic attack on the α-face of the same carbon by the acetyl group, but it is also possible that nicotinamide hydrolysis happens first, followed by rotation of the ribose ring to allow the acetyl group to attack the α-face of carbon 1 (monomolecular nucleophilic substitution, or S$_N$1-like reaction).

Class IV

Most histone deacetylases fall into the preceding three classes. However, phylogenetic studies have indicated the existence of a fourth class of enzymes that are unrelated to the NAD$^+$-dependent class III enzymes and show significant coding sequence variations from the class I/II histone deacetylases. The only well-characterized member of this class is HDAC11, and although bioinformatic analyses have predicted the existence of other members on the basis of DNA sequences present in the human genome, the actual proteins have yet to be isolated.

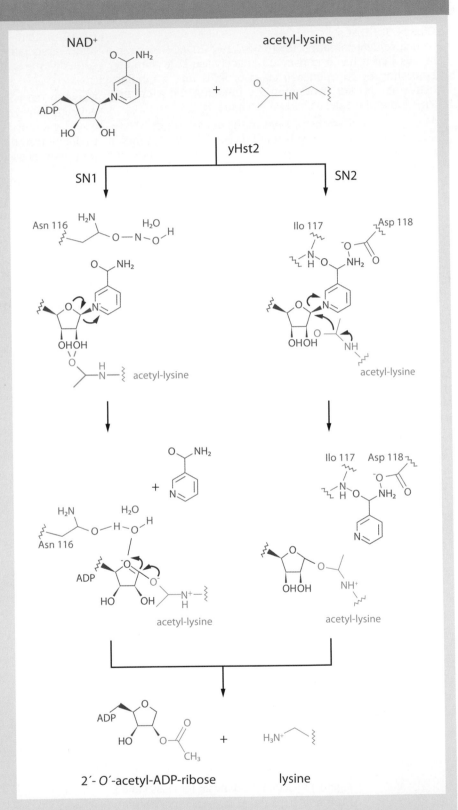

Figure 2 Reaction mechanism for class III HDACs. [Adapted from Hodawadekar SC & Marmorstein R (2007) *Oncogene* 26, 5528. With permission from Macmillan Publishers Ltd.]

by the chemical process needed to react with either lysine or arginine. We will look at the methylation of lysine first, largely because it has been very well studied but also because it provides us with an opportunity to examine a specific catalytic domain whose activities extend into functions well beyond those of histone methylation.

The SET domain

The primary catalytic subunit responsible for histone lysine methylation is the SET domain (Box 7.3). It is named for several *Drosophila melanogaster* phenotypes: Su(Var), Enhancer of Zeste, and Trithorax. The SET domain is universally present in eukaryotic cells and a few prokaryotes. SET is the active component of a wide range of protein methyltransferases; it seems to have arisen relatively late in evolution. Although most experimentally characterized enzymes containing SET are involved in methylating specific lysines of histones H3 and H4, others are known to be involved in methylating nonchromosomal proteins such as cytochrome *c*. Methylation occurs by transfer of methyl groups from the cellular pool of *S*-adenosylmethionine, the same donor as for DNA methylation. The SET domain represents a complex polypeptide fold that is entirely unrelated to previously characterized *S*-adenosyl-L-methionine (AdoMet)-dependent methyltransferases containing the Rossmann fold or SPOUT domains. This observation, combined with the largely eukaryote-specific expression of SET, suggests the relatively "late" evolutionary arrival of the SET domain.

SET 7/9

The SET7/9 enzyme is able to monomethylate lysine 4 of histone H3 (H3K4) but does not seem to catalyze the addition of any more methyl groups to the same ε-amino group of the target lysine. Other SET-domain-containing enzymes are able to do this, so it is likely that substrate specificity in the SET histone methyltransferases is achieved via structural alterations of the fold that directs the histone N-terminal tail substrate into the active site (see Box 7.3). One would expect that a good alignment of the methyl group of AdoMet with the lone pair of electrons on the ε-amino group would lead to an efficient methyl transfer, whereas a bad alignment would make the methyl transfer less likely. This structure–reactivity relationship provides an explanation of why SET7/9 is a monomethyltransferase. Indeed, the crystal structure of SET7/9 complexed with *S*-adenosyl-L-homocysteine (AdoHcy) and a histone H3K4me peptide showed that the K4me side chain in this product complex of monomethylation adopts a conformation that might prevent the lone pair of electrons from further reaction with a second AdoMet molecule. Mutations of the amino acids at critical positions can alter this specificity. It has been observed that the Tyr 3053→Phe mutation in SET7/9 not only led to an increase in the enzyme's monomethylation activity but also changed SET7/9 into a dimethyltransferase.

Another possible mechanism for reducing the possibility of dimethylation arises not from steric hindrance but from the disruption of the formation of the near-attack reactive conformation. As a result of the dynamic nature of the protein structure in solution, the binding pocket of SET7/9 can be enlarged to accommodate the additional methyl group in its active site during the dimethylation reaction, and the binding channel of AdoMet can also open up slightly. The consequence is that solvent water molecules are allowed to penetrate into the channel and form a water chain from the bottom of the channel (lysine nitrogen) to the surface. The nearest water molecule in the active site directly forms a hydrogen bond with the lone-pair electrons of substrate lysine nitrogen, which is very unfavorable for the methylation reaction. During the reaction process,

Box 7.3 The SET domain is the basis of histone methyltransferase activity

Figure 1 Structure of the SET domain is conserved across a range of proteins. [From Yeates TO (2002) *Cell* 111, 5. With permission from Elsevier.]

The structure of SET contains three β-pleated polypeptide sheets (**Figure 1**), and the active site and region for binding the *S*-adenosylmethionine (AdoMet) methyl donor are formed at the junction between the two barrels; not surprisingly, these are regions of very high sequence conservation across species.

The AdoMet-binding pocket, which is 0.95 nm × 1.30 nm in width and 0.85 nm deep, is formed by the interaction of loops and strands from the SET domain and C-terminal polypeptide domain (**Figure 2**). The bottom of the pocket comprises a region of β-pleated sheet, and the overall effect of this structure is to bind AdoMet so that nitrogen 6 of the adenine base moiety faces inward, whereas the hydroxyl group of adenosine points outward into the solvent and the methyl group points toward a narrow channel that is formed by a group of tightly conserved tyrosine residues. This seems to generate a highly polar environment due to the presence of hydroxyl and carbonyl groups whose function may be to favor the access of non-methylated lysine substrates, but more importantly it sterically constrains the target lysine to a position opposite the donor methyl group. In view of this there is good reason to believe that this region is the active site of the SET domain and that the probable reaction mechanism involves an S_N2 reaction in which the substrate lysine, a methyl group, and the leaving thioester group of AdoMet form a linear arrangement from which the unprotonated lysine is able to make a nucleophilic attack on the AdoMet methyl group.

The basic reaction is shown in **Figure 3** and is derived from data obtained from studies of the SET7/9 histone methyltransferase.

(a)

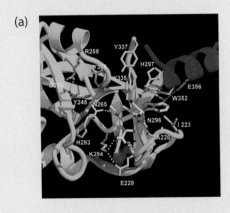

(b)

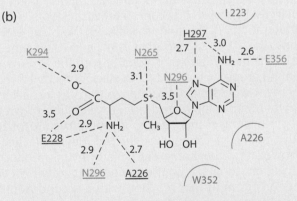

Figure 2 The AdoMet binding pocket for SET7/9.
(a) This structure represents the AdoMet-binding region near the C-terminus of the enzyme. (b) The structure of the AdoMet molecule is shown in relation to specific amino acids of the SET7/9 active site. [a, from Kwon T, Chang JH, Kwak E et al. (2003) *EMBO J.* 22, 292. With permission from Macmillan Publishers Ltd.]

Figure 3 Reaction mechanism for the methylation of histone H3 lysine 4 by SET7/9. [From Guo H-B & Guo H (2007) *Proc. Natl. Acad. Sci. USA* 104, 8797. With permission from the National Academy of Sciences.]

the lone-pair electrons of substrate lysine nitrogen needs to be desolvated and its hydrogen bond network with water molecules needs to be reorganized, which leads to a higher activation barrier for the dimethylation reaction in SET7/9.

EZH2

SET domains are the functional unit of many histone methyltransferases. **Table 7.1** shows the principal histone methyltransferases known to use the SET domain as their catalytic unit. Many of these methylate different target lysines on histone N-terminal tails that may carry other post-translational modifications. Additionally, many of these enzymes function as part of larger multi-subunit protein assemblies, such as the polycomb repressive complex (PRC). A good example of this is the histone methyltransferase EZH2 (short for Enhancer of Zeste 2), which uses its SET domain to give rise to dimethylated and trimethylated versions of lysine 27 within histone H3 (H3K27me2/3). This specificity is clearly different from the H3K4 monomethylation performed by SET7/9, but the basic methylation mechanism is the same despite the considerable differences in sequence and structure between the two enzymes.

EZH2 requires subunits that will allow it to recognize and bind not only to existing H3K27 methylated lysines (because this helps it to propagate the spread of this mark across regions of chromatin) but also to the other components—namely SUZ12 and EED—of PRC2 within which its functions are performed. (PCR2 is a complex consisting of three core proteins: EZH2, SUZ12, and EED.) A sequence between residues 400 and 510 in EZH2 is required for binding of the complex to the H3K27me3 mark, but the C-terminus (residues 1–400) seems to be required for binding to SUZ12 (short for Suppressor of Zeste 12), a core protein of the PRC2 complex that enhances the binding of EZH2 to H3K27me3 but is not strictly necessary for binding to occur. Indeed, EZH2 will only bind this histone modification efficiently when it is bound to two PRC2 components, SUZ12 and EED (Embryonic Ectoderm Development). This enhancement in binding is completely independent of the SET domain of EZH2, which is still able to methylate H3K27, albeit weakly in the absence of the other proteins.

Human SET domain proteins

The first human histone methyltransferase identified was the human SET domain protein SUV39H1, which methylates Lys 9 of histone H3 in all three of its possible states (monomethylation, dimethylation, and trimethylation). The enzyme is a member of a larger group of proteins that are thought to have originated at about the same time as the SET/ESET group of methyltransferases (ESET stands for ERG-associated protein with SET domain, an H3-specific methyltransferase) during evolution. The *SUV39H1* gene was characterized in *Drosophila* long before the recognition of the protein's function in human cells. Genetic screens

TABLE 7.1 SET DOMAIN-CONTAINING HISTONE METHYLTRANSFERASES

Enzyme	Target lysine
G9A	H3K9 (mono and dimethyl)
SUV39H1	H3K9 (trimethyl)
SUV39H2	H3K9 (trimethyl)
GLP	H3K9 (mono and dimethyl)
EZH1	H3K27 (trimethyl)
EZH2	H3K27 (trimethyl)
SUZ12	H3K27 (trimethyl)
MLL	H3K4 (trimethyl)

for suppressors of position effect variegation (PEV) in *Drosophila* and *Schizosaccharomyces pombe* have identified a subfamily of about 30–40 loci, which are referred to as Su(Var)-group genes (standing for suppression of variegation). The Su(Var) group extends over many more proteins than just the histone methyltransferases, and several histone deacetylases, protein phosphatase type 1, and *S*-adenosylmethionine synthetase 16 have been classified as belonging to the Su(Var) group. More specifically, *Su(Var)2–5* (which is allelic to *HP1*) and *Su(Var)3–7* encode heterochromatin-associated proteins, as does *Su(Var)3–9*, which is the *Drosophila* homologue of human *SUV39H1*. This enzyme possesses a chromodomain similar to that of heterochromatin protein 1 (HP1), which we introduced in Section 6.1; this chromodomain is essential for its ability to bind trimethylated H3K9.

Another example of a SET-dependent histone methyltransferase is G9a, an H3K9-methylating enzyme, although it is probably associated more with facultative heterochromatin than with constitutive heterochromatin, a notion supported by its apparent inability to trimethylate H3K9 (addition of this third methyl group is generally only found in constitutive heterochromatin). G9a is expressed ubiquitously and was originally mapped to the class III region of the major histocompatibility complex locus. Mouse G9a is 1263 amino acid residues long, and although the amino half of G9a has very little similarity to any conserved protein domain, the carboxy half has homology with ANK (ankyrin) repeats, and it contains the preSET and SET domains. A second distinguishing feature of G9a is the presence of several ANK repeats implicated in protein–protein interactions in very diverse protein families. Despite this, there is still substantial sequence similarity between G9a and SUV39H1.

MLL-family proteins

One of the chief effectors of H3K4 methylation is the **mixed-lineage leukemia 1 (MLL1)** enzyme, a protein that is known also for being a master regulator of *Hox* and other genes. Core components of the human MLL histone methyltransferase complex were found to form a structural platform, with one component (WDR5) mediating association between the specific histone H3K4 substrate and the methyltransferase complex. This novel regulatory mechanism, which is conserved from yeast to humans, is required for both methylation and downstream target gene transcription.

There are four proteins in the MLL family in humans. Of these, MLL1, which is homologous to the *Drosophila* protein Trithorax, has been the most intensively studied because of its involvement in a variety of acute lymphoid and myeloid leukemias when its gene has been disrupted by chromosomal translocations. Several years ago, the bacterially expressed MLL SET domain was shown to have modest methyltransferase activity itself. However, like many other SET-dependent methyltransferases, MLL1 exists as part of a complex with enhanced methyltransferase activity; this complex includes several other proteins, such as a HAT and the core components WDR5, RbBP5, and Ash2L (**Figure 7.1**).

Interestingly, WDR5, RbBP5, and Ash2L form an independent complex in the absence of an MLL protein, indicating that together they form a structural platform for association with the different Set1 or MLL-family H3K4 methyltransferases. This association is more than just a structural alignment of the proteins. *In vitro* histone methyltransferase (HMT) assays revealed that the absence of RbBP5 or Ash2L from the complex significantly decreases H3K4 trimethylation. In contrast, the absence of WDR5 completely abolishes methylation activity. The central role of WDR5 seems to be linked to its ability to specifically recognize N-terminal polypeptide

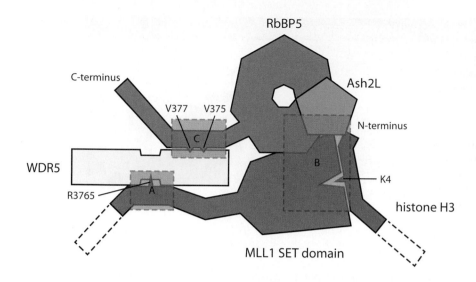

RbBP5

Ash2L

C-terminus

V377 V375

C

N-terminus

B

WDR5

R3765

A

K4

histone H3

MLL1 SET domain

Figure 7.1 A schematic model of the active complex between WDR5, RbBP5, and the MLL1 SET domain. Interactions between the methyltransferase and the WDR5 and RbBP5 domains are summarized. Residues that are critical for the interactions and the target histone lysine have been indicated with arrows. A, represents the site of interaction between MLL1 SET domain and WDR5; B, represents the interaction site between RbBP5, Ash2L, MLL SET domain, and the target Lysine 4 on histone H3; C, represents the interaction between the C-terminus of RbBP5 and WDR5.

chains of H3 and bind them in a conformation that exposes them for methylation. It does so via interaction of the histone's N-terminal tail with seven WD40 repeats—a protein motif comprising seven β-pleated sheets of around 40 amino acid residues each (see Figure 6.7)—that form a depression in the structure of WDR5.

Non-SET-dependent methyltransferases

Placing the methyl mark on histones is not always performed by a SET-domain-containing protein, and one of the best-characterized examples of this is DOT1L, which normally methylates Lys 79 on H3 (**Box 7.4**). Lys 79 occupies a position in the ordered core domain of histone H3 and resides in a short turn connecting the first and second helices of the conserved histone fold. Both H3 Lys 79 residues in the histone octamer are exposed and do not interact with DNA or other histones, so it is perhaps not surprising that a different type of enzyme is needed to methylate a structure that is very different from the N-terminal polypeptide tail.

Dot1 is an evolutionarily conserved histone methyltransferase that was originally identified in *Saccharomyces cerevisiae* as a disruptor of telomeric silencing. Sequence analysis revealed that it possesses characteristic AdoMet-binding motifs, similar to those in protein arginine methyltransferases. However, both yeast Dot1 and its human counterpart hDOT1L show intrinsic histone methyltransferase activity toward Lys 79 of histone H3 in the nucleosome core.

The histone arginyl methyltransferases

Protein arginine methyltransferases (PRMTs) are AdoMet-dependent enzymes that catalyze the monomethylation and dimethylation of peptidyl arginine residues. They are classified into four classes, I–IV. Class I, II, and III enzymes methylate the terminal guanidino nitrogens of arginine, whereas class IV enzymes catalyze the monomethylation of the internal (or δ) guanidino nitrogen. Class I PRMTs are the most numerous and have a common seven-stranded β-pleated sheet structure. Class II enzymes are basically SET-domain-containing lysine methyltransferases that are also capable of methylating arginine; class III comprises membrane-bound PRMTs. Most arginine methylation is performed by class I and II enzymes, so most of our discussion will concern enzymes of these types. They act on a wide range of cellular proteins. Although there are several enzymes capable of performing this reaction, the predominant arginine methyltransferase in human cells is PRMT1, which is responsible for approximately 85% of all arginine methylations.

Box 7.4 DOT1—an atypical histone methyltransferase

The full-length hDOT1L consists of 1537 amino acid residues, but only the 360 amino acids at the N-terminal share significant sequence similarity with yeast Dot1. Residues 1–416 contain the active HMTase catalytic domain, whose core has an elongated structure of approximately 8 nm × 4 nm × 4 nm (**Figure 1**). The structure comprises an N-terminal domain with five α-helices, two pairs of short β-strand hairpins, and a C-terminal domain with a seven-strand central β-pleated sheet and five α-helices. The binding site for *S*-adenosylmethionine is formed between the polypeptide loop linking the N-terminal and C-terminal domains, where it interacts with specific amino acid residues, and lysine 79 of H3 can access the site via a channel formed in the D1 region.

The structure of the channel is shown in more detail in **Figure 2** and the relative position of the *S*-adenosylmethionine group is ideal for an "in-line" methyl transfer reaction probably resulting from nucleophilic attack by a deprotonated ε-amino group of lysine in a similar manner to that adopted by the SET domain. The key difference is that the putative lysine-binding channel and the *S*-adenosylmethionine-binding domain are separated by a greater distance in the DOT1L structure, but this may well be an essential feature because the enzyme must achieve a closer interaction with the nucleosome surface than the SET-containing enzymes that act on histone N-terminal tails. The enzyme interacts with all four histone residues on the same side of the nucleosome surface, and the area surrounding lysine 79 makes contact in all of these models. In addition, some regions of the C-terminal region of DOT1L make contact with the DNA wrapped around the nucleosome, but perhaps more importantly the C-terminus is also positioned close to a C-terminal α-helix component of histone H2B. The presence of ubiquitin on lysine 123 of H2B seems to be a prerequisite for the methylation of H3 lysine 79, and the molecular models suggested above indicate that the catalytic core of DOT1L may be able to interact with the ubiquitin group covalently attached to the C-terminal helix of H2B. This observation is consistent with the proposal that ubiquitination of histone H2B may serve as a spacer between adjacent nucleosomes to allow access of histone modification enzymes such as DOT1L.

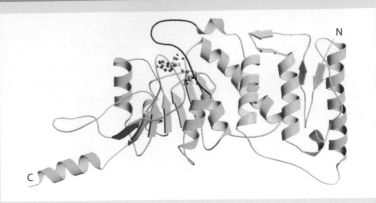

Figure 1 Structure of DOT1L. The N-terminal region (residues 1–126) is colored yellow, the open structure (residues 141–332) is shown in cyan, and the loop L–EF connecting the two regions is shown in dark red. The bound *S*-adenosyl methionine molecule is shown in a ball-and-stick model (gray, carbon; blue, nitrogen; red, oxygen; yellow, sulfur). [From Min J, Feng Q, Li Z et al. (2003) *Cell* 112, 711. With permission from Elsevier.]

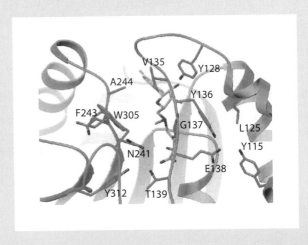

Figure 2 Interaction of polypeptides in DOT1 to form the HhK79 specific access channel. Lysine 79 is shown as the orange and red stick structure. [From Min J, Feng Q, Li Z et al. (2003) *Cell* 112, 711. With permission from Elsevier.]

PRMT1 is an essential enzyme during development. The evidence for this comes from experiments showing that mouse embryos lacking the *Prmt1* gene die shortly after implantation. However, the enzyme seems to be dispensable for many cellular processes, because embryonic stem cells from such *Prmt1*-knockout mice are viable in culture.

The basic methylation reactions of PRMTs are shown in **Figure 7.2**. Individual PRMT isoenzymes use AdoMet as the methyl donor and catalyze the transfer of a methyl group to the ω-nitrogen of a peptidyl arginine residue, yielding AdoHcy and ω-N^G-monomethylarginine (MMA). Specific isoenzymes can also catalyze the transfer of a second methyl group from AdoMet either to the same nitrogen atom or to the other ω-nitrogen to generate ω-N^G,N^G-asymmetrically dimethylated arginine (ADMA) and ω-N^G,$N^{G'}$-symmetrically dimethylated arginine (SDMA), respectively.

Because of this difference in methylation specificity, peptidyl-ADMA-generating PRMTs are classified as class I enzymes (PRMTs 1, 3, 4, 6, and 8), whereas those isoenzymes that catalyze the formation of peptidyl-SDMA are classified as class II (PRMTs 5 and 7) enzymes. Note that no methyltransferase activity has been observed for PRMTs 2 and 10; thus these enzymes cannot be classified. Also note that two other PRMTs have been identified: PRMTs 9 and 11. However, these isoenzymes may not represent genuine PRMTs, consistent with the fact that catalytic domains within these enzymes lack significant homology to the catalytic core domain present in other PRMT family members. PRMT1 and PRMT4/CARM1 (coactivator-associated arginine methyltransferase 1) are the best-characterized PRMTs, and these enzymes methylate histones H3 at Arg 2, Arg 17, and Arg 26 (by PRMT4/CARM1) and H4 at Arg 3 (by PRMT1).

Although class I PRMTs are known to catalyze the monomethylation and asymmetric dimethylation of arginine residues, it is unclear whether these two events are coupled; in other words, does the enzyme (1) release the monomethylated species before rebinding it to facilitate the second methylation event (a distributive mechanism), or (2) bind to its substrates in such a way that monomethylation and dimethylation occur sequentially, without the release of the monomethylated species (a processive mechanism)? If the first situation predominates, one would expect the monomethylated species to possess a significantly higher affinity for the enzyme; in contrast, if the second situation predominates, one would expect an obligate order of product release that results solely in the production of the dimethylated species.

Figure 7.2 Reactions catalyzed by PRMT isoenzymes. All PRMT isoenzymes catalyze the formation of monomethylarginine (MMA) and *S*-adenosyl-L-homocysteine (AdoHcy). Type I isoenzymes go on to generate asymmetric dimethylarginine (ADMA) after the second round of methylation (by reaction with *S*-adenosylmethionine, AdoMet), whereas type II isoenzymes produce symmetric dimethylarginine (SDMA). [Adapted from Smith BC & Denu JM (2009) *Biochim. Biophys. Acta* 1789, 45. With permission from Elsevier.]

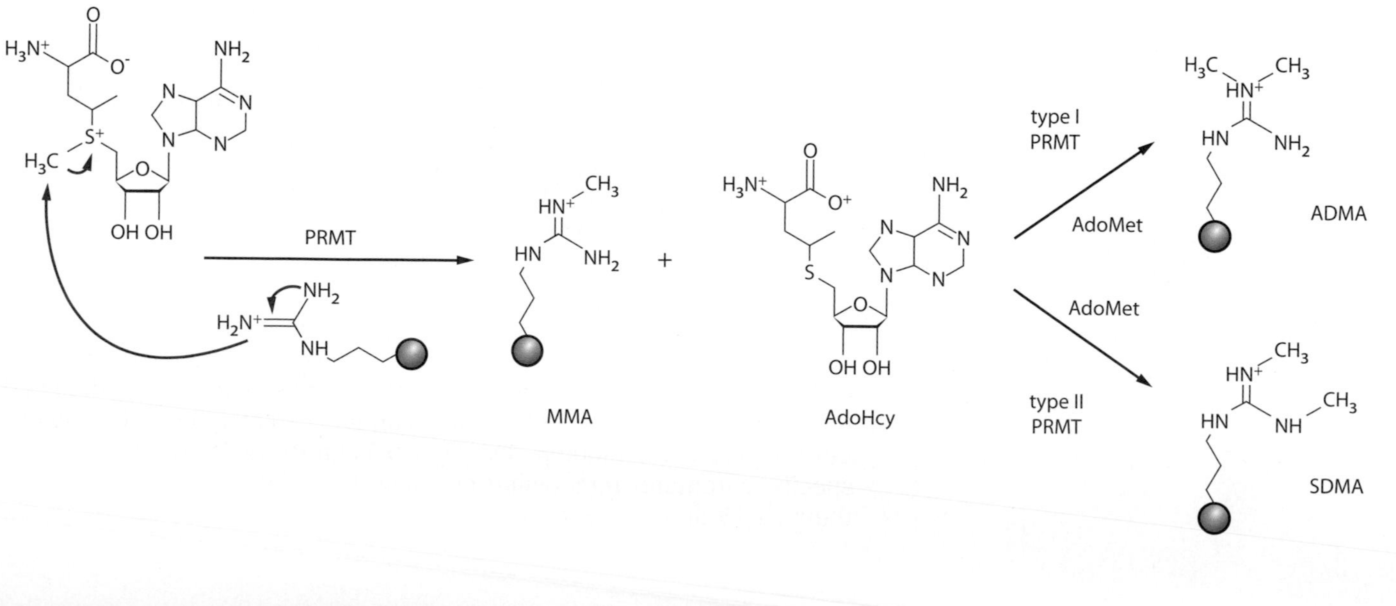

Once more, studies with synthetic peptides have helped to shed light on the nature of the catalytic mechanism of PRMT1. They show that peptides containing monomethylated arginines have lower catalytic constants than the unmethylated analogs, which is inconsistent with the idea that monomethylated substrates possess an intrinsically higher affinity for PRMT1. In addition, mass spectroscopic studies of the PRMT1-catalyzed reaction show that roughly equal amounts of monomethylated and dimethylated arginines are produced even if there is a large excess of the unmodified substrate present. Together, these results are most consistent with a partly processive mechanism in which the release of AdoHcy and the rebinding of AdoMet occur on a timescale similar to that of the release of monomethylarginine-containing products. Although similar results have been observed for PRMT6 and interpreted as being consistent with a processive mechanism, the fact that monomethylated arginine and dimethylated arginine are produced in similar quantities rules out such a purely processive mechanism for PRMT1.

Histone methylation is reversible using histone demethylases

Lysine-specific demethylase 1

Until recently, it was unclear whether methylation of lysine residues was a permanent post-translational histone modification. Although it was clear that cells need to be able to remove methylated histones, it was uncertain whether this occurred via an enzymatic mechanism that removed the methyl groups themselves or simply replaced the whole histone molecule with a new, unmethylated one. In 1973, Paik and Kim were the first to observe the demethylation of calf thymus histones, but they were unable to isolate the enzyme that might have been responsible for this activity. It took another three decades for the first histone demethylase to be isolated and characterized. In 2004, a lysine-specific demethylase (LSD1) was shown to demethylate monomethylated and dimethylated Lys 4 of histone H3, with the concomitant formation of formaldehyde.

LSD1 belongs to the amine oxidase superfamily; it oxidatively demethylates lysine residues by a mechanism that is dependent on the cofactor flavin adenine dinucleotide (FAD) (**Box 7.5**). The FAD cofactor is essential to the demethylation reaction because its heterocyclic rings (the flavin component) can act as an electron acceptor in the type of oxidation reaction performed by monoamine oxidases.

As the first step of the process, FAD undergoes oxidation by molecular oxygen and produces stoichiometric hydrogen peroxide along with formaldehyde as the demethylation by-products (see Box 7.5). Because both of these products are potentially very harmful to the cell (hydrogen peroxide can damage both DNA and proteins, and formaldehyde is a known mutagen), there must be a strong evolutionary pressure to retain this demethylation system, particularly as it operates in the nucleus—the perfect site to induce DNA damage. Nevertheless, control of the demethylation system must be very tight.

The structure of LSD1 is very different from that of the histone methyltransferases. This at first glance might seem to be a rather obvious requirement were it not for the documented reversibility of many enzyme-catalyzed reactions. An important feature in the LSD1-mediated demethylation process is the mechanism for substrate recognition. LSD1 requires at least the first 20 N-terminal amino acid residues of the histone tail for productive binding. This is a peculiarity of LSD1 because site-specific chromatin remodeling enzymes typically do not require more than 10–15 amino acids for efficient substrate binding. Such a specific

Box 7.5 The mechanism of action of the flavin adenine dinucleotide-dependent monoamine oxidases

Flavin adenine dinucleotide (FAD) (**Figure 1**) undergoes oxidation by molecular oxygen as the first step of the process producing stoichiometric hydrogen peroxide (H_2O_2) along with formaldehyde as the demethylation by-product (**Figure 2**). Because both of these products are potentially very harmful to the cell (hydrogen peroxide can damage both DNA and proteins, and formaldehyde is a known mutagen), there must be a strong evolutionary pressure to retain this demethylation system particularly because it operates in the nucleus—the perfect site to induce DNA damage. Nevertheless, control of the demethylation system must be very tight. The FAD-dependent demethylation of dimethylated Lys 4 of histone H3 proceeds through the hydrolysis of an iminium ion after a two-electron oxidation of the amine by FAD (see Figure 2).

The sequence of amino acids that form the histone N-terminal tail binding site of LSD1 induces conformational changes that fold and position H3K4 in proximity to FAD, in the appropriate position to undergo flavin-dependent oxidative demethylation (**Figure 3**). Biochemical evidence indicates that substrate recognition by LSD1 requires a peptide comprising at least the first 20 N-terminal residues of H3—that is, peptides of shorter length bind poorly to the enzyme and are not demethylated.

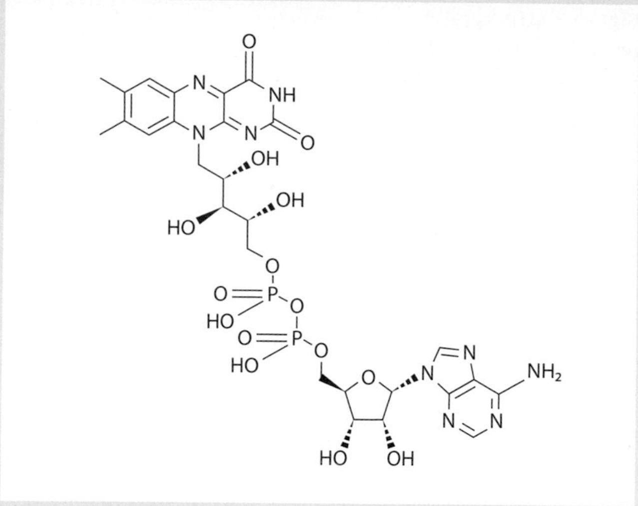

Figure 1 Structure of flavin adenine dinucleotide.

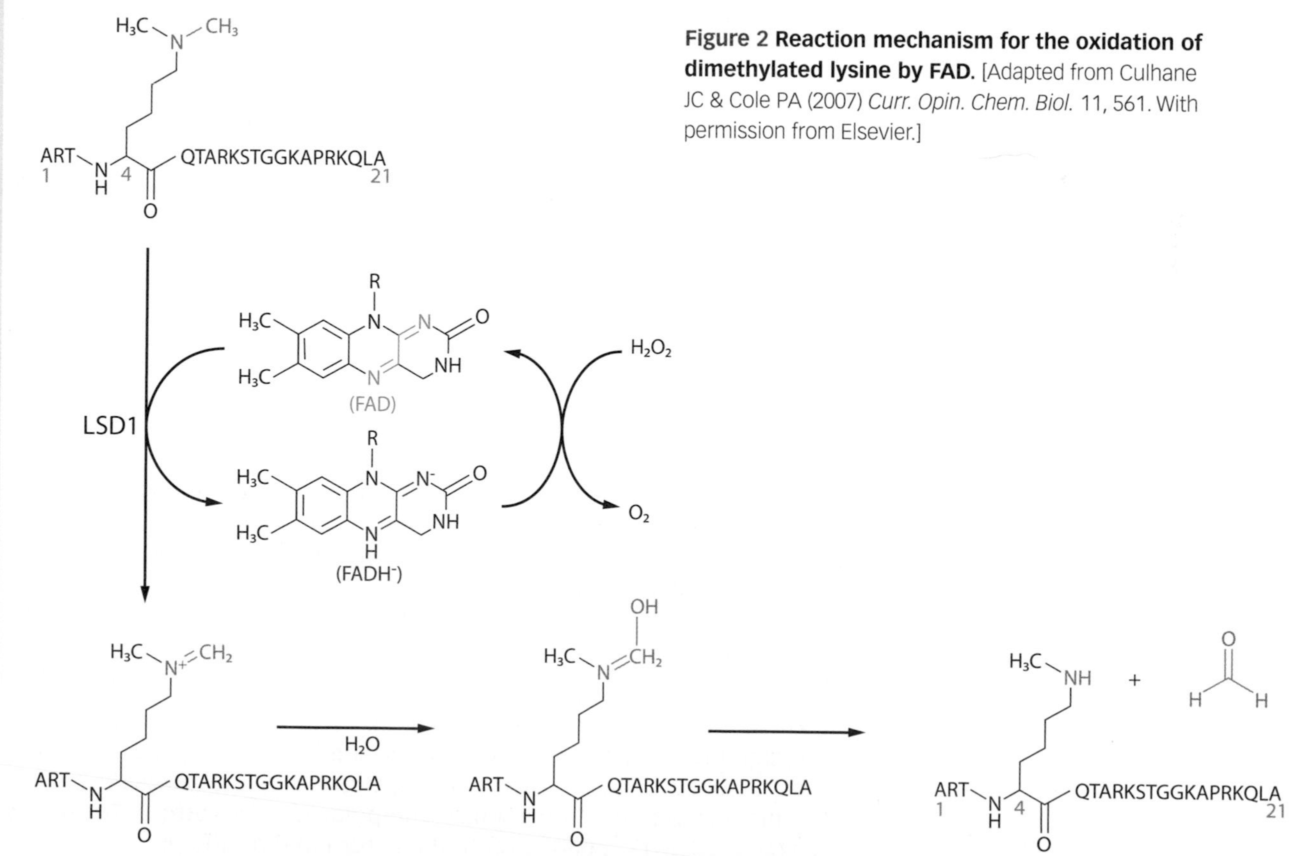

Figure 2 Reaction mechanism for the oxidation of dimethylated lysine by FAD. [Adapted from Culhane JC & Cole PA (2007) *Curr. Opin. Chem. Biol.* 11, 561. With permission from Elsevier.]

Continued

Box 7.5 *(continued)*

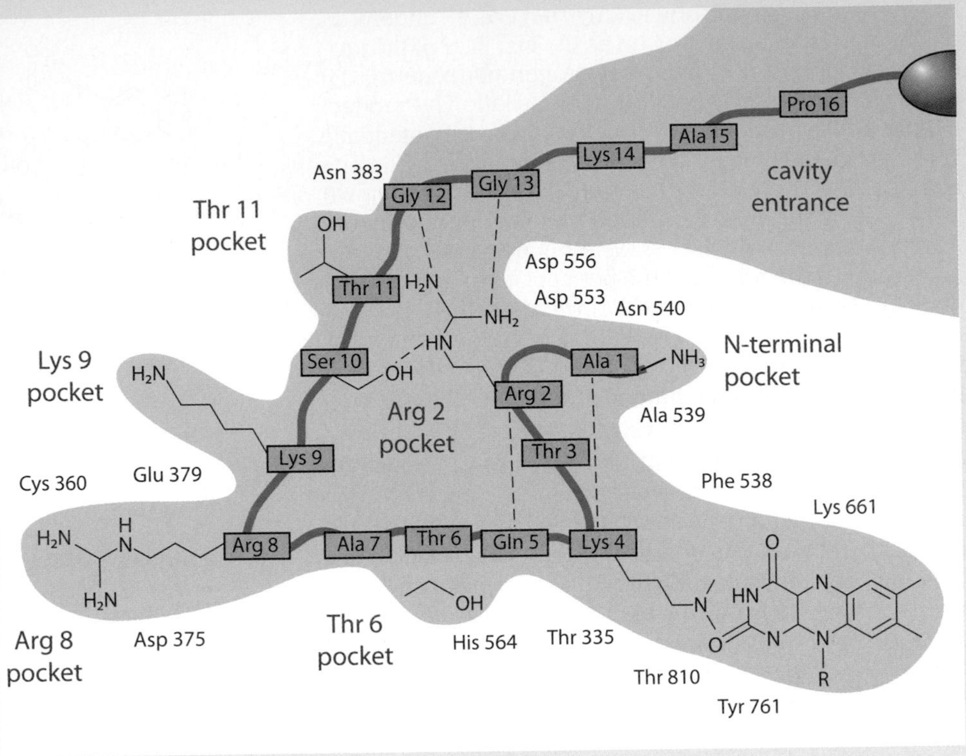

Figure 3 Structural deformation of the histone N-terminal tail by LSD1 to place lysine 4 in the correct position for demethylation. [Adapted from Forneris F, Binda C, Battaglioli E & Mattevi A (2008) *Trends Biochem. Sci.* 33, 181. With permission from Elsevier.]

recognition mechanism enables LSD1 to sense the epigenetic message encoded by the histone tail. Indeed, the presence of other epigenetic markers on H3 affects LSD1, decreasing its catalytic activity and, in some cases (such as lysine hyperacetylation or Ser 10 phosphorylation), completely hampering the reaction. This implies that the true LSD1 substrate is the H3 tail stripped of all other epigenetic modifications.

Looking at the arrangement of domains in this protein, we see that LSD1 is an asymmetric molecule consisting of three distinct structural domains. Two of the domains, the N-terminal SWIRM domain and the C-terminal FAD-binding amine oxidase domain, are closely packed together to form a globular core structure (**Figure 7.3**).

We encountered the equivalent of SWIRM domains in Section 4.1. The SWIRM domain is one of the functional units of the SWI/SNF family of chromatin remodeling complexes. This similarity implies a degree of nucleosome recognition and binding ability in the LSD1 enzyme. A third, or "tower," domain protrudes from the globular structure as an extended α-helix, which seems to be needed for the interaction of LSD1 with other members of the protein complex within which the enzyme functions.

Like many histone-modifying enzymes, LSD1 does not act alone. The architecture of the substrate-binding site of LSD1 is characterized by the presence of various niches that accommodate the side chains of the histone peptide. In particular, the polar residues Arg 2, Thr 6, Arg 8, Lys 9, and Thr 11, in addition to the N-terminal amino group, lie in well-defined pockets, thereby establishing specific interactions with the surrounding protein residues. The addition of other epigenetic modifications introduces steric and electrostatic perturbations that alter this network of interactions, thus explaining the negative effect that nearly all epigenetic modifications have on LSD1–H3 binding.

Demethylating trimethylated lysine 4 on H3

The mechanism used by LSD1 is unlikely to demethylate trimethylated H3K4 because trimethylated lysine is an unlikely substrate for a hydride exchange reaction. This is because it is already a positively charged trimethylammonium species that cannot undergo imine formation, as this would violate the valency requirement of the amino nitrogen. Demethylation of trimethylated H3K4 clearly occurs, so alternative mechanisms have been proposed to account for it, and LSD1 is unlikely to be able to catalyze these reaction mechanisms. In contrast, the iron(II) 2-oxoglutarate oxygenases can hydroxylate alkyl groups directly through a hydroxyl radical, providing a chemically plausible path to the demethylation of quaternary ammonium salts (**Box 7.6**). This mechanism seems to be the preferred option for a wide range of histone-modifying enzymes that are characterized by the possession of variants of the so-called Jumonji domain (the Jumonji histone demethylases), in addition to other protein components that mediate their ability to bind to methylated lysines in the context of the histone N-terminal tail. Jumonji histone demethylases are specific to individual methylation states and positions of the methyl-lysines that they can demethylate. For example, on histone H3, the JARID1 class of four enzymes is capable of removing methyl groups from Lys 4 but is inactive against trimethylated Lys 9 (see Box 7.6). This implies that specific amino acids from the Jumonji domains interact very precisely with those of the histone tails.

Demethylating methylated arginine

Neither the FAD-dependent monoamine oxidases nor the Jumonji domain-containing enzymes will remove methyl groups from monomethylarginine or dimethylarginine. Despite the discovery of histone arginine methylation more than a decade ago, corresponding arginine demethylases were unknown until recently. This changed with the discovery of a novel mechanism by which **peptidyl arginine deaminase 4 (PAD4)** was shown to also convert histone tail H3 and H4 methylarginine residues into citrulline in the regulation of estrogen-responsive gene transcription. The subsequent structure determination of the PAD4 enzyme has told us a great deal about how this enzyme might work (**Box 7.7**).

The substrate preference for PAD4 seems to be restricted to monomethylarginine; indeed, modeling studies suggest that, although monomethylarginine can be accommodated, asymmetrically and symmetrically dimethylated arginine cannot because of steric occlusion. PAD4 directs the side chain of the target arginine residue deep into the active-site cleft, and five normally unstructured amino acids of the histone peptide form an ordered β-turn-like conformation. PAD4 does not recognize a defined amino acid sequence, but instead seems to require an unstructured sequence surrounding the target arginine residue, which could account for the broad range of arginine positions it is able to demethylate (H3R2, H3R8, H3R17, H3R26, and H4R3).

Notably, the conversion of histone monomethylated arginines to citrulline by PAD4 does not represent the second half of an on/off switch, unlike the relationship we have observed between HATs and HDACs or histone methyltransferases and their corresponding demethylases. Demethylation by PAD4 poses a problem in that citrulline is a nonstandard amino acid whose structure does not represent a direct substitute for arginine. Citrulline is therefore unlikely to be able to replicate arginine's function within the histone N-terminal tail and would be expected to alter

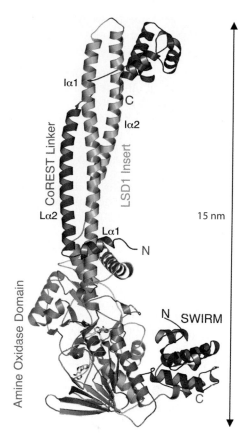

Figure 7.3 Structure of LSD1. CoREST, co-repressor of RE1 silencing transcription factor; C, C-terminus; N, N-terminus. [From Yang M, Culhane JC, Szewczuk LM et al. (2007) *Biochemistry* 46, 8058. With permission from American Chemical Society.]

Box 7.6 The Jumonji domain acts as an iron(II)-dependent 2-oxoglutarate oxygenase that can demethylate trimethylated lysine

The Jumonji domain is another example of a widely used protein module that was first identified in a transcription factor needed for neural development in mouse embryos. Mutants for this gene develop an abnormal neural-plate phenotype that looks a bit like a cross, hence the naming of the gene by its Japanese discoverers (Jumonji means cruciform in Japanese); however, variants of this domain seem to be present in huge numbers of proteins. Sequence analyses show that even the Jumonji gene can be subdivided into several variants, and the gene originally discovered from the neural malformation experiments has become known as Jumonji C or JmjC (Table 1). Even so, this is probably one of the most widely occurring protein building blocks, because the JmjC sequence can be identified in more than 11,000 proteins currently in protein databases such as Uniprot or INTERPRO. The JmjC domain has a very characteristic topology. It forms a double-stranded β-helix (DSBH) fold, also known as a jelly-roll motif, that normally consists of eight anti-parallel β-strands. The DSBH fold in proteins containing a JmjC domain can form an enzymatically active pocket by coordinating Fe(II) and the co-substrate 2-oxoglutarate. In almost all cases, the two-electron oxidation of the "prime" substrate is coupled to the conversion of 2-oxoglutarate into succinate

and CO_2 (**Figure 1**). This class of enzymes is therefore also known as nonheme-Fe(II)-2-oxoglutarate-dependent dioxygenases. Other Jumonji-type domains exist that result from variation within the basic sequence identified from experiments with the mouse neural plate. These alternative domains are frequently combined in the Jumonji-based histone demethylases in humans: about 20 JmjC (usually multidomain) proteins have been identified and grouped into seven distinct subfamilies: JHDM1, JHDM2, JMJD2, PHF2, PHF8, Jumonji (A+T)-rich interactive domain (JARID), ubiquitously transcribed tetratricopeptide repeat X/Y-linked (UTX/UTY), and JmjC domain only.

The problem of removing methyl groups from a quaternary ammonium moiety such as the trimethylated ε-amino group of lysine seems to be largely the preserve of the JARID subdivision, and several examples of enzymes performing such demethylations are listed in Table 1, although our knowledge of these is still incomplete. We know slightly more about the JMJD2 class of enzymes such as JMJD2A, which comprises the Jumonji N domain, the Jumonji C domain, a C-terminal domain, and a zinc-finger motif; together these domains create a potential catalytic center in the core of the protein. This enzyme is able to recognize and demethylate lysines 9 and 36 on histone H3.

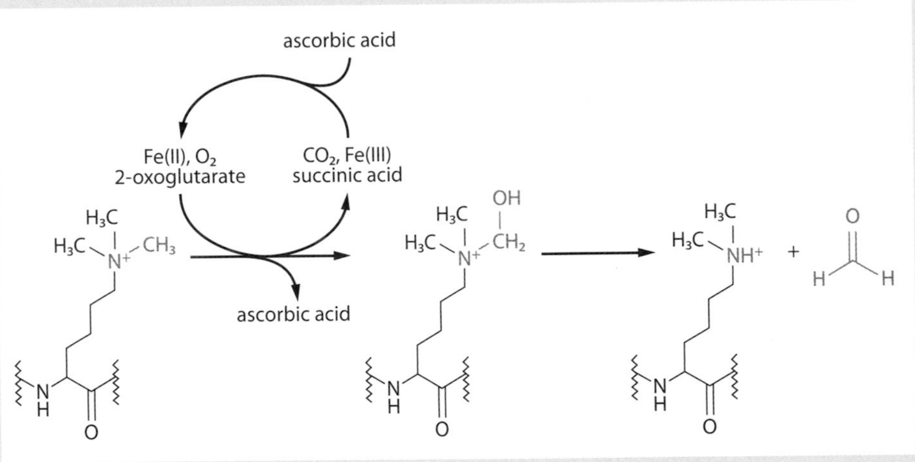

Figure 1 Reaction mechanism used by the Jumonji domain.

TABLE 1 HISTONE DEMETHYLASES CONTAINING JUMONJI DOMAINS

Demethylase	Alternative name	Site specificity	Demethylates
(Flavin-dependent) LSD1	BHC110, AOF2	H3K4	Mono, di
(Fe(II) 2-oxoglutarate dependent) JHDM1A	FBXL11	H3K36	Mono, di
JMJD1A	JHDM2A, TSGA	H3K9	Mono, di
JMJD2A	JHDM3A	H3K9, H3K36	Tri, di
JMJD2B	KIAA0876	H3K9	Tri, di
JMJD2C	GASC1	H3K9, K3K36	Tri, di
JMJD2D	KIAA0780	H3K9	Tri, di, mono
JARID1A	RBP2	H3K4	Tri, di
JARID1B	PLU-1	H3K4	Tri, di
JARID1C	SMCX	H3K4	Tri, di
JARID1D	SMCY	H3K4	Tri, di

Box 7.7 Mechanism of PAD4

The mechanism for converting methylarginine to citrulline relies on the following reaction and the active site of PAD4 achieves this by allowing the deprotonated thiol group of Cys 645 to act as a nucleophile to generate a tetrahedral adduct; this collapses to form a stable amidinocysteine intermediate and ammonia (see **Figure 1**). Hydrolysis of the intermediate leads to the formation of citrulline, which seems to be a characteristic feature of the PAD4 enzyme because other amidinotransferases allow the tetrahedral adduct formed from other methylarginine-containing non-histone proteins to collapse to form ornithine.

Proposed PAD4 catalytic mechanism

A low-pK_a active-site cysteine residue is the nucleophile in this reaction, leading to the formation of an amidinocysteine intermediate (steps 1 and 2). Although His 471 is drawn as the general base, a plausible alternative mechanism involves product-assisted catalysis, namely deprotonation of a water molecule by ammonia (steps 3

and 4). Asp 350 is appropriately positioned to accept a proton from the tetrahedral intermediate and thereby promote the anti-elimination of the Cys nucleophile (step 5). Asp 350 could donate this proton to bulk solvent to regenerate the initial form of the enzyme (step 6). Note that this mechanism has unmethylated arginine as the substrate for the sake of clarity, but the reactions are equally applicable to methylated arginine.

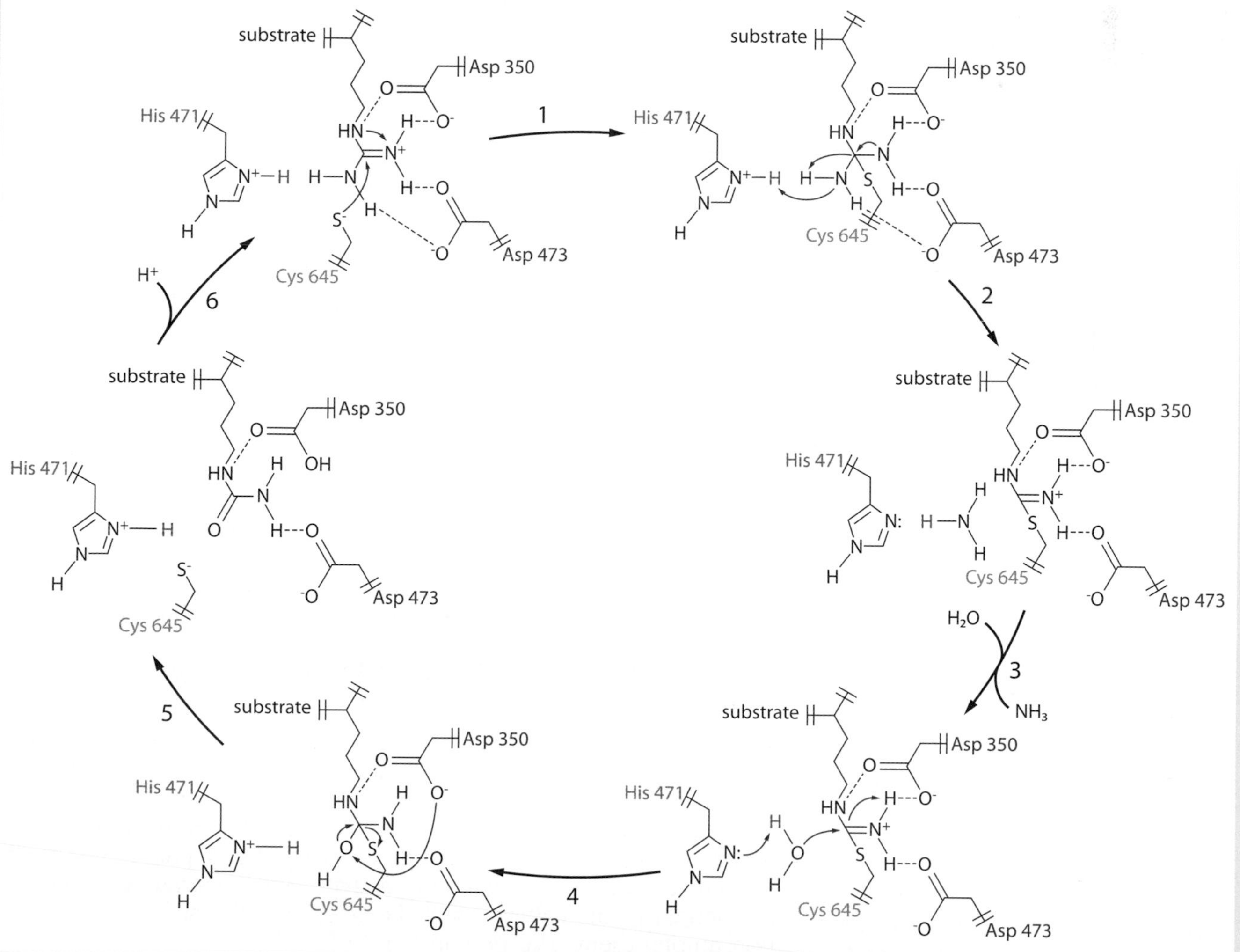

Figure 1 Proposed catalytic mechanism of PAD4. [Adapted from Arita K, Hashimoto H, Shimizu T et al. (2004) *Nat. Struct. Mol. Biol.* 11, 777. With permission from Macmillan Publishers Ltd.]

or impede the histone's function. Theoretically, the citrulline would have to be removed and replaced by arginine to recover the structure–function relationship, and although this could be performed by an aminotransferase enzyme, none of these have been shown to perform any functions related to histone modification.

The other possibility for reinstating full function is simply to remove the whole histone molecule and replace it with a new one, but this seems an enormous effort to remove a histone N-terminal tail modification. Furthermore, because the induction of PAD4 activity in cultured cells results in a global increase in citrullinated histones and decreased levels of arginine methylation, such total histone replacement may not occur rapidly.

There is no doubt that histone replacement can and does occur *in vivo*, as is demonstrated by actively transcribed regions of chromatin undergoing a dynamic replacement of histone H3.1 with the histone variant H3.3. However, it seems unlikely that this process is specifically designed to counteract histone methylation, because it ultimately relies on disassembly of the nucleosome and replacement of the histone molecule to regenerate an intact unmodified nucleosome. This process causes not only a loss of the histone methylation marks but also a loss of other important histone modifications found on the same molecule. That notwithstanding, the fact that levels of citrulline do increase in the presence of active intracellular PAD4 is supportive of a role for this enzyme in antagonizing histone arginine methylation *in vivo*, but how the cell deals with the citrulline that PAD4 creates is still open to question.

7.3 ENZYMES THAT PHOSPHORYLATE OR DEPHOSPHORYLATE HISTONES

It is generally accepted that phosphorylation is one of the most common methods of altering protein structure and/or function, and as we saw earlier, histones are no exception. Phosphorylation is generally performed by kinase enzymes, but the number and variety of kinases normally active in the cell are so large that a brief digression into their classification is valuable at this point. Protein kinases transfer the γ phosphate from nucleotide triphosphates (often ATP) to one or more amino acid residues in a protein substrate side chain, resulting in a conformational change that affects protein function. These enzymes fall into two broad classes, characterized with respect to substrate specificity: serine/threonine-specific kinases and tyrosine-specific kinases. The serine/threonine kinases phosphorylate the hydroxyl groups on the side chains of these two amino acids, which are broadly similar in structure. This class probably represents the bulk of the roughly 500 kinase enzymes currently known, but the tyrosine kinases are an important subclass that is more often involved in signal transduction mechanisms within the cell.

Kinases catalyze the phosphorylation of specific amino acids on histones

The protein kinases are assembled in a modular fashion, in common with proteins in many other classes. However, the only module that is common to all and that shares extensive sequence similarities is the one corresponding to the catalytic subunit. The noncatalytic modules are very probably important for determining other functional parameters such as subcellular location, regulation of activity, and substrate specificity, and in contrast to the catalytic unit there is very little sequence conservation among them. The protein kinase fold is highly conserved among both tyrosine and serine/threonine kinases. It contains an N-terminal lobe involved in ATP binding, and a C-terminal lobe that is thought to be involved in protein substrate recognition.

Several lines of evidence suggest that the transition state for **protein tyrosine kinase (PTK)** reactions is dissociative. In a dissociative (as opposed to an associative) transition state (**Box 7.8**, **Figure 1**), there is little bond formation between the entering nucleophile (in this case

Box 7.8 Structures and mechanism of kinase-catalyzed phosphorylation

Many of the interactions made by the phosphate groups of the nucleotide are mediated by bound metal ions, usually Mg^{2+} or Mn^{2+}. The activation loop immediately adjacent to the catalytic centre has been implicated in defining the catalytic activity of the protein kinases and the association of regulatory subunits. This region of the protein is often subjected to phosphorylation as a means to control its kinase activity. Most protein kinases are known to be regulated by the phosphorylation or dephosphorylation of equivalent residues in this region; for instance, the cell cycle kinase CDK2 is activated by a combination of phosphorylation of the equivalent residue, Thr 160, and binding of cyclin (see **Figure 1** for (a) the reaction mechanism and (b) enzyme structure).

A portion of the active site of protein kinase A (PKA) shows the conserved elements of active protein kinase structures (**Figure 2**). Asp 184 is positioned to support the ATP phosphate tail by a hydrogen bond to the main chain of the Asp 184-Phe-Gly (DFG) motif. The same peptide bond is part of a three-turn to the DFG + 2 residue. This orients the peptide bond between Phe 185 and Ala 188 so that it stabilizes the position of Arg 165, which in turn is the primary interaction partner of the phosphorylated Thr 197. This provides a path for the activating phosphorylation to stabilize ATP in a favorable position for transfer of the phosphate to a substrate.

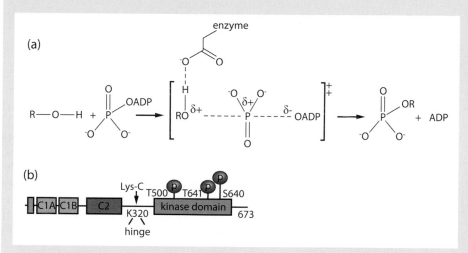

Figure 1 Reaction mechanism for phosphorylation of serine or threonine. Part (a) represents the reaction mechanism and part (b) represents the enzyme structure.

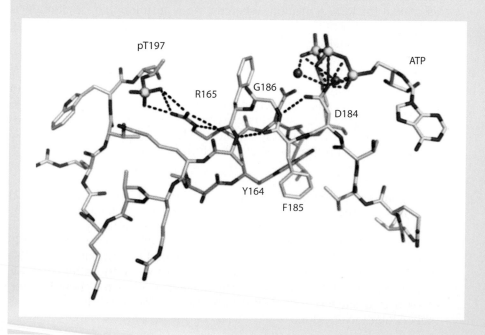

Figure 2 Structure of the catalytic centre of protein kinase A. Cyclic AMP, the DFG motif, and phosphorus atoms are colored yellow, the rest of the activation loop is colored gray, and the HRD motif from the catalytic loop is colored tan. [Adapted from Ten Eyck LF, Taylor SS & Kornev AP (2008) *Biochimica et biophysica acta. Proteins and proteomics* 1784, 238. With permission from Elsevier Limited.]

tyrosine) and the attacked phosphorus before substantial departure of the leaving group (in this case ADP). Reaction rates for PTK interaction with a series of fluorotyrosine-analog-containing substrates show minimal dependence on the basicity of the nucleophile, supporting a dissociative transition state. Moreover, the form of the attacking substrate phenol is required to be neutral rather than the chemically more reactive phenoxide anion, arguing against the importance of nucleophilicity in the reaction.

Having learned the basic structures of kinases, we can now examine those members of this class of enzymes that are responsible for histone phosphorylation. There are quite large numbers of enzymes that perform this function, but unlike many of the other post-translational N-terminal tail modifications, the histone kinases do not have only histones as their target substrates. Most of the following examples show multiplicity of function, and although they are intended to be as broad a representation of the histone kinases as possible, we cannot include all the enzymes of this type owing to their very large numbers.

A variety of serine kinases phosphorylate serine 10 on histone H3

Ribosomal S6 kinases

The **ribosomal S6 kinases (RSKs)** are a family of broadly expressed serine/threonine kinases that are activated by extracellular signal-regulated protein kinases, such as ERK1 and ERK2, in response to growth factors, hormones, and neurotransmitters. Ribosomal S6 kinases are an important effector system in the regulation of cell division and survival. These enzymes form part of a much broader class referred to as the growth-factor-activated protein kinases that have major roles in the mediation of transduction of growth factor signals. Collectively, these kinases phosphorylate a large array of cellular proteins and thereby regulate cellular division, survival, metabolism, transmembrane ion flux, migrative behavior, and differentiation.

The RSK enzymes comprise two kinase domains connected by a linker peptide of approximately 100 amino acid residues. The C-terminal kinase domain (CTD) is responsible for autophosphorylation at Ser 386, which is critical for RSK activation, whereas the N-terminal kinase domain is believed to phosphorylate exogenous substrates of RSK. The latter are quite broad in nature, because RSKs have been suggested to phosphorylate and activate transcription factors such as the estrogen α-receptor and cyclic AMP response-element-binding protein (CREB). RSK2 is known to phosphorylate Ser 10 on histone H3, although knowledge of how it recognizes and phosphorylates H3S10 is still lacking, and mutations of the gene encoding this enzyme are responsible for the human disease Coffin–Lowry syndrome, which is characterized by severe mental retardation and progressive skeletal deformations.

MSK1 and MSK2

Mitogen- and stress-activated protein kinases 1 and 2 (MSK1 and 2) are also members of the growth-factor-activated protein kinase family and have been shown to phosphorylate histone H3, although their mode of activation by signal transduction differs slightly from that of RSK2. MSK1 is activated by the Ras/mitogen-activated protein kinase (MAPK) and p38 pathways, whereas RSK2 does not respond to the latter pathway. Like RSK2, MSK1 is a predominantly nuclear protein, but there is some evidence to suggest that its function may be controlled by altering its subcellular localization.

There has been some controversy as to the true identity of the histone H3S10 kinases, because some evidence supports the possibility that MSK1 alone is capable of adding this histone modification, yet other studies show that RSK2 cannot phosphorylate Ser 10 (or at least have failed to demonstrate that it *can* phosphorylate Ser 10). Once more, however, and despite substantial evidence for the involvement of MSK1 and MSK2 in H3S10 phosphorylation, structural and mechanistic studies of their modes of action are lacking at the time of writing.

Aurora kinases

The involvement of this group of centrosome-localized serine/threonine kinases in histone H3S10 phosphorylation was first noted in *S. cerevisiae* in which knockouts of the gene encoding this protein were deficient in histone phosphorylation. Subsequent studies in mammalian cells showed that Aurora B is able to phosphorylate Ser 28 on H3. Aurora activities seem to be associated predominantly with progression of the mitotic and meiotic cell cycles, in which these kinases have highly significant roles. In this context, immunocytochemical analyses have revealed that Aurora B co-localizes with H3 phosphorylated not only at Ser 10 from late G2 phase to metaphase, but also at Ser 28 from prophase to metaphase (**Figure 7.4**). In addition, the transfection of a dominant-negative mutant of Aurora B in mitotic HeLa cells resulted in a decrease in H3 phosphorylation, not only at Ser 10 but also at Ser 28.

The Aurora kinases have excited considerable interest owing to their apparent involvement in multiple human cancers. Because they fall into the more general classification of **cAMP-, cGMP-dependent protein kinase C-like kinases (AGC kinases)**, there is quite a lot of information about their three-dimensional structures. They seem to be controlled by autophosphorylation in the same way as many other kinases, and we have data concerning their interaction with many other non-histone-related proteins.

Human Auroras A–C are kinases of a size ranging from 309 to 403 amino acid residues that have a relatively high sequence divergence among species. For example, the overall differences in sequence identities between human and rodent proteins are 82% for Aurora A, 84% for Aurora B, and 78% for Aurora C. Their domain and three-dimensional structures (**Figure 7.5**) show them to be broadly similar to kinases in general. Despite this wealth of information, we do not have a clear picture of how they might interact with the histone H3 N-terminus and select serines 10 or 28 for phosphorylation.

MST1 kinase phosphorylates Ser 14 on histone H2B

The MST proteins belong to a large family of serine/threonine kinases called **sterile 20 proteins (Ste20)** that contains at least 35 members. All members of the Ste20 family contain a conserved kinase domain and a structurally diverse region, which is implicated in regulation. On the basis of their domain architecture, the Ste20 group is further divided into two

Figure 7.4 Scheme showing mechanisms by which Aurora B and type 1 protein phosphatase (PP1) regulate histone H3 phosphorylation at serine 10 (Ser 10) or at serine 28 (Ser 28). (a) Aurora B phosphorylates H3 at both Ser 10 and Ser 28 during late G2 phase. However, because PP1 dephosphorylates H3 phosphorylated at Ser 28 (P-Ser 28) exclusively, P-Ser 28 is undetectable before mitosis. (b) On entry into mitosis (from prophase to metaphase), Ser 28 phosphorylation is increased to a detectable level, probably through inactivation of PP1 by Cdc2 kinase.

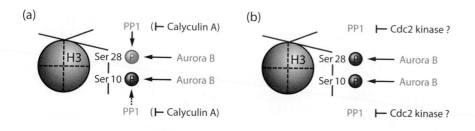

Figure 7.5 Domain organization of Aurora kinases A–C. As shown here, Aurora kinases present three domains: the N-terminal and the C-terminal domains contain most of Aurora's regulatory motifs, and the central region contains the catalytic domain. In addition to the kinase activity, this central domain also presents regulatory motifs, as the crystal structure of the Aurora A–TPX2 complex has shown. [Adapted from Bolanos-Garcia VM (2005) *Int. J. Biochem. Cell Biol.* 37, 1572. With permission from Elsevier.]

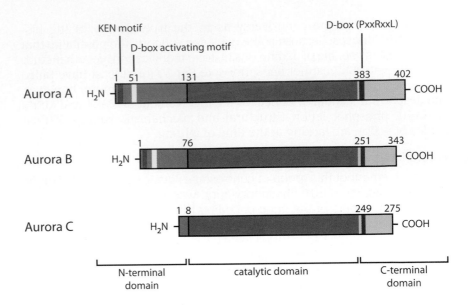

distinct families, **p21-activated kinase (PAK1)** and **human germinal-center kinase (GCK)**. The two families are distinguished by the relative positions of their kinase domains, with the PAKs containing a C-terminal kinase domain and the GCKs, including MST1, containing an N-terminal kinase domain. MST1 is 487 amino acid residues long and also contains a C-terminal regulatory domain and an extreme 56-residue C-terminal α-helical domain implicated in homodimerization.

MST1 is a proapoptotic cytosolic kinase. Activation of MST1 via the caspase-3 pathway leads to apoptosis, and overexpression of MST1 results in the induction of apoptosis in a variety of cell backgrounds through pathways that involve the activation of stress-activated protein kinase (SAPK). Allis and co-workers demonstrated that MST1 phosphorylates histone 2B at Ser 14, which specifically correlates with the onset of apoptosis in human cells. Involvement of histone modification in the apoptotic process is probably necessary to bring about chromatin condensation, which is the most recognizable nuclear hallmark of apoptosis.

Histone phosphatases remove phosphates from histone residues

Protein dephosphorylation occurs by hydrolysis of the phosphate esters of amino acids such as serine and tyrosine. The reaction liberates a phosphate ion and the free hydroxyl group of the amino acid side chain. Phosphatases can be divided into two main categories: **cysteine-dependent phosphatases (CDPs)** and **metallophosphatases (MMPs)**, which are dependent on metal ions in their active sites for activity.

The metallophosphatases are probably the most common or at least the best-characterized group (**Box 7.9**). There are four subgroups of enzymes in this classification, termed phosphoprotein phosphatases 1, 2A, 2B, and 2C, and these account for a substantial proportion of the phosphatase activity toward phosphoproteins involved in controlling cellular function. The catalytic subunits of the phosphoprotein phosphatase (PPP) family show high sequence homology, and the active sites of, for example, PP1 and PP2B are all very similar (although the control of these catalytic subunits depends on regulatory subunits in a complex interplay of inhibition and allosteric effects). These enzymes are serine/threonine-specific protein phosphatases.

Cysteine-dependent phosphatases encompass the protein tyrosine phosphatases (PTPs), of which there are 107 members encoded in the human genome, with the class I cysteine-based PTPs constituting the largest group. This group can be further subdivided into 61 dual-specificity phosphatases and 38 tyrosine-specific PTPs. All members of the PTP family catalyze metal-independent dephosphorylation of phosphoamino acids, using a covalent phosphocysteine intermediate to facilitate hydrolysis.

Box 7.9 Structures and mechanism of the histone phosphatases

Highly conserved residues in these enzymes serve to bind two metal ions within ~0.3 nm of each other, and further cationic residues (histidine and arginine) line the shallow depression on the surface to form the active site. The two metal ions are held in octahedral coordination sites with a bridging hydroxyl group and bridging acetate oxygen atom. Inorganic phosphate (product) or inhibitors (such as tungstate) form a 1,3 bridge between the two metal ions (**Figure 1**).

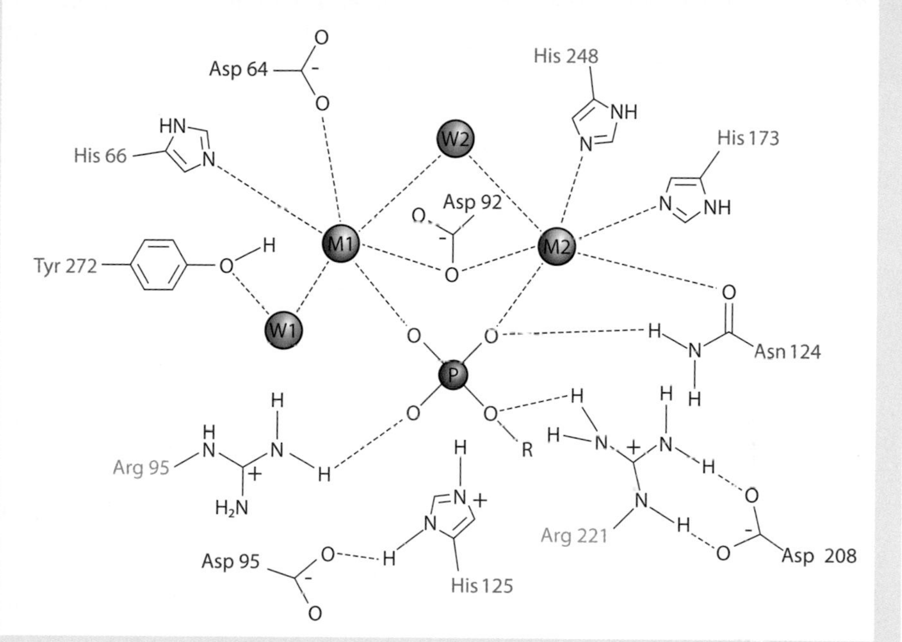

The 280-residue catalytic domain of protein tyrosine phosphatase (PTP) consists of an α-helix/β-pleated sheet structure, and biochemical and structural studies have led to a detailed understanding of the catalytic mechanism. Key features of the domain include the PTP signature motif, the mobile Trp-Pro-Asp "WPD" loop that in the closed conformation positions the conserved and catalytically important aspartate residue, and the phosphotyrosine recognition loop.

Figure 1 Binding of metal ions by specific amino acids of the histone phosphatases brings the polypeptides into the correct conformation to form the catalytic center.

PTPs show exceptional substrate specificity *in vivo*, which is conveyed both by the catalytic domain and other regulatory mechanisms including restricted subcellular localization, post-translational modification events (for example phosphorylation), specific tissue distribution, and accessory or regulatory domains. Within the catalytic domain, the phosphotyrosine recognition loop (between α1 and β1) contributes to the selective recognition of phosphotyrosine over phosphoserine/phosphothreonine, and two nonconserved residues following the conserved tyrosine in this loop are important in substrate interactions (**Figure 2**).

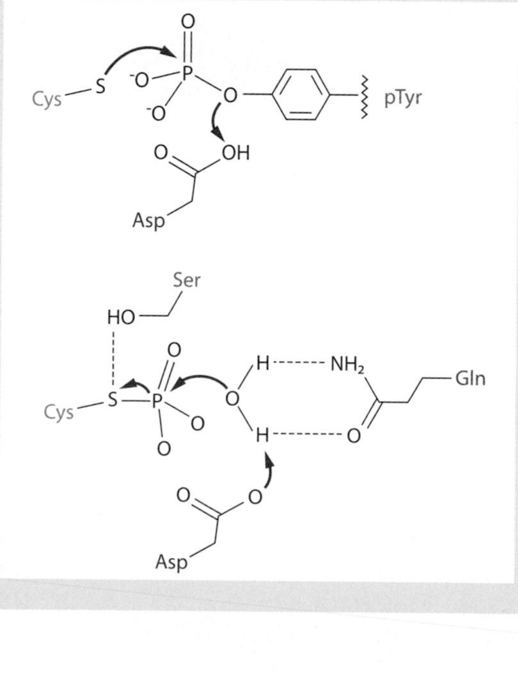

Figure 2 Specific amino acids are brought into close proximity to the phosphate groups on tyrosine to permit precise removal.
(Courtesy of David Briggs, University of Manchester.)

7.4 ENZYMES THAT ADD AND REMOVE UBIQUITIN ON HISTONES

E3 ubiquitin ligases add ubiquitin to lysine

The principal proteins that catalyze the addition of ubiquitin to specific lysine residues in the histone N-terminal tails and on the nucleosome core belong to the E3 ubiquitin ligase family. These enzymes operate in conjunction with an E1 ubiquitin-activating enzyme and an E2 ubiquitin-conjugating enzyme (note that E1 and E3 are the protein modules central to the assembly of these enzymes). There is one major E1 enzyme, shared by all ubiquitin ligases; it uses ATP to activate ubiquitin for conjugation and transfers the ubiquitin to an E2 enzyme. The E2 enzyme interacts with a specific E3 partner and transfers the ubiquitin to the target protein. The E3 ligase, which may be a multiprotein complex, is in general responsible for targeting ubiquitination to specific substrate proteins.

The principal E3 ligases involved in lysine ubiquitination are the polycomb repressive complex (PRC) components RING1B and BMI1. These are both essential to the process, because although only RING1B has the necessary catalytic activity, it is only sufficiently active when bound to BMI1. RING1B encompasses the region required for the histone H2A monoubiquitination E3 ligase activity, and BMI1 is sufficient for the stable interaction with RING1B and stimulation of its E3 ligase activity. There is also some evidence to suggest that BMI1 is required for nucleosomal interaction, especially because the BMI1–RING1B complex adds ubiquitin only to Lys 119 of histone H2A. This lysine residue is situated close to the globular histone core of the nucleosome; hence a degree of nucleosome interaction capacity would be required to position RING1B in close proximity to its target.

A variety of enzymes remove ubiquitin from lysine

Cleavage of the amide bond between lysine and ubiquitin requires the ubiquitin-processing protease enzymes, of which several examples have been characterized in human cells. De-ubiquitinating enzymes can act at various points in the ubiquitin pathway, including polyubiquitin chain processing to remove ubiquitin from substrates to rescue substrates from degradation, or the removal of residual ubiquitin to assist in the proteasomal degradation of substrate. Many novel candidate de-ubiquitinating enzymes have been identified, but only a few of them are necessary components of the histone ubiquitination/de-ubiquitination system.

There are now approximately 100 human de-ubiquitinating enzymes that have been identified by *in silico* efforts and activity-based profiling. They can be divided into five subclasses: (1) ubiquitin C-terminal hydrolases, (2) ubiquitin-specific processing proteases (USPs), (3) Machado–Joseph disease protein domain proteases, (4) ovarian tumor proteases, and (5) JAMM (JAB1/MPN/Mov34 metalloenzyme) motif proteases.

The largest subclass is the USPs, with more than 50 members. These are cysteine proteases that contain highly divergent sequences and show strong homology mainly in two regions that surround the catalytic Cys and His residues—the so-called Cys box and the His box. The structure of USP14 shown in **Figure 7.6** is typical of these enzymes. Shown in the figure are the relative positions of the cysteine and histidine residues central to the catalytic function of this active site.

Recently, several USPs that recognize the monoubiquitinated histones H2A and/or H2B have been identified. Among these enzymes, three USPs contain a zinc-finger ubiquitin-specific protease (ZnF-UBP) domain,

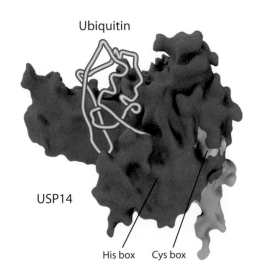

Figure 7.6 Structure of the de-ubiquitinating enzyme USP14. The relative positions of the Cys and His boxes together with the ubiquitin recognition site are shown. [From Hu M, Li P, Song L et al. (2005) *EMBO J.* 24, 3747. With permission from Macmillan Publishers Ltd.]

indicating that this domain has a crucial role in regulating their activity. The ZnF-UBP domain—also known as deacetylase and ubiquitin-specific protease (DAUP), polyubiquitin-associated zinc-finger (PAZ), or binder of ubiquitin zinc-finger (BUZ)—is an additional domain that consists of approximately 100 residues and is thought to regulate protease activity. Examples of these zinc-finger-type enzymes include USP22, which was recently identified as a de-ubiquitinase for monoubiquitinated histones H2A and H2B, and USP16 (also known as UBP-M), whose activity is specific to monoubiquitinated histone H2A.

Interestingly, neither free recombinant USP22 nor Ubp8p (the yeast homologue of USP22) can efficiently de-ubiquitinate H2B or H2A (in the case of USP22) *in vitro*, indicating that interaction with other cellular components is needed to permit this activity. These enzymes become part of the Spt-Ada-Gcn5-acetyltransferase (SAGA) complexes *in vivo* and seem to be active only in this complex. This interaction seems also to remove any necessity for USP22 to recognize and bind to the ubiquitin moiety on the histone. Indeed, in complete contrast to many other ubiquitin-specific proteases, no interaction between USP22 and ubiquitin can be demonstrated experimentally.

7.5 ENZYMES THAT ADD AND REMOVE THE SUMO GROUP ON HISTONES

E3 SUMO ligases add the SUMO group to lysine

SUMOylation means the addition of a small protein, SUMO (small ubiquitin-related modifier protein). SUMOylation of lysine is performed with a mechanism that is somewhat analogous to that of ubiquitination, in that catalysis of the final reaction in the pathway is brought about by an E3 ligase. SUMO is first activated for conjugation by the E1 enzymes Aos1/Uba2, subsequently transferred to the E2 conjugation enzyme Ubc9 (ubiquitin-conjugating enzyme 9), and finally conjugated to target proteins by an E3 SUMO ligase.

Different classes of E3 SUMO ligase proteins have been reported. Protein inhibitors of activated STAT (PIAS), RanBP2, and polycomb group protein Pc2 have been shown to act as SUMO E3 ligases because they are able to interact with both the Ubc9 enzyme and the subtstrate protein (in this case the histone). PIAS proteins comprise at least four different types, namely 1, 3, x, and y, all of which contain two principal protein domains: an N-terminal SAP domain and a central RING-finger motif similar to that of the RING1B enzyme responsible for histone ubiquitination. More recently, it was shown that the nucleoporin protein RANBP2 (Ran-binding protein 2) functions as a SUMO E3 ligase for substrates such as the PML body protein SP100 and histone deacetylase 4. Polycomb repressive complex protein 2 (PRC2, also known as CBX4) was shown to be a SUMO E3 ligase for the transcriptional co-repressor CtBP.

Since then, other substrates have been identified for Pc2, but in comparison with other E3s, such as RanBP2 or the PIAS family, relatively few are known for Pc2. It is possible that Pc2 has a more limited repertoire of substrates, perhaps as a result of its specific subnuclear localization within polycomb bodies. This may allow it to promote SUMOylation of a specific pool of proteins, one that is localized to polycomb bodies, whereas other E3s may regulate SUMOylation of the same or similar substrates at other cellular locations.

As was noted for the enzymes in the ubiquitin pathway, there is very little (or even less) information concerning the precise structures and mechanisms of the E3 SUMO ligases. Therefore, at present, we cannot accurately describe how they recognize and add SUMO groups to specific lysine residues on histones.

SUMO-specific proteases remove the SUMO group from lysine

The SUMO modification pathway is an example of a reversible system that is controlled by a series of on and off enzymes. The removal step is quite simple because only two SUMO-deconjugating enzymes have been discovered in simpler organisms such as yeast, and a total of only six in humans. The SUMO removal system makes use of the SUMO-specific proteases. These have a catalytic domain characterized by the catalytic triad (histidine, aspartate, and cysteine), as shown in the *S. cerevisiae* SUMO-specific protease Ulp1. The secondary structure of Ulp1 possesses seven α-helices and seven β-strands and shares similarity with other cysteine proteases in the active site. The active site includes the central α-helix, three β-strands, and the catalytic triad that presents the cysteine and histidine residues in the correct orientation relative to the substrate to catalyze the reaction.

The six human SUMO proteases are human sentrin/SUMO-specific proteases (SENP), which can be divided into three families. The first family consists of SENP1 and SENP2, which have broad specificity for the three mammalian SUMOs (SUMO1 to SUMO3). The second family includes SENP3 and SENP5, which favor SUMO2 and SUMO3 as substrates and are localized in the nucleolus. The third family contains SENP6 and SENP7, which have an additional loop inserted in the catalytic domain and also seem to prefer SUMO2 and SUMO3. From an evolutionary standpoint, SENP1–SENP3 and SENP5 are more closely related to Ulp1, whereas SENP6 and SENP7 are related to Ulp2. The crystal structure of SENP2 shows that it resembles other cysteine proteases in that its active site comprises Cys 548, His 478, and Asp 495.

Comparative proteolysis assays show that SENP2 hydrolyzes SUMO2 better than SUMO1 or SUMO3 precursors. This C-terminal hydrolase activity is apparently dependent on the respective C-terminal tails of these SUMOs. However, SENP2 efficiently hydrolyzes SUMO1, SUMO2, or SUMO3 from RanGAP1. Furthermore, the isopeptidase activity of SENP2 is twentyfold stronger than its C-terminal hydrolase activity, suggesting that SENP2 interacts more readily with SUMO conjugates than with SUMO precursors.

7.6 ENZYMES THAT ADD AND REMOVE BIOTIN ON HISTONES

Biotinidase and biotin holocarboxylase synthetase can biotinylate histones

Post-translational modification of histones by addition of the vitamin biotin to specific lysine residues is another mechanism for the epigenetic regulation of genome activity. It has been suggested that the addition mechanism relies on the cleavage of biocytin (biotin conjugated to the ε-amino group of lysine in a non-histone carrier protein) by the enzyme biotinidase. This forms a thioester bond between biotin and a cysteine residue near the active site of biotinidase, and this conjugate is capable of transferring the biotinyl moiety to a lysine residue of another protein, such as a histone (**Figure 7.7**).

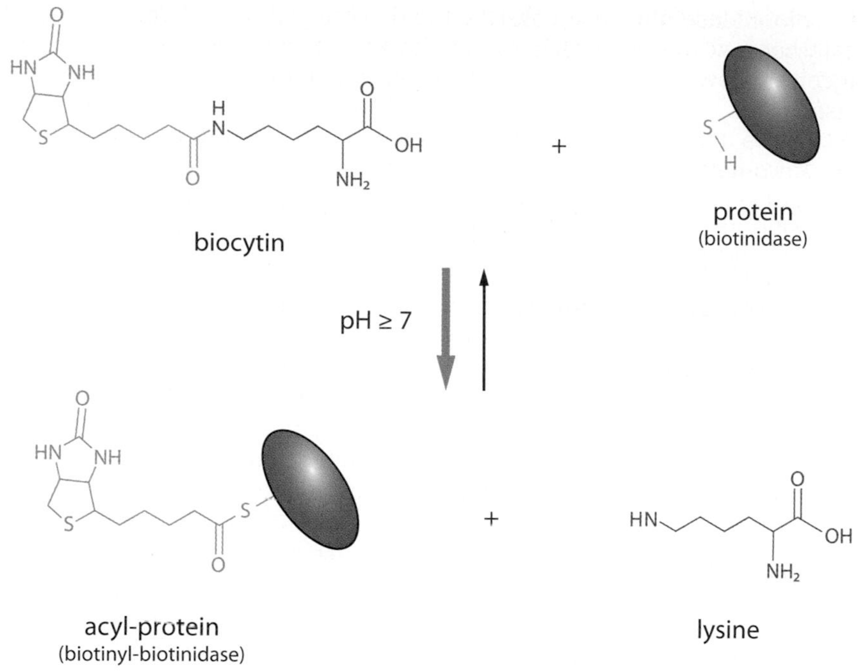

biocytin

+

protein
(biotinidase)

pH ≥ 7

acyl-protein
(biotinyl-biotinidase)

+

lysine

Figure 7.7 Mechanism of biotinylation of lysine. "Free" biotin exists as a conjugate with a non-histone carrier protein and is referred to as biocytin. To transfer the biotin moiety to a histone lysine residue, the enzyme biotinidase transfers biotin to one of its free thiol groups, from which it is able to react with lysine residues present on histone proteins.

Biotinidase has biotinyltransferase activity at neutral to alkaline pH. Therefore, under physiological conditions, biotinidase probably transfers biotin to suitable nucleophilic acceptor molecules, such as histones (Figure 6.12).

The amino acid sequence of human biotinidase is homologous with bacterial aliphatic amidases and some bacterial and plant nitrilases. One of the regions of homology contains a cysteine residue involved in the active site of aliphatic amidases and nitrilases. This suggests that Cys 245 of biotinidase is the cysteine residue in the active site of the enzyme and is the site of thioester formation that is integral for enzyme function. It is also possible that biotin holocarboxylase synthetase may perform biotinylation of histone lysine residues.

Biotin is the essential coenzyme for four biotin-dependent carboxylases in humans. **Biotin holocarboxylase synthetase (HCS)** covalently binds biotin to the ε-amino group of a lysine residue of each of the apocarboxylases, thereby forming biotinylated, active holocarboxylases. If this enzyme can do this as part of its normal biological function, it may be able to perform a similar reaction with lysine when it is part of a histone molecule.

The open reading frame of the gene for full-length HCS encodes a 726-residue protein having a predicted molecular weight of 81 kDa. HCS has been detected in cytoplasm, mitochondria, cell nuclei, and nuclear lamina. Both HCS transcript abundance and the nuclear translocation of HCS protein depend on cellular biotin supply. Nuclear localization is consistent with a role of HCS in chromatin remodeling.

Enzymes that remove biotin from histone lysine residues

The classical role of biotinidase in metabolism is to hydrolyze biocytin (biotinyl-ε-lysine), a degradation product of biotin-dependent carboxylases. Hydrolysis of biocytin releases free biotin, which is recycled in the synthesis of new holocarboxylases. In view of this, it has been proposed that biotinidase may also catalyze the debiotinylation of histones, especially because this reaction has been shown to occur *in vitro*, but it remains uncertain how the same enzyme could act as a biotinyl transferase in one case and as a debiotinylase in another. It has been proposed that if biotinidase is a dual-action enzyme, its regulation is achieved by interactions with chromatin-associated proteins, post-translational modifications and alternative splicing, and substrate availability. However, there is still the possibility that histone debiotinylases other than bitoinidase remain to be discovered.

The enzyme systems detailed in this chapter represent the current state of knowledge concerning the mechanism by which covalent post-translational modifications can be introduced and removed from the amino acid residues comprising the N-termini of histones. However, this tells us little about how these mechanisms are directed to act at some regions of the genome while leaving other regions alone. The next chapter attempts to clarify the locus-specific control of histone-modifying enzyme activity.

KEY CONCEPTS

- Histone acetylation and deacetylation are catalyzed by the histone acetyltransferases and histone deacetylases, respectively. These enzymes form part of multiprotein complexes that recognize specific genomic loci and are able to establish epigenetic marks in a precise manner at these positions.

- Histone methylation is more complex because the target lysine residues can be subject to monomethylation, dimethylation, or trimethylation. The primary enzyme unit responsible for histone lysine methylation is the SET domain, and most histone methyltransferase enzymes incorporate this protein. Histone demethylase enzymes remove methyl groups from lysine residues. Both the histone methyltransferases and histone demethylases function as part of multiprotein complexes that can direct epigenetic modifications to specific gene loci.

- Arginine can also be subjected to monomethylation and dimethylation through the activities of histone methyltranferases distinct from those that methylate lysine residues.

- Serine, threonine, and tyrosine residues can be phosphorylated, and this introduces a further level of epigenetic gene expression control. Protein kinases phosphorylate these residues, again as part of multiprotein complexes that can identify specific loci, and histone phosphatases dephosphorylate them.

- Small proteins such as ubiquitin and SUMO can modify specific lysine residues to introduce further structural epigenetic controls.

- Biotin can be added to specific lysine residues to introduce an additional epigenetic modification.

FURTHER READING

Agger K, Christensen J, Cloos PA & Helin K (2008) The emerging functions of histone demethylases. *Curr Opin Genet Dev* 18:159–168 (doi:10.1016/j.gde.2007.12.003).

Allis CD, Berger SL, Cote J et al. (2007) New nomenclature for chromatin-modifying enzymes. *Cell* 131:633–636 (doi:10.1016/j.cell.2007.10.039).

Avvakumov N & Côté J (2007) The MYST family of histone acetyltransferases and their intimate links to cancer. *Oncogene* 26:5395–5407 (doi:10.1038/sj.onc.1210608).

Bedford MT & Clarke SG (2009) Protein arginine methylation in mammals: who, what, and why. *Mol Cell* 33:1–13 (doi:10.1016/j.molcel.2008.12.013).

Black JC, Van Rechem C & Whetstine JR (2012) Histone lysine methylation dynamics: establishment, regulation, and biological impact. *Mol Cell* 48:491–507 (doi:10.1016/j.molcel.2012.11.006).

Cedar H & Bergman Y (2009) Linking DNA methylation and histone modification: patterns and paradigms. *Nat Rev Genet* 10:295–304 (doi:10.1038/nrg2540).

Chang B, Chen Y, Zhao Y & Bruick RK (2007) JMJD6 is a histone arginine demethylase. *Science* 318:444–447 (doi:10.1126/science.1145801).

Cheung P & Lau P (2005) Epigenetic regulation by histone methylation and histone variants. *Mol Endocrinol* 19:563–573 (doi:10.1210/me.2004-0496).

Cloos PA, Christensen J, Agger K & Helin K (2008) Erasing the methyl mark: histone demethylases at the center of cellular differentiation and disease. *Genes Dev* 22:1115–1140 (doi:10.1101/gad.1652908).

de Ruijter AJ, van Gennip AH, Caron HN et al. (2003) Histone deacetylases (HDACs): characterization of the classical HDAC family. *Biochem J* 370:737–749 (doi:10.1042/BJ20021321).

Di Lorenzo A & Bedford MT (2011) Histone arginine methylation. *FEBS Lett* 585:2024–2031 (doi:10.1016/j.febslet.2010.11.010).

Iberg AN, Espejo A, Cheng D et al. (2008) Arginine methylation of the histone H3 tail impedes effector binding. *J Biol Chem* 283:3006–3010 (doi:10.1074/jbc.C700192200).

Kooistra SM & Helin K (2012) Molecular mechanisms and potential functions of histone demethylases. *Nat Rev Mol Cell Biol* 13:297–311 (doi:10.1038/nrm3327).

Kothapalli N, Camporeale G, Kueh A et al. (2005) Biological functions of biotinylated histones. *J Nutr Biochem* 16:446–448 (doi:10.1016/j.jnutbio.2005.03.025).

Kouzarides T (2002) Histone methylation in transcriptional control. *Curr Opin Genet Dev* 12:198–209 (doi:10.1016/S0959-437X(02)00287-3).

Lau PN & Cheung P (2013) Elucidating combinatorial histone modifications and crosstalks by coupling histone-modifying enzyme with biotin ligase activity. *Nucleic Acids Res* 41:e49 (doi:10.1093/nar/gks1247).

Lee KK & Workman JL (2007) Histone acetyltransferase complexes: one size doesn't fit all. *Nat Rev Mol Cell Biol* 8:284–295 (doi:10.1038/nrm2145).

Marmorstein R (2003) Structure of SET domain proteins: a new twist on histone methylation. *Trends Biochem Sci* 28:59–62 (doi:10.1016/S0968-0004(03)00007-0).

Marmorstein R (2004) Structural and chemical basis of histone acetylation. *Novartis Found Symp* 259:78–98.

Marmorstein R & Trievel RC (2009) Histone modifying enzymes: structures, mechanisms, and specificities. *Biochim Biophys Acta* 1789:58–68 (doi:10.1016/j.bbagrm.2008.07.009).

Thomas T & Voss AK (2007) The diverse biological roles of MYST histone acetyltransferase family proteins. *Cell Cycle* 6:696–704 (doi:10.4161/cc.6.6.4013).

Utley RT & Cote J (2003) The MYST family of histone acetyltransferases. In Workman JL (ed.). Protein complexes that modify chromatin, pp 203–236. Springer.

Wysocka J, Allis CD & Coonrod S (2006) Histone arginine methylation and its dynamic regulation. *Front Biosci* 11:344–355.

Xiao B, Wilson JR & Gamblin SJ (2003) SET domains and histone methylation. *Curr Opin Struct Biol* 13:699–705 (doi:10.1016/j.sbi.2003.10.003).

Zhang Y & Reinberg D (2001) Transcription regulation by histone methylation: interplay between different covalent modifications of the core histone tails. *Genes Dev* 15:2343–2360 (doi:10.1101/gad.927301).

8

LOCUS-SPECIFIC CONTROL OF HISTONE-MODIFYING ENZYME ACTION

It will have become clear from the previous chapters that although the enzyme systems that add and remove the post-translational modifications, to either the histone N terminal tails or the globular histone core of the nucleosome, recognize specific structures around their target amino acids, they probably have very little ability to recognize the specific areas of the genome that must be subjected to modification in response to cellular events such as differentiation. The key issue is how these systems are able to distinguish one genomic locus from another, given that the loci may have very similar patterns of epigenetic modifications. The latter reality implies that the only means to differentiate such loci is to use the information contained in the underlying DNA sequence.

In Section 5.2 we read how DNA methyl-binding domains are able to interact with histone acetyltransferases and methyltransferases that can modify the structure of chromatin and prevent access of the transcriptional machinery to the DNA. This is only one method by which DNA sequence information can be used to position histone-modifying enzymes at the desired loci, and that is the subject of this chapter.

8.1 HISTONE ACETYLATION AND DEACETYLATION AS A PROTEIN COMPLEX ACTIVITY

NURD is a well-known deacetylation complex

We introduced one of the better-known epigenetic modification protein complexes, the large nucleosome remodeling and histone deacetylation complex (NURD), in our discussion of DNA methylation (see Section 5.2). As the name suggests, NURD is able to mediate ATP-dependent chromatin remodeling and deacetylation of histones. However, although we alluded to the ability of NURD to deacetylate histones, and thereby repress gene transcription by altering the chromatin architecture to restrict access of the transcriptional machinery, we neglected a more detailed discussion of this activity until the chemical basis of histone modification had been covered. Histone deacetylases HDAC1 and HDAC2 are integral components of the NURD complex, as shown in **Figure 8.1**. The ATP-dependent chromatin-remodeling activity of NURD resides in one or both of the two Mi-2 proteins, α and β (also known as CHD3/CHD4).

In addition to the two subunits with enzymatic activity, the Mi-2–NURD complex contains several other proteins of importance. The smallest subunit is a member of the methyl-CpG-binding domain (MBD) family of

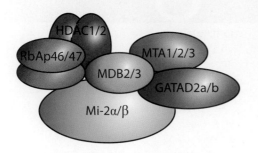

Figure 8.1 Some of the proteins that comprise the NURD complex.

proteins. The MBD subunit, like all other known subunits of the complex, is an interchangeable component varying between the family members MBD2 and MBD3. One of these two proteins, MBD2, has the capacity to selectively recognize methylated DNA. The other, MBD3, apparently lost this function during vertebrate evolution owing to a critical amino acid change in the MBD fold.

An additional subunit of the NURD complex consists of RbA p46 and/or RbA p48. These proteins were originally identified by their interaction with the retinoblastoma (Rb) tumor suppressor protein and are presumed to be structural subunits of Mi-2–NURD that act as protein interaction surfaces assisting complex assembly. Some versions of the complex are also thought to incorporate a second structural and/or regulatory subunit, p66a or p66b, which interacts with the MBD2 subunit and may be involved in interactions of the complex with methylated DNA. Both p66 isoforms have the capacity to interact directly with core histones.

The ability of NURD to recognize methylated DNA is important to its function in gene repression, because the promoters of nontranscribed genes need to lose their histone acetylation. However, locus specificity of NURD deacetylation is unlikely to be achieved simply by the presence or absence of methylated DNA. As we saw in Section 5.3, where the methylation of specific DNA sequences by DNMT3A was described, the NURD complex is probably recruited to a defined sequence of a gene promoter region through the binding of a specific transcription factor, or even a group of such proteins. The interaction of NURD with the transcription factor need not be direct.

An example of NURD-mediated activity is the repression of the target genes of GATA-1 by the transcriptional repressor FOG-1 during erythroid cell development. FOG-1 binds to the NURD co-repressor complex *in vitro* and *in vivo*. The interaction is mediated by a small conserved domain at the extreme N-terminus of FOG-1 that is necessary and sufficient for NURD binding, but FOG-1 does not have any intrinsic DNA-binding capacity. Previous work indicated that FOG-1 assists GATA-1 in associating with select GATA elements *in vivo*, and its mode of action seems to be that of a binding partner that provides GATA-1 with the ability to recruit repression complexes such as NURD. This observation would seem straightforward were it not for contradictory data indicating that FOG-1 also functions as co-activator, because it has been detected (by chromatin immunoprecipitation) at genes *activated* by GATA-1.

This raises the critical question as to how FOG-1 can function as activator despite its strong association with NURD. The most likely possibility is that FOG-1 might associate with NURD only at genes repressed by GATA-1/FOG-1. At activated genes, NURD might be displaced by a co-activator complex, analogous to what occurs with nuclear hormone receptors upon ligand binding.

SIN3A acts as a scaffold on which repressor proteins may assemble

The SIN3 protein is a prototypical transcriptional co-repressor that is found in large multiprotein complexes comprising histone deacetylases. Besides functioning as a molecular scaffold for complex assembly, the SIN3 co-repressor is recruited through direct protein–protein interactions by a large variety of sequence-specific DNA-binding factors (**Figure 8.2**). Underscoring the protein's important role in cellular physiology, constitutive SIN3 knockouts are characterized by embryonic lethality, whereas conditional knockouts lead to developmental defects caused by the deregulation of genes that control proliferation, homeostasis, and apoptosis.

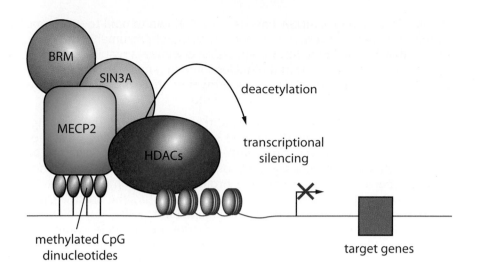

methylated CpG
dinucleotides

deacetylation

transcriptional
silencing

target genes

Figure 8.2 SIN3A acts as a scaffold to coordinate binding of transcription factors, components of the NURD complex, and factors recognizing methylated DNA. The individual proteins are shown. MeCP2 recognizes methylated DNA; through the recruitment of histone deacetylases (HDACs) coordinated by SIN3A, the complex of proteins can recognize repressed genes and maintain their inactive state by ensuring that histone acetylation at their loci is minimal. [From Bienvenu T & Chelly J (2006) *Nat. Rev. Genet.* 7, 415. With permission from Macmillan Publishers Ltd.]

At the molecular level, the 150 kDa SIN3 co-repressor harbors six regions that are highly conserved from yeast to human. Two closely related forms called SIN3A and SIN3B are found in higher organisms, and the paralogs seem to have both common and unique functions. The SIN3 proteins contain four imperfect copies of the PAH (paired amphipathic helix) domain that serve as primary interaction sites for transcription factors.

Unlike the NURD complex, SIN3A is only one component of a variety of repression complexes that form around defined promoter regions when required by the cell. This greatly complicates any attempt to describe these complexes because potentially they all contain different proteins. However, there are many references in the scientific literature to SIN3A coordinating the interactions needed to assemble such complexes, and mention of a few examples is worthwhile.

That the transcription factors OCT4, NANOG, and SOX2 associate with SIN3A in embryonic stem cells has been shown by immunopreciptation studies in which antibodies against NANOG protein pull down substantial amounts of SIN3A, MTA1/MTA2 (more MTA1 than MTA2), and HDAC2. However, there is currently a lack of data about the specific genes that are repressed by the three transcription factors using SIN3A. A more precise example may be repression of the E2F1 and androgen receptor genes by the transcription factor EBP1 in breast and prostate cancer cell lines. This particular activity came to light from the observation that ectopic overexpression of EBP1 in these types of cancer cells inhibited their proliferation and caused them to differentiate, suggesting that EBP1 was able to alter the gene expression profile of the cells. Chromatin immunoprecipitation assays showed the binding of EBP1 to the E2F1 promoter region, and additional immunoprecipitations using antibodies directed against other components of potential SIN3A-containing complexes (such as HDAC2), and against SIN3A itself, showed that these proteins form a direct association with EBP1. These interactions were investigated in more detail using each of the individual component proteins made by *in vitro* translation of cDNA vectors containing their coding sequences. These experiments indicated that EBP1 binds SIN3A directly. However, *in vitro*-translated HDAC2, which co-immunoprecipitated with EBP1 from cell lysates, failed to interact with the EBP1 translated *in vitro*, suggesting that these two were unable to enter into a direct binding and required SIN3A to act as a "bridge" between them.

The NURD complex is associated with a distinct chromatin remodeling activity promoted by the MI-2 protein, which is analogous to the ISWI

ATPases. Interestingly, SIN3A has also been shown to bind to ISWI-type proteins (at least in *Drosophila melanogaster*), and presumably the presence of this protein in the SIN3A repression complex is required to cause the DNA to "loop" away from the nucleosome in a manner analogous to that observed for the NURD complex. However, there is some debate as to why this looping away should be required.

A possible rationale is that NURD acts in concert with methyl-binding-domain proteins, such as MBD2, and some relaxation of the histone–DNA interaction may be necessary to allow MBD2 to access the 5-methylcytosine-enriched segments of the DNA. There is less evidence for this type of interaction with the proteins that comprise the SIN3A assembled complexes, but some data suggest that the methyl-CpG-binding proteins MECP2 and MBD4 are able to bind to SIN3A and HDAC1; some relaxation of the DNA–histone interaction in the nucleosome may therefore be required to allow these methyl-binding domains to access their targets.

Surprisingly, ISWI seems to be enriched in areas of hypoacetylated chromatin. However, this could result from its functional association with the SIN3A histone deacetylase complex, which would have already led to a loss of acetyl groups from the associated chromatin. One would predict that changes in the level of ISWI expression should affect the global levels and distribution acetylation of histones H3 and H4. This is indeed true, as is evident from ISWI overexpression in *Drosophila*, in which significant decreases of H3 and H4 acetylation are observed, possibly because the overloading of wild-type ISWI on polytene chromosomes could recruit SIN3A HDAC activity to a great number of chromatin loci.

Protein complexes containing histone acetyltransferases promote transcription

The association of histone acetyltransferases (HATs) with active transcription means that they are often found as part of the protein complex that drives the assembly of the transcription machinery at gene promoters. We saw in Section 2.2 how this assembly occurs in stages, with the delivery of the TATA box binding protein (TBP) to the core promoter as part of the TFIID protein complex. This complex also contains HAT activity because gene activation requires chromatin remodeling, which in turn is presumed to require an increase in histone acetylation. This also seems to be a general feature of other complexes that deliver the TBP to promoters.

There are several well-characterized examples of this type of complex in addition to TFIID, the most common probably being the SAGA (SPT–ADA–GCN5 acetyltransferase) complex. Both the SAGA complex and the TFIID complex are shown in **Figure 8.3**, where we can see that SAGA contains the protein GCN5 in addition to TBP and several other components. GCN5, a protein conserved from *Saccharomyces cerevisiae* to humans, is one of the most studied HATs. This protein carries two conserved domains: an acetyltransferase domain that is required for its catalytic activity, and a bromodomain that binds acetylated lysine residues.

Recombinant yeast GCN5 preferentially modifies histone H3 K14 and histone H4 K8 and K16 *in vitro*. However, the incorporation of yeast GCN5 into multisubunit complexes expands its substrate specificity, enabling it to acetylate histone H3 K9 and K18, in addition to K14. Mammalian GCN5 shows broadly similar characteristics to its yeast counterpart but has also been identified in complexes that do not deliver TBP to the promoter, such as SPT3–TAF9–GCN5 acetyltransferase (STAGA) and the TATA-binding protein (TBP)–free TAF complex (TFTC) (see **Figure 8.4**).

(a)   (b)

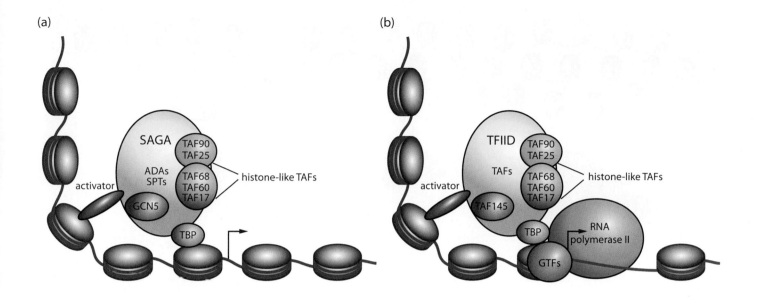

Both SAGA and STAGA can use an alternative HAT, the p300–CBP-associated factor (PCAF), although the precise reasons behind this selection are not clear. It is possible that PCAF interacts with other proteins needed to control promoter activity. For example, PCAF is known to acetylate p53 at lysine 382, allowing p53 to be recruited to the promoter of one of its target genes (*p21*). So the choice of HAT to incorporate into such complexes may depend on the target gene to be up-regulated and its downstream application in the cell.

The PCAF-containing complex actually leads us to look at another quite different HAT, namely p300, which competes with PCAF for binding at the *p53* promoter. p300 is actually involved in controlling a much wider range of genes. Tri-iodothyronine affects gene expression in a wide range of cellular pathways and functions, including gluconeogenesis, lipogenesis, insulin signaling, adenylate cyclase signaling, cell proliferation, and apoptosis. The ability to do this relies on thyroid hormone response elements (consensus sequence AGGTCAnnnnAGGTCA) in the promoters of genes involved in these processes, to which the hormone can bind after transport into the nucleus from its receptor on the cell membrane. This binding recruits the p300 HAT and/or PCAF (see **Figure 8.5**), so that transcription can begin. It is interesting that this occurs at a site distal to the TATA box binding site, which even in the repressed promoter can still be occupied by a TBP delivery system that can even load RNA polymerase II in readiness for transcription. Transcription does not actually begin until p300 or PCAF has completed histone acetylation to allow the transcription complex to proceed along the gene's coding sequence.

p300 is quite a versatile HAT: it seems to do things other than simply acetylate histones. Although strictly speaking this is not an example of direct epigenetic control of genome function, one of the known activities of p300 is to acetylate various lysines in the MAX protein. This inhibits the ability of MAX to activate *c-myc* target genes. A possible mechanism for this inhibitory action may be to interfere with the ability of MAX to form the MYC–MAX heterodimer, which in turn may establish a preference for forming the opposing MAD–MAX dimer, leading to gene repression through HDAC recruitment. This implies that p300 is able to control the level of histone acetylation at the promoters of *c-myc* target genes, albeit in an indirect manner and one that is completely opposite to its normal gene-activating mechanism.

Figure 8.3 Protein complexes containing subunits with histone acetylation activity that acetylate histones in promoter proximal nucleosomes. (a) The transcription adaptor complex (SAGA) and (b) general transcription factor (TFIID). Subunits with histone acetylation activity (pink), nucleosomes (purple), histone-like TAFs (dark green), and additional TAFs (light green) are shown. SAGA also includes ADA and SPT proteins, and TFIID contains additional TAFs. Both complexes can interact with transcriptional activators, and bind or contain the TATA-binding protein (TBP). TFIID is also crucial in the formation of transcription complexes, including additional general transcription factors (GTFs) and RNA polymerase II. [From Grant PA & Workman JL (1998) *Nature* 396, 410. With permission from Macmillan Publishers Ltd.]

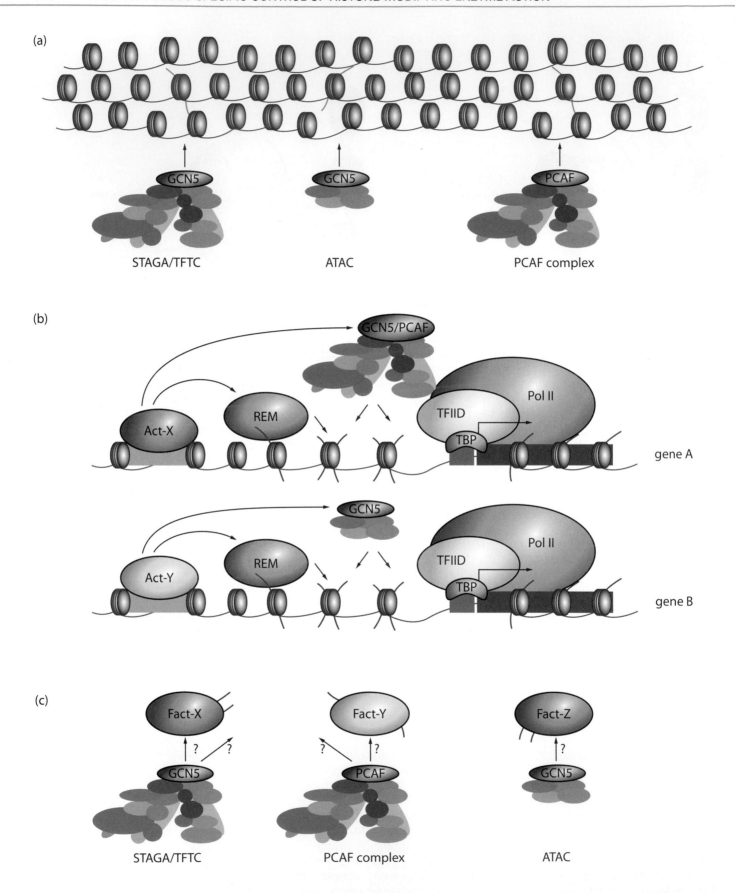

Figure 8.4 The distinct GCN5- and PCAF-containing complexes possess global histone acetylation activity and locus-specific co-activator functions, together with acetyltransferase activity on non-histone substrates. (a) Global acetylation. STAGA/TFTC, PCAF, and/or ATAC complexes contribute to structural integrity of the chromosomes with a preference for global acetylation of histones H3 or H4 on different nucleosomes (brown ovals). (b) GCN5- and PCAF-containing complexes also function as co-activators at specific genes. Once a specific activator (Act-X or Act-Y) has bound to its binding site or enhancer region (light gray box) chromatin-modifying HAT complexes (STAGA/TFTC/PCAF-type, upper panel) and ATP-dependent chromatin-remodeling complexes (REM) are recruited to the vicinity of the core promoter (purple box). Nucleosomes are acetylated, ATP-dependent chromatin-remodeling complexes are binding to the acetylated histone tails (short green lines), and nucleosomes are mobilized as a result of these activities (short arrows). Nucleosomes covering the core promoter get evicted; thus, TFIID and other general transcription factors including RNA polymerase II (Pol II) can bind to the core promoter region and transcription will start from the given gene (dark gray box). (c) GCN5- and PCAF-containing complexes regulate transcription by directly acetylating transcription factors. The GCN5- or PCAF-dependent acetylation (short green lines) on these nuclear factors (Fact-X, Fact-Y, Fact-Z) can have positive and negative effects on transcription regulation. [From Nagy Z & Tora L (2007) *Oncogene* 26, 5341. With permission from Macmillan Publishers Ltd.]

8.2 COMPLEXES OF THE HISTONE METHYLTRANSFERASES

The histone methyltransferase (HMTasc) enzymes described in Section 7.2 also depend on being included in multiprotein complexes for their specificity of action. We have already discussed an important example of this dependence in the MLL–WRD5-containing complex (see Section 7.2). Other SET domain-containing HMTases form complexes such as COMPASS and the polycomb repressive complexes.

The yeast SET1–COMPASS complex is the sole histone H3 K4 methyltransferase in that organism and consists of the catalytic SET1 protein and seven other, noncatalytic proteins. Mammalian SET1-like complexes are more widely expressed and provide nonredundant functions *in vivo*. These latter complexes are referred to as SET1A and SET1B, and their SET1-containing proteins show considerable sequence similarity to their yeast counterparts, despite the fact that other components of the complexes, such as CFP1 and WRD82, seem to be unique to mammals.

Polycomb repressive complexes (PRCs) have the HMTase EZH2 as an essential component (see Section 7.2), and this allows a considerable range of histone modifications to be performed. Human EZH2, in complex with different isoforms of EED, another polycomb-group protein, exhibits HMTase activity *in vitro* toward H3 K27 and H1 K26, as well as limited activity toward H3 K9. Another component of PRC, SUZ12, is also necessary for these activities. SUZ12 is also able to bind to heterochromatin protein 1 (HP1), suggesting that SUZ12 may have a role in constitutive heterochromatin formation in mammalian cells. Levels of murine SUZ12 remain constant during mouse embryonic stem (ES) cell differentiation, whereas EED and EZH2 levels decrease, providing a second line

Figure 8.5 A molecular model for basal repression by co-repressors in the absence of tri-iodothyronine and transcriptional activation by co-activators in the presence of tri-iodothyronine. CBP, cAMP-response-element binding protein; DRIP, Vitamin D receptor interacting protein; GTF, general transcription factor; HDAC, histone deacetylase; P/CAF, p300/CBP-associated factor; RXR, retinoid X receptor; SRC, steroid receptor co-activator; TAF, TATA-binding-associated factor; TBP, TATA-binding protein; TF, transcription factor; TR, thyroid hormone receptor; TRAP, TR-associated protein; TRE, thyroid hormone-response element. [From Yen PM, Ando S, Feng X et al. (2006) *Mol. Cell. Endocrinol.* 246, 121. With permission from Elsevier.]

(a)

(b)

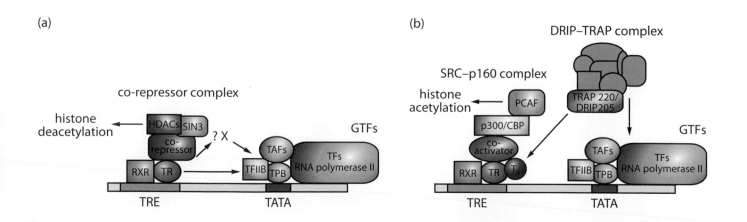

of evidence that SUZ12 may function outside the EED–EZH2 complexes. In addition, SUZ12 also seems to be able to recruit the histone H3K4 demethylase LSD1. So not only are PRCs capable of adding repressive modifications, they are also able to remove activating ones to ensure that silencing of their target genes is wholly effective. A similar interaction exists between RING6a and JARID1D.

How PRCs are directed to their target genes remains a mystery, but it is thought that this involves binding to polycomb binding elements (PBEs) within their promoters. Components of some polycomb-group complexes, such as the Pleiohomeotic protein of *Drosophila*, are known to have PBE-binding activity and to be able to recruit many of the remaining polycomb-group proteins into a functional complex. The function of similar PBE-binding proteins in mammals has yet to be elucidated.

Acting in a fashion opposite to that of the polycomb-group complexes are proteins belonging to the trithorax group. These also exist in several different multiprotein complexes; intriguingly, two of those complexes also possess HMTase activity. However, their HMTase activity is, as expected, specific to methylation of H3 K4, in accordance with the generally gene-activating function of trithorax-type complexes. For many of the HMTase complexes, specificity of action is achieved through transcription factor binding, just as it is for the complexes performing acetylation. An example is shown in **Box 8.1**, which describes the recruitment of histone-modifying enzymes by the RUNX family of transcription factors.

Gene-specific recruitment of an HMTase-containing complex does not necessarily require a DNA-binding transcription factor, because many of the accessory proteins that regulate cellular activities such as cell proliferation and cell cycle can bind to several histone-modifying enzymes and in doing so target them to multiple gene loci. HCF-1 is just such a regulator of cell proliferation, and **Figure 8.6** shows a diagram of the domain structure of HCF-1, together with the types of proteins known to bind to it.

This suggests that HCF-1 acts as a scaffold for the assembly of histone-modification complexes. However, the complex that HCF-1 forms with VP16 in HSV-transfected cells is rather unusual in that it contains H3K4 methyltransferase activity (in the form of a SET1–ASH2 HMTase) and HDAC activity (introduced into the complex through association with the SIN3A protein). At first glance, the presence of activating and repressive histone modifiers in the same complex seems somewhat contradictory. However, a detailed analysis of this protein "supercomplex" seems to indicate that although HCF-1 can bind the SIN3 HDAC and SET1–ASH2 HMTase complexes simultaneously, the HSV protein VP16 does not stably associate with this supercomplex. Instead, it is found to selectively bind to HCF-1 only when this is bound to the activating SET1–ASH2 HMTase complex. Therefore, as a transcriptional activator, VP16 distinguishes between the chromatin-modifying activities of proteins bound to a co-regulator and associates with the protein complex in a manner that permits VP16 to perform its function. Thus the introduction of VP16 into the cell by the virus activates several genes associated with cell proliferation. It is interesting, however, that HCF-1 does also bind the opposing repressive complex, an act that must have some function. It has been suggested that tethering the HDAC and HMTase functions simultaneously to the gene that HCF-1 regulates provides a platform that can lead to oscillating on and off states of transcription, and that some as yet unknown system alters the balance between these states to activate or repress cell proliferation.

Box 8.1 Recruitment of histone-modifying enzymes by the RUNX family of transcription factors

The RUNX family are DNA-binding transcription factors that regulate the expression of genes involved in cellular differentiation and cell cycle progression. The family includes three mammalian RUNX proteins (RUNX1, RUNX2, and RUNX3) and two homologues in *Drosophila*. RUNX-mediated repression involves the recruitment of co-repressors such as mSIN3A and Groucho, as well as histone deacetylases. Furthermore, RUNX1 and RUNX3 associate with SUV39H1.

RUNX family members were initially implicated in transcriptional activation, and indeed these factors bind to core elements of enhancers and promoters that are required for the activation of a large number of genes. RUNX proteins bind DNA as heterodimers with core binding factor β (CBFβ), a non-DNA-binding subunit that increases the affinity of RUNX factors for DNA. The RUNX proteins are poor activators on their own, even though they are able to interact with transcriptional co-activators such as histone acetyltransferases. However, when placed in the appropriate context, RUNX family members act as organizing factors at the promoters and enhancers of target genes, where they associate with cofactors and other DNA-binding transcription factors that are required for gene regulation. In contrast, the RUNX proteins are potent repressors of transcription even when expressed alone, suggesting that their default function may be one of transcriptional repression. **Figure 1** shows the extent of protein interactions possible for the RUNX1 proteins, several of which are enzymes that modify histones when incorporated into multiprotein complexes.

Although many genes contain binding sites for the various RUNX family members, repression and activation by RUNX family members is cell-type specific and depends on associated cofactors. One of the first co-repressors that was found to bind to RUNX factors was the human Groucho homolog TLE. Studies involving mutant versions of the RUNX1 protein demonstrated binding to SIN3A, HDACs 1, 2, 5, and 6, p300, and SUV39H1. The presence of the last of these indicates that RUNX family members are required not only for short-term transcriptional repression by changes in histone acetylation, but also for more permanent gene silencing, because SUV39H1 trimethylates H3K9 to permit the binding of HP1 as a first step toward heterochromatinization.

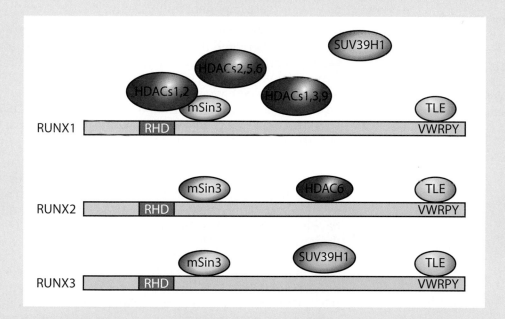

Figure 1 RUNX family members recruit co-repressors, HDACs, and SUV39H1 to repress transcription. RUNX family members recruit the Groucho homologue TLE to their C-terminal domains that contain the Groucho-binding motif VWRPY. They also associate with the mSin3 co-repressors that recruit HDACs 1 and 2. Binding of mSin3 is required for RUNX1-mediated repression of the p21Waf1/Cip1 promoter in NIH 3T3 cells. RUNX1 associates strongly with HDACs 1, 3, and 9 and weakly with HDACs 2, 5, and 6. RUNX2 binds HDAC6. RUNX1 and RUNX3 recruit SUV39H1 to repress transcription. RHD, runt homology domain. [From Durst KL & Hiebert SW (2004) *Oncogene* 23, 4220. With permission from Macmillan Publishers Ltd.]

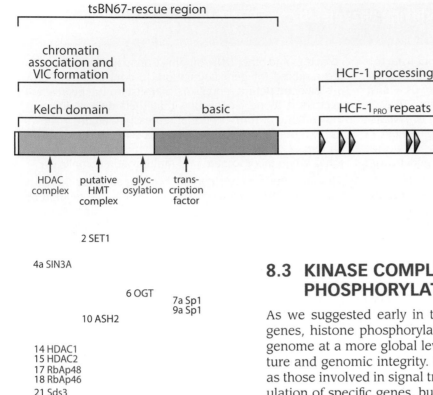

Figure 8.6 Protein binding to HCF-1. The regions of the HCF-1 protein that the proteins bind to are indicated. NLS, nuclear localization signal. [From Wysocka J & Herr W (2003) *Trends Biochem. Sci.* 28, 294. With permission from Elsevier.]

8.3 KINASE COMPLEXES FOR HISTONE PHOSPHORYLATION

As we suggested early in this book, rather than regulating individual genes, histone phosphorylation may be responsible for controlling the genome at a more global level, one associated with chromosome structure and genomic integrity. Phosphorylation of accessory proteins, such as those involved in signal transduction, are probably integral to the regulation of specific genes, but phosphorylation of histones at specific loci may not be very important for this activity.

A possible exception to this is the repair of double-strand DNA breaks. Cells must overcome the constant threat of damage to their DNA for healthy survival. Double-strand breaks (DSBs) are the most deleterious form of DNA damage. DSBs are generated by exposure to ionizing radiation or genotoxic chemicals, and can also occur during normal cellular processes such as DNA replication. If left unrepaired, DSBs can result in cell death or in loss of genetic information, chromosomal translocations, and genome instability, all potentially leading to cancer development.

In eukaryotes, DSBs and their detection and repair occur in the context of chromatin in which DNA is tightly bound to histones. The relaxation of these interactions through ATP-dependent chromatin remodeling is an essential part of the repair process of DSBs, but histone modifications also have a role. For example, the phosphorylation of H2AX (γ-H2AX) is one of the best-characterized histone modifications that occurs on the chromatin at the sites of DSBs; it forms visible nuclear foci and is important for efficient DSB repair. This phosphorylation event promotes the interaction of H2AX with SWI/SNF-type chromatin remodeling complexes, such as BRG1, that are needed to temporarily remove the DNA from the nucleosome surface so that it becomes accessible to the DNA repair machinery.

The cell checkpoints that permit cell cycle progression also rely in part on phosphorylation of H2AX. Mediator of DNA-damage checkpoint 1 (MDC1) is known to bind directly to the phosphorylated form of H2AX and also to the anaphase-promoting complex/cyclosome (APC/C) protein, which is an E3 ubiquitin ligase involved in cell cycle control. APC/C is active from early mitosis through late G1. During this time, APC/C targets many critical regulators of the cell cycle for degradation and is essential for proper cell cycle progression. APC/C acquires its substrate specificity via two co-activators, CDC20 and CDH1. CDC20 is active during the metaphase–anaphase transition, whereas CDH1 is active during later stages of mitosis and G1.

8.4 COORDINATION AMONG CHROMATIN-MODIFYING COMPLEXES

HDAC complexes respond to other histone modifications

In accordance with the histone code hypothesis, there should be some means of interrogating the pattern of post-translational modifications at a particular N-terminal histone tail to coordinate the efforts of the various epigenetic modification complexes that need to act there. Acetylation of histones is normally associated with methylation of adjacent, or at least nearby, lysine residues. This enrichment of acetyl groups at a particular locus may be thought of as a means to reinforce the message given to the cell's transcriptional machinery. Does this imply that there is any form of interaction between HAT complexes and those that either detect or add the methyl groups to the adjacent lysine residues?

The CREB-binding protein (CBP) (which is known to have an intrinsic HAT activity) participates in the transcriptional activation of several target sequences by physically interacting with sequence-specific transcription factors. This seems to occur in parallel with the highly similar protein p300, which as we saw earlier is recognized as having HAT activity. Immunoprecipitation experiments show that p300–CBP co-precipitates with other proteins possessing strong histone methyltransferase activity that preferentially methylates lysines 4 and 9 on histone H3. Similarly, HDAC1 and HDAC2 are known to associate with the endogenous human protein MMSET, an active HMTase both *in vivo* and *in vitro*, that targets histones H3 and H4 (although it shows a preference for H4 K20). These observations emphasize the apparent role of MMSET in the repression of several genes, because methylation of H4 K20 is generally associated with transcriptional repression and/or heterochromatin formation. It is important to note that MMSET is able to function as part of a complex that also deacetylates chromatin, this being the other immediately essential function to allow heterochromatinization.

Noncoding RNA can regulate histone-modifying complexes

The evidence that RNA-directed processes help to orchestrate chromatin architecture and epigenetic memory is growing rapidly and is already compelling. We saw earlier (Section 5.5) how small noncoding RNAs (ncRNAs) can regulate the activity of DNA methyltransferases by using the RNA interference pathways and also by direct binding to proteins, such as MECP2, that are central to the regulation of DNA methylation. Additionally, there is a wealth of information that substantiates the ability of ncRNAs to regulate histone-modifying complexes (**Figure 8.7**). It is known, for instance, that RNA is an integral component of chromatin and that many of the proteins involved in chromatin modifications have the capacity to bind RNA or complexes containing RNA. These proteins include not only the DNA methyltransferases and DNA methyl-binding domain proteins, but also many other proteins involved in histone methylation and acetylation.

An example is the yeast SET1 H3K4 methyltransferase, which contains, in addition to its catalytic SET domain, a conserved RNA-recognition motif (RRM1). The structure of this motif has been determined and shown to bind RNAs only when it is associated with the whole protein complex. In *S. cerevisiae*, SET1 catalyzes monomethylation, dimethylation, and trimethylation of H3K4, and a good example of this activity is the *GAL10* gene. The extensively studied yeast *GAL1–GAL10* gene cluster is tightly regulated by the availability of sugar in the environment. Unexpectedly, under repressive conditions (that is, low sugar availability), the 3′ region

Figure 8.7 Long noncoding RNA-mediated chromatin remodeling. Some long ncRNAs that are transcribed by RNA polymerase II recruit transcriptional repressive complexes including PcGs and G9a to silence specific genomic regions, both *in cis* (top) and *in trans* (bottom). [From Chen L-L & Carmichael GG (2010) *Curr. Opin. Cell Biol.* 22, 357. With permission from Elsevier.]

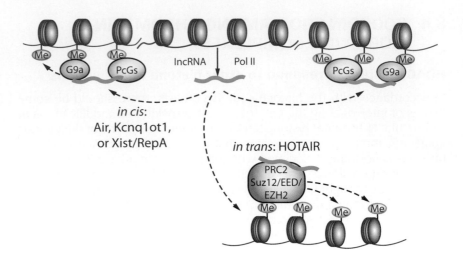

of the *GAL10* coding sequence is trimethylated by SET1 on histone H3K4, a modification normally characteristic of 5′ regions of actively transcribed genes. It seems that this trimethylation is controlled by the expression of a long ncRNA whose sequence lies within the 3′ end of the *GAL10* gene.

Long ncRNAs are also involved in many epigenetic processes, with increasing reports that RNAs can direct and regulate both chromatin activator complexes (CACs) and chromatin repressor complexes (CRCs). Good examples of these are the well-known role of RoX ncRNAs in *Drosophila* dosage compensation (wherein the RoX ncRNAs guide the generic MSL complex to specific sites on the X chromosome to promote global gene activation in males) and the involvement of ncRNAs in parental imprinting. Perhaps the best-characterized example of control by long ncRNA is the XIST/TSIX-mediated X-chromosome inactivation in mammals, which has recently been shown to intersect with the RNA interference (RNAi) pathway. The mechanisms behind these processes are as yet far from clear, but substantial progress has been made toward gaining a greater understanding of RNA-based chromatin control. For example, at the *KCNQ1* and *IGF2R* imprinted clusters of mammals, the long antisense ncRNAs Kcnq1ot1 and Air have both been shown to coat target chromatin regions in a manner similar to the activity of XIST, and to interact with HMTase complexes (G9a and polycomb) to direct the imprinting of specific genes in placental tissues.

Polycomb and trithorax are examples of chromatin activator and repressor complexes controlled by noncoding RNA

Chromatin regulation by the polycomb and trithorax complexes also seems to involve ncRNA; that is, functional RNA not translated into a protein. In many of the genes regulated by polycomb or trithorax, specific response elements for these complexes exist within the promoter regions of the target genes. Polycomb gene (*PcG*) and trithorax gene (*trxG*) regulators are recruited to specific chromosomal elements that are present in the *cis*-regulatory region of target genes. The same element can act as an activating or a silencing module. In the repressed state, the elements represent polycomb response elements (PREs) and facilitate the recruitment of polycomb-group proteins. In the activated state, the DNA elements function as trithorax response elements (TREs) and recruit trithorax proteins. However, there is evidence to suggest that transcription of ncRNAs from PRE/TRE elements switches silent PREs into TREs, which indicates that transcription as ncRNA is important in epigenetic activation.

The mcRNA transcripts of three TREs located in the *Drosophila* homeotic gene Ultrabithorax (*Ubx*) mediate transcriptional activation by recruiting the epigenetic regulator ash1 to the template TREs. TRE transcription coincides with *Ubx* transcription and the recruitment of ash1 to TREs. The SET domain of ash1 binds all three TRE transcripts, with each TRE transcript hybridizing with and recruiting ash1 only to the corresponding TRE in the chromatin of the *Ubx* gene, where it methylates H3K4.

Quite how a short sequence of RNA is able to direct a protein complex to a precise region of the DNA has not been completely clarified, but it is possible that a small segment of antisense RNA generated from the DNA sequence of the polycomb or trithorax response elements is able to anneal back to a longer nascent transcript as it emerges from the ribosome. This may recruit chromatin-modifying proteins, such as ash1, either directly through their ability to bind to the short RNA sequence, or indirectly via guidance by the argonaute protein.

This is one possible method by which ncRNAs may be able to direct chromatin-modification protein complexes to precise genomic loci, but it is also one that seems to be largely restricted to yeast, because there is little evidence that small RNAs have a polycomb/trithorax (PRE/TRE) targeting role in vertebrates. In contrast, there is accumulating evidence for the involvement of longer ncRNAs in vertebrate PRE/TRE regulation. **Table 8.1** lists the noncoding transcripts that have been mapped to PRE/TRE binding sites in flies and mammals. These noncoding transcripts are typically several hundred base pairs to several kilobases long, and those that have been studied in detail are subject to complex and dynamic regulation.

For example, one of the best-studied fly PRE/TREs is bxd, a long, 26 kb embryonic RNA polymerase II (Pol II) transcript that runs through the PRE/TRE and other nearby regulatory sequences and undergoes complex differential splicing. In addition, three shorter larval Pol II transcripts of 350 to 1108 nucleotides, arising from within the PRE/TRE element, have also been mapped. Strikingly, *in situ* hybridization analysis of most of the transcripts shown in Table 8.1 reveals that, in both flies and mammals, their spatial and temporal expression correlates precisely with the global domain of activation of the gene that is regulated by the PRE/TRE, suggesting that they are under similar regulation to their cognate genes. For the *Hox* clusters in both flies and vertebrates, the noncoding transcripts appear slightly earlier than the mRNA of their cognate *Hox* genes. This has led to the proposal that they have a role in early activation by opening regulatory chromatin domains, thereby switching PRE/TRE elements to an active state in the cells where the cognate gene is activated.

There are several lines of evidence indicating that both polycomb and trithorax proteins possess RNA-binding activities, and these are outlined in the diagram in **Figure 8.8**.

Immunoprecipitation of a given protein followed by use of the reverse transcription polymerase chain reaction (RT-PCR) for specific RNAs suggests that each protein can pull down several different RNAs and may engage in different interactions in different cell types. For example, the Kcnq1ot1 ncRNA, involved in silencing imprinted genes, was found to be associated with SUZ12 and EZH2 in mouse placenta, where its target genes are silenced, but not in fetal liver, where they are active. The fact that these RNA interactions can occur is interesting, but it does not give us much information about the mechanism needed to recruit polycomb/trithorax to one locus and not to another.

TABLE 8.1 NONCODING RNAS IN POLYCOMB/TRITHORAX REGULATION IN FLIES AND MAMMALS

Name of polycomb/ trithorax response element	Gene regulated	Genomic location relative to gene	Coding direction	Correlation to regulated gene	Cell type/tissue
Flies					
Iab8	*Abd-B*	Downstream	Sense	Co-expressed	Kc and SF4 cells
Fab-7	*Abd-B*	Downstream	Sense	Spatial	2–6-hour *Drosophila* embryos
Iab6	*Abd-B*	Downstream	Sense	Spatial	2–6-hour embryos
MCP	*Abd-B*	Downstream	Sense	Spatial	2–6-hour embryos
Iab4	*Abd-A*	Upstream	Sense/antisense	Spatial	2–6-hour embryos
Iab3	*Abd-A*	Promoter	Sense	Spatial	2–6-hour embryos
Iab2	*Abd-A*	Promoter	Sense	Spatial	2–6-hour embryos
Bxd embryonic	*Ubx*	Upstream	Sense	Spatial	2–6-hour embryos
Bxd larval	*Ubx*	Upstream	Sense	Spatial	Larval 3rd leg disc
Dfd	*Dfd*	Intron	Antisense	Co-expressed	Kc and SF4 cells
En	*En*	Upstream	Sense/antisense	Spatial	Embryos
Salm	*Salm*	Upstream	Sense/antisense	Spatial	Embryos
Slou	*Slou*	Upstream	Sense/antisense	Spatial	Embryos
Til	*Til*	Upstream	Sense/antisense	Spatial	Embryos and larval brain
Mammals					
HOTAIR	*HoxD* cluster	Intergenic	Antisense	Opposite	Human adult fibroblasts
Hox transcripts	*Hox* clusters	Intergenic	Antisense	Co-expressed	Adult tissues and placenta
Hoxb5/b6as	*HoxB5/b6*	Overlaps promoter	Antisense	Co-expressed	Mouse ES cells/mouse embryos
Evx1as	*Evx1*	Overlaps promoter	Antisense	Co-expressed	Mouse ES cells/mouse embryos
Rep A	X-chromosome	Within *Xist* RNA	Sense	Transcribed before Rep A up-regulated	Mouse embryonic fibroblasts
Kcnq1ot1	Imprinted genes in kcnq1 region	Within imprinted cluster	Antisense to Kcnq1	Opposite	Mouse embryonic fibroblasts, placenta, fetal liver

Several theoretical models have been proposed to account for this specificity of interaction with polycomb/trithorax (see **Figure 8.9**). The models seem to require the chromatin to be in a conformation that permits loading of the transcriptional machinery onto the DNA, because transcription of the RNA is always required. This requirement for transcription suggests that ncRNA transcribed through PRE/TREs can both activate and silence the transcription of the associated coding gene, depending on the nature and level of the transcript that is expressed.

One study in which the human *HOX* gene complexes were ectopically activated suggests that ncRNA transcription through polycomb-group-binding sites is involved in *HOX* gene activation. In this study, induction of differentiation in human embryonic carcinoma cells by retinoic acid was used to provoke progressive up-regulation of the *HOXA* cluster genes. Noncoding RNAs adjacent to each gene within the cluster behaved similarly to their adjacent genes and, indeed, were activated

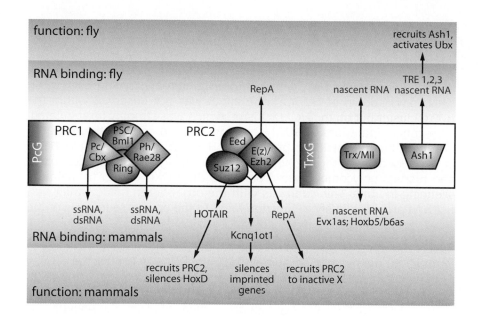

Figure 8.8 Binding of RNA by PcG and TrxG proteins. The PRC1, PRC2, and TrxG complexes are shown. Where fly and vertebrate homologs have different names, the fly protein name is given first. The interactions of fly proteins with RNA are shown above the drawing, and the vertebrate interactions are shown below the drawing. RNA names given in red indicate that the interaction has been tested *in vitro* with recombinant protein or protein domains, demonstrating direct binding. For RNA names in black, the interaction was demonstrated by chromatin immunoprecipitation, regular immunoprecipitation, or RNA pulldown from protein extracts, so these interactions may be indirect. Where functions have been investigated, these are indicated in the top and bottom panels. [From Moazed D (2011) *Cell* 146, 510. With permission from Elsevier.]

more rapidly in response to treatment with retinoic acid. It was apparent that mammalian ncRNA transcription precedes *HOX* transcription. Chromatin immunoprecipitation experiments demonstrated that SUZ12 and EZH2 binding to the cluster decreased after treatment with retinoic acid, indicating a correlation between the transcription of ncRNAs and the loss of polycomb-group binding. However, overexpression of a transgenic ncRNA from the region between *HOXA3* and *HOXA4* genes did not have any effect on any of the *HOX* genes. This suggests that the RNAs themselves are not sufficient for activation and that the act of transcription rather than the RNA molecule itself may be the important activating signal. However, this result does not rule out the possibility that the ncRNA does have a structural role, but one that only functions *in cis* at the locus from which it is transcribed.

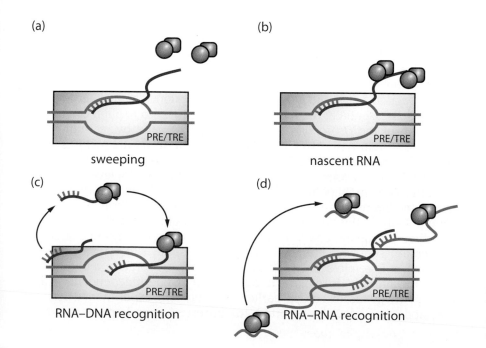

Figure 8.9 Models for noncoding RNA in PcG/TrxG regulation and recruitment. The PRE/TRE is shown as a melted DNA. Noncoding RNA transcripts from the top ("sense") strand are shown in red; those from the antisense strand are shown in blue. The protein complexes may be either PcG or TrxG complexes. (a) Sweeping: the act of transcription removes chromatin-bound complexes from DNA. (b) Nascent RNA: PcG or TrxG complexes bind to the nascent RNA and/or to single-stranded DNA that is being transcribed. (c) RNA–DNA recognition: the noncoding RNA is bound by PcG or TrxG complexes independently of chromatin, and is recruited back to a DNA site of complementary sequence by DNA–RNA pairing. (d) RNA–RNA recognition: a single strand of the noncoding RNA binds to PcG or TrxG proteins and guides them to a nascent RNA transcribed from the opposite strand, the one with complementary sequence.

Activation of *HOX* genes is not the only contribution made by ncRNAs. There is also recent evidence that other ncRNAs are important for silencing polycomb-group targets. A 2.2 kb antisense ncRNA named HOTAIR is expressed from the human *HOXC* cluster at the boundary between the active and silent *HOX* genes in foreskin fibroblasts. Surprisingly, on siRNA knockdown of HOTAIR, the effect was not observed within the *HOXC* cluster but on the *HOXD* cluster, which was strongly up-regulated, with concomitant loss of SUZ12 binding. This, in combination with the observed binding of PRC2 proteins to HOTAIR, suggests that HOTAIR is expressed from the *HOXC* cluster specifically in certain cell types and recruits polycomb-group proteins to the *HOXD* cluster to maintain silencing *in trans*. This contrasts with the observation made earlier that ncRNAs are more likely to function *in cis* to the coding sequences of the *HOXA* genes.

Such contradictory observations suggest that we still have a great deal to learn about the mechanisms employed by ncRNAs to control the activity of individual genes through polycomb or trithorax complex recruitment. The models depicted in Figure 8.9 are largely theoretical at the time of writing. Because they all have supporting evidence from the scientific literature, it is difficult to assess which of them is more likely to be a general mechanism. The concept of "sweeping" chromatin proteins away from the DNA by passage of the transcription complex (Figure 8.9a) is one we have seen before in relation to chromatin remodeling, but this requires that the gene (or cluster of genes in the case of *HOX*) will have already been remodeled for activation by other mechanisms. This type of mechanism could probably contribute best to the maintenance of gene activity after an initial transcription event by recruiting histone-modifying agents, such as trithorax, to establish active marks, such as histone H3K4 methylation. However, this will only work if a transcription factor has already bound to its target site to initiate transcription in the first place.

The second model relies on binding of the polycomb or trithorax complex to a strand of nascent RNA (Figure 8.9b), and this could be important for both the initiation and maintenance of expression. However, once again, the process that leads to transcription of the nascent RNA is not clear, and neither is a reason why this activation method should be preferred over transcription factor binding.

The final two models propose mechanisms by which regulation and recruitment may occur *in trans*, meaning that the transcription of the ncRNA and its regulatory effects are uncoupled. Figure 8.9c shows the RNA–DNA recognition model, in which the ncRNA associates with polycomb-group or trithorax proteins independently of chromatin and is guided back to a site with a complementary sequence by RNA–DNA pairing. This is essentially the model proposed earlier to explain the trans-activation of the *Drosophila* Ultrabithorax (*Ubx*) gene by recruiting the trithorax protein Ash1. This RNA–DNA recognition model requires only one strand of the recruiting RNA to be transcribed, as is the case for many of the PRE/TRE transcripts that have been studied. However, for several PRE/TREs, noncoding transcription is detectable from both strands, raising the interesting possibility of an RNA–RNA recognition mechanism (Figure 8.9d). In this model, a free RNA strand binds to polycomb-group or trithorax proteins and anneals back to a nascent transcript of complementary sequence, transcribed from the same site. This proposed mechanism is reminiscent of fission yeast heterochromatin recruitment. If this mechanism does exist, it is very likely to be independent of the RNAi pathway, as discussed earlier. Although no experimental test of this model has yet been reported, the observed strand-specific binding of some polycomb-group or trithorax proteins, and the observation that several PRE/TREs are transcribed from both strands means that this idea certainly merits further investigation.

KEY CONCEPTS

- Epigenetic modifying enzymes are probably incapable of recognizing specific DNA sequences, so they need to form complexes with other proteins that can direct them to specific loci. A good example of this is the NURD complex, which combines histone deacetylases and transcription factors with other proteins that recognize methylated DNA. The whole complex is capable of repressing genes and remodeling nucleosome positions.

- There are several other examples of complexes of this type. Some promote gene repression, whereas others activate genes (the latter usually contain histone acetyltransferases and/or histone methyltransferases). Complexes of this type often include the scaffold protein SIN3A.

- Several of these protein complexes are able to recognize other histone modifications already present at a genomic locus. This is necessary to promote interaction between histone-modifying activities such as the coordination between acetylation and the types of histone methylations found in actively transcribed genes.

- Noncoding RNAs are often responsible for controlling the activities of histone-modifying protein complexes.

FURTHER READING

Brockdorff N (2013) Noncoding RNA and Polycomb recruitment. *RNA* 19:429–442 (doi:10.1261/rna.037598.112).

Chan HM & La Thangue NB (2001) p300/CBP proteins. HATs for transcriptional bridges and scaffolds. *J Cell Sci* 114:2363–2373.

Ciferri C, Lander GC, Maiolica A et al. (2012) Molecular architecture of human polycomb repressive complex 2. *Elife* 1:e00005 (doi:10.7554/eLife.00005).

Denslow SA & Wade PA (2007) The human Mi-2/NuRD complex and gene regulation. *Oncogene* 26:5433–5438 (doi:10.1038/sj.onc.1210611).

Feng Q & Zhang Y (2003) The NuRD complex: linking histone modification to nucleosome remodeling. *Curr Top Microbiol Immunol* 274:269–290.

Hekimoglu B & Ringrose L (2009) Non-coding RNAs in polycomb/trithorax regulation. *RNA Biol* 6:129–137.

Icardi L, Mori R, Gesellchen V et al. (2012) The Sin3a repressor complex is a master regulator of STAT transcriptional activity. *Proc Natl Acad Sci USA* 109:12058–12063 (doi:10.1073/pnas.1206458109).

Imoberdorf RM, Topalidou I & Strubin M (2006) A role for gcn5-mediated global histone acetylation in transcriptional regulation. *Mol Cell Biol* 26:1610–1616 (doi:10.1128/MCB.26.5.1610-1616.2006).

Kogo R, Shimamura T, Mimori K et al. (2011) Long noncoding RNA HOTAIR regulates polycomb-dependent chromatin modification and is associated with poor prognosis in colorectal cancers. *Cancer Res* 71:6320–6326 (doi:10.1158/0008-5472.CAN-11-1021).

Leeb M, Pasini D, Novatchkova M et al. (2010) Polycomb complexes act redundantly to repress genomic repeats and genes. *Genes Dev* 24:265–276 (doi:10.1101/gad.544410).

Lempradl A & Ringrose L (2008) How does noncoding transcription regulate *Hox* genes? *Bioessays* 30:110–121 (doi:10.1002/bies.20704).

Luo M, Ling T, Xie W et al. (2013) NuRD Blocks reprogramming of mouse somatic cells into pluripotent stem cells. *Stem Cells* 31:1278–1286 (doi:10.1002/stem.1374).

Marango J, Shimoyama M, Nishio H et al. (2008) The MMSET protein is a histone methyltransferase with characteristics of a transcriptional corepressor. *Blood* 111:3145–3154 (doi:10.1182/blood-2007-06-092122).

Mattick JS, Amaral PP, Dinger ME et al. (2009) RNA regulation of epigenetic processes. *Bioessays* **31**:51–59 (doi:10.1002/bies.080099).

Mercer TR & Mattick JS (2013) Structure and function of long noncoding RNAs in epigenetic regulation. *Nat Struct Mol Biol* **20**:300–307 (doi:10.1038/nsmb.2480).

Morey L & Helin K (2010) Polycomb group protein-mediated repression of transcription. *Trends Biochem Sci* **35**:323–332 (doi:10.1016/j.tibs.2010.02.009).

Nagy Z & Tora L (2007) Distinct GCN5/PCAF-containing complexes function as co-activators and are involved in transcription factor and global histone acetylation. *Oncogene* **26**:5341–5357 (doi:10.1038/sj.onc.1210604).

Petruk S, Sedkov Y, Riley KM et al. (2006) Transcription of *bxd* noncoding RNAs promoted by trithorax represses *Ubx* in *cis* by transcriptional interference. *Cell* **127**:1209–1221 (doi:10.1016/j.cell.2006.10.039).

Tang L, Nogales E & Ciferri C (2010) Structure and function of SWI/SNF chromatin remodeling complexes and mechanistic implications for transcription. *Prog Biophys Mol Biol* **102**:122–128 (doi:10.1016/j.pbiomolbio.2010.05.001).

Tsai MC, Manor O, Wan Y et al. (2010) Long noncoding RNA as modular scaffold of histone modification complexes. *Science* **329**:689–693 (doi:10.1126/science.1192002).

Wang KC, Yang YW, Liu B et al. (2011) A long noncoding RNA maintains active chromatin to coordinate homeotic gene expression. *Nature* **472**:120–124 (doi:10.1038/nature09819).

Wilusz JE, Sunwoo H & Spector DL (2009) Long noncoding RNAs: functional surprises from the RNA world. *Genes Dev* **23**:1494–1504 (doi:10.1101/gad.1800909).

Xie W, Ling T, Zhou Y et al. (2012) The chromatin remodeling complex NuRD establishes the poised state of rRNA genes characterized by bivalent histone modifications and altered nucleosome positions. *Proc Natl Acad Sci USA* **109**:8161–8166 (doi:10.1073/pnas.1201262109).

Yuan W, Wu T, Fu H et al. (2012) Dense chromatin activates Polycomb repressive complex 2 to regulate H3 lysine 27 methylation. *Science* **337**:971–975 (doi:10.1126/science.1225237).

Zhang Y, Ng HH, Erdjument-Bromage H et al. (1999) Analysis of the NuRD subunits reveals a histone deacetylase core complex and a connection with DNA methylation. *Genes Dev* **13**:1924–1935.

EPIGENETIC CONTROL OF CELLULAR FUNCTION

SECTION 2

9

EPIGENETIC CONTROL OF CELL-SPECIFIC GENE EXPRESSION

Epigenetic modification of the genome really has one overarching purpose: to control the output of information from the DNA content of the cell's nucleus in a manner that allows the cell to perform a defined set of functions in a reproducible manner. The greater part of this control is aimed at the repression or activation of specific genes, and there are several mechanisms by which this may be achieved. Naturally, as one would expect, direct control of the gene by methylation and/or acetylation of specific amino acids on the N-termini of the histones at its locus coupled to an absence of DNA methylation in the gene's control sequences is an important means of maintaining the gene in a state in which transcription factor binding is permitted. However, there are other influences on gene activity, such as the genes' relative positions in the nucleus, proximity to the nuclear membrane, and incorporation into transcription "factories" such as the nucleolus. All of these factors are subject to epigenetic control in some way.

9.1 EPIGENETIC CONTROL OF CHROMOSOME ARCHITECTURE

The position of DNA within separate subnuclear compartments reflects the expression or repression of genes

The cell nucleus is the most prominent compartment in the eukaryotic cell, and for a long time it was considered to be a static and scarcely structured compartment. The nucleus was thought to be substantially altered only during cell division, after the formation of metaphase chromosomes and partitioning of the chromosome complement into daughter cells. Another common image of the nuclear compartment has been that of a contingent cellular region in which nuclear components are randomly located. These static views of the cell nucleus have changed drastically in recent years, mainly as a result of technical advances in the field of microscopy. The cell nucleus is currently considered to be a highly complex and organized compartment in which nuclear components tend to occupy nonrandom positions, leading to a precise definition of the nuclear architecture concept.

Nuclear architecture is the result of the morphological and functional heterogeneity generated by the positioning of different subnuclear compartments inside the nucleus (**Figure 9.1**). A subnuclear compartment has been defined as a macroscopic region within the nucleus that is morphologically and/or functionally distinct from its surroundings. Two types

Figure 9.1 Compartmentalization of the mammalian cell nucleus. The mammalian cell nucleus contains chromatin in the form of chromosome territories (CTs). CTs may overlap at their touching borders (intermingling) or create the so-called interchromatin space (white). Constitutive heterochromatin (dark brown) is mainly found as pericentromeric chromatin in patches throughout the nuclear volume, at the nuclear periphery, and around nucleoli. Nuclear pore complexes, the double-layered nuclear membrane, and the meshwork-like nuclear lamina are structural hallmarks in the periphery of the nucleus. Chromatin loops with associated transcription factories may extrude out of chromosome territories within the nucleolus and also throughout the nucleoplasm. Transcription (green), replication (yellow), and DNA repair processes (light blue) usually occur in small domains with a diameter of less than 100 nm. A diverse set of nuclear bodies, such as speckles, paraspeckles, the perinucleolar compartment, Cajal bodies, and PML bodies, are found in the interchromatin space. [Adapted from Hemmerich P, Schmiedeberg L & Diekmann S (2011) *Chromosome Res.* 19, 131. With permisson from Springer.]

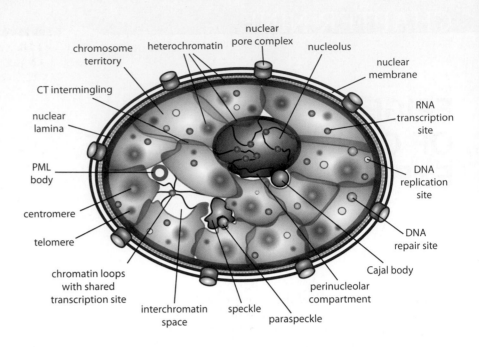

of subnuclear compartment are usually considered: nuclear bodies and chromosome territories, including associated chromatin domains.

Nuclear bodies are distinct subnuclear regions of different sizes lacking a lipid membrane and usually characterized by a definite protein composition. The most prominent nuclear body is the **nucleolus**, a factory for ribosome biogenesis and the site of RNA Pol II-independent ribosomal DNA (rDNA) gene transcription. There are many other nuclear bodies that are exclusively characterized by the presence of one or more specific proteins. Of special relevance are the **Cajal bodies**, which are the proposed sites of assembly of small nuclear ribonucleoprotein (snRNP), and **PML bodies**, which are of unknown function apart from being the main repositories of the promyelocytic leukemia protein (PML). The existence of thousands of RNA Pol II transcription factories dispersed in the nucleoplasm of mammalian cells has also been well documented.

Chromosome territories are slightly different structures (see Figure 9.1). The chromatin contained within each chromosome is not randomly distributed in the nucleus but occupies a specific location known as the **chromosome territory** or **domain**, a feature that constrains the whole spatial organization inside the nuclear compartment. It has been proposed that chromosome territories are not compact, inaccessible structures. Instead they are thought to be permeated by nucleoplasmic channels, creating a porous entity of enlarged surface area that is accessible to different nuclear factors. Large-scale chromatin domains, belonging to one or more chromosome territories, are also morphologically defined as heterochromatic, highly condensed, genomic regions or as euchromatic, less-condensed regions (this corresponds to the definitions given in Chapter 3).

Several reports also indicate a precise positioning of chromosome territories and chromatin domains relative to a radial orientation in mammalian cells. In some cases the radial position has been correlated to the chromosome's gene density, and in other cases to chromosome size.

Positioning of chromosome territories relative to other chromosome territories has been also reported, and certain chromatin domains, such as heterochromatic regions, tend to associate with the nuclear membrane and with the nucleolus. The above observations question the importance of the role of the nuclear architecture in gene positioning and function. Many gene loci tend to localize inside their corresponding chromosome territories and have strong preferential positioning with respect to the nuclear center. This positioning is not a direct consequence of gene activity and probably reflects the nonrandom location of the corresponding chromosome territory.

The nuclear skeleton is central to subnuclear organization

For a substantial period there has been biochemical evidence that the synthetic activities of the nucleus were attached to large intranuclear structures. The earliest indication of such structures came from the ability to extract apparently quite large particles from cells that not only contained DNA or RNA polymerase activity but were also clearly associated with many other component molecules. Initial experiments with HeLa cells synchronized in early S phase showed dense, morphologically discrete, ovoid bodies strung out along what appeared to be a three-dimensional structure—the nucleoskeleton—within the nucleus.

The **nucleoskeleton** is composed of many interacting structural proteins that provide the framework for DNA replication, transcription, and a variety of other nuclear functions. The attachment of chromosomes to the nuclear lamina is a characteristic of this structure (**Figure 9.2**). The principal protein components of the nuclear lamina are the lamins, of which there are two classes of proteins, the A-type and B-type lamins; these are expressed from three genes: *LMNA*, *LMNB1*, and *LMNB2*. Somatic cells express at least one B-type lamin, and A-type lamins are expressed in differentiated cells, with *LMNA* expressing two major A-type lamins—A and C—that are known to co-localize (in addition to many other proteins) with the DNA replication regions of the nucleus.

Lamin B1 seems to be particularly important for nucleoskeleton integrity. Alteration in the structure of the lamin network was evident in nucleoids prepared from lamin-B1-depleted cells, in which loss of stable DNA loops suggests that lamin-B1-dependent interactions are a key determinant of the organization of chromatin loops into so-called functional domains within the nucleus. Most notably, lamin B1 is essential for maintaining natural levels of gene expression, whereas loss of lamin A/C has no obvious functional implications. The major role of A-type lamins, by contrast, seems to be linked to the structural maintenance of the nuclear compartment and the regulation of nuclear mechanics. However, A-type lamins have been implicated in the functional organization of the nuclear interior, possibly by providing a structural scaffold for nuclear compartments. Mutations in *LMNA* (that is, lamin A/C) have been shown to cause a whole range of human disorders associated with disruption of chromatin structure. The defects in chromatin organization caused by lamin deficiency can be imitated in experiments by eliminating two lamin-associated proteins such as MAN1 and emerin, which are responsible for tethering chromatin at the nuclear periphery.

The precise ability of the nuclear lamina to restrict the movement of chromatin is most probably dependent on the structure of individual chromatin domains. Because transcriptionally active chromatin tends to be located away from the nuclear periphery, this location may be related to the pattern of histone modifications present at the promoters and coding sequences of active genes.

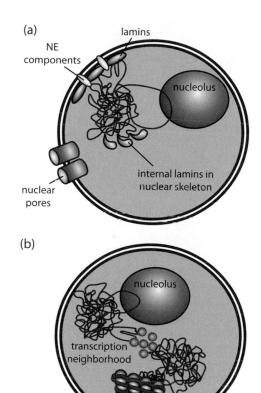

Figure 9.2 Tethering of chromosome territories to nuclear structures restricts their movement and contributes to the maintenance of chromosome positioning. The positions of chromosomes could result in activation or silencing of the genes they contain. (a) Attachment of chromosomes to the nuclear envelope (NE) involves lamins (orange) and frequently results in the repression of gene transcription. (b) Chromosome territories are separate from the nuclear envelope and are often associated with transcription neighborhoods (also known as transcription factories), as indicated by the green circles. [Adapted from Parada LA & Misteli T (2002) *Trends Cell Biol.* 12, 425. With permission from Elsevier.]

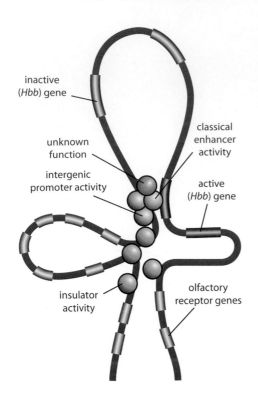

Figure 9.3 A chromatin hub. A hub that is formed by long-range interactions between the hemoglobin β-chain complex (*Hbb*) genes and the locus control region is shown. Transcriptional activation involves the physical association of genes and their regulatory elements. Looping out of the DNA between enhancers (green circles) and promoters (orange circles) brings these regions into contact with one another. The contacts, which are stabilized by specific interactions between transcription factors bound to both elements, are thought to be critical for the activating effects of enhancers. [Adapted from Chakalova L, Debrand E, Mitchell JA et al. (2005) *Nature Rev. Genet.* 6, 669. With permission from Macmillan Publishers Ltd.]

9.2 SPATIAL ORGANIZATION OF GENE TRANSCRIPTION IN THE NUCLEUS

The complexity of the transcriptional machinery means that it has significant potential for organizing the local environment around the transcribed gene. Although much of this organization revolves around the activity of promoters and enhancers and the assembly of the pre-initiation complex (see Chapter 2, Section 2.3), there is now a considerable body of evidence indicating that the activating effect of enhancers is the result of direct contact with promoters through looping out of the intervening sequences. The recruitment of general transcription factors (GTFs) to the promoters of some genes seems to be accompanied by a parallel recruitment to enhancer elements before transcription can be activated. This suggests a model in which the binding of sequence-specific factors to distal elements (enhancers) recruits GTFs, which are then transferred to the promoter by direct contact and looping out of the intervening sequence.

In support of this notion, functional dissection of mammalian gene expression domains has revealed that they contain multiple elements that combine to activate transcription. These elements frequently coincide with deoxyribonuclease I (DNase I)-hypersensitive sites that are created by the binding of specific combinations of tissue-specific and ubiquitous transcription factors to clustered binding sites. Recent studies have also suggested that several DNase I-hypersensitive sites in a domain can come together to form a complex structure containing multiple loops, which has been termed an **active chromatin hub (ACH)**.

The formation of contacts between regulatory elements through local DNA looping is directly involved in regulating gene expression (**Figure 9.3**). The size of these loops depends on the spacing between gene sequences, and we might therefore expect that organisms with smaller or more densely packed genomes would produce correspondingly smaller loops than those of higher organisms such as mammals. Yeasts, for example, seem to form loops of only a few kilobases, whereas humans or mice produce loops over several hundred kilobases.

The ability to form chromatin loops is the first step toward organization of the genome into the functional and nonfunctional domains, and this differentiation probably occurs through the regulation of large-scale chromatin organization when control sequences such as promoters and enhancers come into contact during loop formation. The complexes that form between the transcription factors and RNA polymerases at transcribed gene loci can also create structures that influence the overall anatomy of the nucleus.

The nucleolus is formed from multiple chromatin loops

A well-characterized example of multiple chromatin loops—that is, one created by the interaction of gene regulatory elements—is the nucleolus, in which the ribosomal RNA (rRNA) genes are located in clusters. The nucleolus is an interesting construction because the rRNA genes are not all located in the same regions of the genome. Approximately 40 rRNA genes exist, but they are arranged into tandem arrays at five different chromosomal locations. Thus the nucleolus seems to take its form and to function by bringing all of these separate genes into one transcription "factory," to produce the large amounts of structural RNAs needed to construct the ribosomes of the cell. RNA polymerase I is responsible for the transcription of these genes.

The recruitment of rRNA genes to the nucleolus seems to rely on their arrangement into the aforementioned tandem repeats, which are also referred to as **nucleolar organizers**; this definition is somewhat uncertain, because there are other chromosomal regions that go by this name that do not contain ribosomal RNA genes (such as the one found on the long arm of human chromosome 1). The rRNA genes containing nucleolar organizers associate with a subnucleolar structure known as the **fibrillar center**, which is a multiprotein structure believed to form part of an intranucleolar "skeleton" that is contiguous (and possibly identical) with the nuclear skeleton connecting to the nuclear envelope; the possible link between the intranucleolar and the nuclear skeletons is conjectural, because the morphologically defined nucleolar skeleton has not been characterized biochemically.

It is not currently clear how the nucleolar organizer regions (NORs) are recognized and recruited, but there are two clues that may help us to understand this process. First, their locations on the short arms of their chromosomes tend to place them in regions distinct from the protein-coding genes that are transcribed by RNA polymerase II. Second, this isolation is reinforced by the presence of substantial blocks of heterochromatinized repetitive satellite DNA around the nucleolar organizers, and it is believed that this also has a role in maintaining the structure of nucleoli and the integrity of rDNA repeats.

The notion that chromatin may have a role in recruiting rRNA genes to the nucleolus is supported by some experimental evidence. For example, typically, nucleoli stain poorly with the fluorescent dye 4',6-diamidino-2-phenylindole (DAPI) (which is surprising, given the amount of DNA present) but are surrounded by a shell of intensely stained heterochromatin. This so-called **perinucleolar heterochromatin** is composed of satellite DNA that surrounds NORs and silent rDNA clusters located on either active or silent NORs. A link between heterochromatin formation and the nucleolus is further strengthened by the observations that heterochromatin from non-NOR-bearing chromosomes associates with nucleoli. Moreover, if the ability to maintain such heterochromatic regions is lost, for example by knockout of the histone methyltransferase SuV39h or the NAD^+-dependent histone deacetylase SIR2, rDNA becomes unstable, leading to profound disorganization of the nucleolus. This suggests that the presence of heterochromatic histone modifications such as H3K9me/me2 and/or H3K27me3 on the nucleosomes surrounding the nucleolar organizer may be responsible, at least in part, for the localization of the nucleolar organizer region to the nucleolus. Such localization may involve some form of binding to the nucleolar skeleton, although at the time of writing there are no data to indicate how this might function.

rRNA genes are clustered for transcription in the nucleus

Because the transcription of rDNA to create new ribosomes places an enormous demand on the energy and raw material resources of the cell, it is highly desirable to restrict its transcription when new ribosomes are not needed. Given the repetitive nature of rRNA genes, two strategies for regulating rRNA synthesis are conceivable. RNA polymerase I transcription may be controlled either by changing the rate of transcription from each active gene or by adjusting the number of genes that are involved in transcription. Although there is evidence for both options, most short-term regulation affects the rDNA transcription cycle, namely, pre-initiation complex assembly, initiation, promoter escape, and transcription elongation or termination. The number of active rRNA genes seems to vary between cell types, suggesting that it is possible to alter the fraction of active genes in the rDNA repeats, whose structure, although large, has features that lend themselves to epigenetic mechanisms of control. Sequences encoding pre-rRNA (13–14 kb) are separated by long

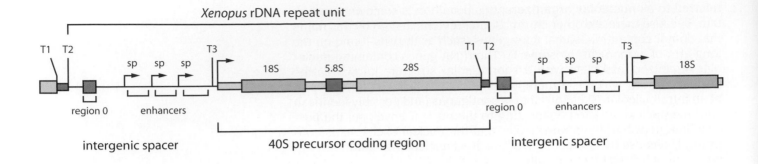

Figure 9.4 Structural organization and regulatory elements of the rDNA repeat unit. The 40S precursor coding region is shown with boxes representing the sequences coding for the 18S, 5.8S, and 28S rRNA. The sites of transcription initiation of the 40S pre-rRNA and transcripts from the intergenic spacer promoter are indicated by short arrows. T1 and T2 represent the 3′ ends of the 28S and 40S rRNA respectively, and T3 is a terminator element upstream of the next gene in the repeat unit. Spacer promoter regions (sp), repetitive region 0, and repetitive enhancer elements (enhancers) are also indicated. [Adapted from Maric C, Levacher B & Hyrien O (1999) *J. Mol. Biol.* 291, 775. With permission from Elsevier.]

intergenic spacers (IGSs) of approximately 30 kb. Regulatory elements, including gene promoters, spacer promoters, repetitive enhancer elements, and transcription terminators, are located in the intergenic spacers (**Figure 9.4**). However, the major part of the IGS seems to be devoid of regulatory elements, comprising a high density of simple sequence repeats and transposable elements.

rRNA gene structure

Interestingly, there are seven variants of this basic rRNA gene structure, differing in terms of sequence polymorphisms both in the variable region of the 28S rRNA and in the 5′-terminal part of the transcription unit. Despite the relatively small differences in sequence, it seems that these variants are regulated independently and, in some cases, in a cell-type-specific manner. Three variant rDNA types are expressed in all tissues (constitutively active), two are expressed in some tissues (selectively active), and two are not expressed at all (silent). The finding that rDNA can exist in genetically distinct subdomains that can be regulated individually in different tissues, suggests a hitherto unappreciated complexity in mammalian rDNA structure and regulation. Regardless of the type of variant in the repeats, they all must be assembled into the nucleolar organizer regions because the nucleolus seems to be the exclusive rRNA transcription zone.

Until quite recently, rRNA genes were thought to adopt a simple head-to-tail tandem array configuration, but recent data have shown this assumption to be incorrect. The actual arrangement seems to be more chaotic, with as many as 30% of the genes being aligned in a non-canonical fashion as palindromic repeats—that is, they are arranged head-to-head and tail-to-tail. It has been presumed that these repeats have to be silenced in some way by the cell to avoid the possibility that a transcript from a gene oriented head-to-tail might undergo base pairing with one from a head-to-head gene, which could cause serious problems for rRNA synthesis.

Regulation of rRNA gene transcription

The level of rRNA gene transcription is tightly controlled by relegating nucleolar organizers to active or inactive states. The active forms comprise relatively uncondensed chromatin (approximately one-tenth as dense as the surrounding satellite DNA) that is bound by RNA polymerase I-specific transcription factors such as upstream binding factor (UBF), whose depletion in the cell represses rRNA synthesis. On inactive organizers, rDNA is packaged in a form that is indistinguishable from the surrounding heterochromatin. Silent organizers can be visualized as condensed foci of rDNA that lack associated RNA polymerase I and its specific transcription factors, as well as the secondary constriction that characterizes the decondensed, open state of rDNA.

It is interesting that both the regulatory and coding sequences of rDNA are rich in CpG islands, but as we might expect from our discussion of DNA methylation in Chapter 5, the inactive genes can be distinguished from the active ones by the degree of methylation (**Figure 9.5**). The level of control imposed by this epigenetic modification is exquisite: a single 5-methylcytosine at position –133 in the upstream control element of the mouse rDNA gene is sufficient to exclude UBF from its nearby binding site, and that alone is sufficient to prevent assembly of the transcription complex. For this methylation to have such a pronounced effect on UBF binding suggests that cytosine 133 is in a favored position, perhaps as a result of being located at the nucleosome surface. Such an exposed site may allow the steric effects of methylation to be more acute than they would be at another, less exposed, cytosine.

In human cells, the methylation status of rDNA, with 25 CpGs residing within the promoter, is more complex. Usually, human rDNA promoters have a mosaic methylation pattern; that is, they are neither completely methylated nor unmethylated but show methylation of a few to most CpGs.

It is also known that the histone modification pattern differs between active and inactive rDNA repeats, with higher levels of histone acetylation (H3K9 ac) and H3K4 methylation in active genes. It also seems that chromatin remodeling complexes have a very significant role in the regulation of rDNA transcription. The Cockayne syndrome protein B (CSB) is a candidate chromatin remodeler that is able to establish euchromatin-like features at active rDNA repeats. CSB has DNA-dependent ATPase activity in common with many of the chromatin remodeling complexes introduced in Chapter 4 and is known to localize to the nucleolus at sites of active rDNA transcription. It is also part of a larger multiprotein complex containing RNA polymerase I, transcription factor IIH (known to bind to rDNA), and transcription initiation factors for RNA polymerase I. In the absence of CSB, RNA polymerase I seems to be unable to complete its function, indicating that the chromatin modeling promoted by CSB is needed to expose the DNA for use by the transcription complex.

Importantly, CSB-mediated activation of rDNA transcription requires association with G9a, a histone methyltransferase. G9a is responsible for monomethylation and dimethylation of H3K9 and facilitates the binding of heterochromatin protein 1γ (HP1γ), a protein containing a chromodomain

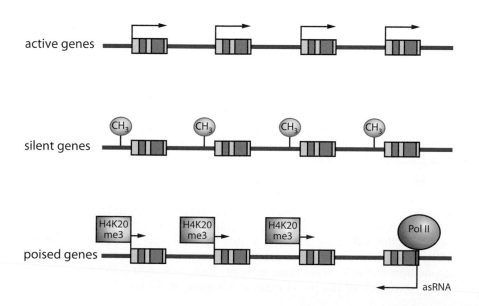

active genes

silent genes

poised genes

asRNA

Figure 9.5 rDNA can be active, silent, or poised. The promoter of active genes is unmethylated, marked by euchromatic histone modifications, and associated with components of the transcription initiation complex. Silent rRNA genes are characterized by DNA methylation of the promoter and are marked by heterochromatic histone modifications. Genes that are ready to undergo rapid activation are described as "poised" and are repressed by the trimethyl H4K20 histone modification, which can be removed more rapidly than DNA methylation. asRNA, indicates antisense RNA. [Adapted from Grummt I & Längst G (2013) *Biochim. Biophys. Acta* 1829, 393. With permission from Elsevier.]

(a)

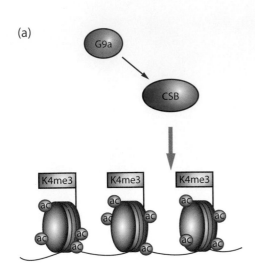

(b)

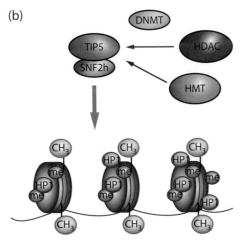

Figure 9.6 CSB and NoRC establish a specific chromatin structure at active and silent clusters of rRNA genes. (a) Active rDNA promoter; (b) silent rDNA promoter. TTF-I recruits either the transcription activator CSB or the silencing complex NoRC (TIP5 and SNF2h) to rDNA. CSB and NoRC interact with distinct chromatin-modifying enzymes that establish epigenetic features that characterize active and silent genes, respectively. The promoter of active, euchromatic genes is unmethylated and marked by acetylated histones H3 and H4 (ac) and trimethylated lysine 4 on histone H3 (H3K4me3). CH$_3$ indicates methylated DNA, "me" indicates methylated histones. The interaction of NoRC with histone deacetylase (HDAC), H3K9 histone methyltransferase (HMT) SET-DB1, and DNA methyltransferase (DNMT) establishes a heterochromatic compact chromatin structure that characterizes silent rRNA genes. [Adapted from Grummt I & Längst G (2013) *Biochim. Biophys. Acta* 1829, 393. With permission from Elsevier.]

that recognizes H3K9 methylation. Methylation of histone H3 at lysine 9 and association with HP1 have well-established roles in heterochromatin formation (as we have seen in Chapter 3, Section 3.2). The finding that G9a is associated with CSB and is required for Pol I transcription suggested additional, surprising functions for H3K9 methylation and HP1γ recruitment in chromatin-based processes. Notably, H3K9me2 and HP1γ are present within the transcribed region of active rDNA repeats, and both H3K9 methylation and association of HP1γ with rDNA are dependent on continuing Pol I transcription.

The presence of an apparently repressive histone modification at an active rDNA gene does not tell us much about how the active state of rRNA genes is maintained and how the cell can distinguish an active gene from an inactive one. DNA methylation seems to be a more important factor in rDNA transcription than it is in genes transcribed by RNA polymerase II. The difference between the two may hinge on an interesting possibility, namely the involvement of specific proteins that either protect the promoters of rDNA from methylation or target them for demethylation.

Proteins that protect or target rDNA for methylation and demethylation

A protein that may serve in this manner is MBD3, a member of the methyl-CpG-binding domain (MBD) family of proteins, which bind to methylated DNA. However, in contrast with the other proteins in its family, MBD3 has two amino acid substitutions in the MBD domain that abolish binding to methylated DNA. Interestingly, MBD3 is associated with the rDNA promoter, and bisulfite mapping revealed that the fraction of rDNA bound to MBD3 is unmethylated. Overexpression of MBD3 decreased methylation of the rDNA promoter, whereas knockdown of MBD3 increased methylation and decreased pre-rRNA synthesis. These results suggest that MBD3 has an important role in maintaining rDNA promoters in an unmethylated and therefore active state, but does DNA methylation affect the overall chromatin structure of the inactive rDNA repeats?

A key component in the establishment and inheritance of a given epigenetic state at specific subsets of rDNA repeats is TTF-I. It is a multifunctional protein that binds to specific terminator elements downstream of the rDNA transcription unit and mediates transcription termination and replication fork arrest (see Figure 9.4 for the positions of terminator elements). TTF-I is able to interact with a wide variety of proteins, but one of its main activities is to recruit a chromatin remodeling complex termed NoRC (nucleolar remodeling complex). NoRC induces nucleosome sliding in a manner that depends on ATP and on the histone H4 tail, but it also seems to do a lot more than simply move nucleosomes along to expose DNA. In a manner similar to that of many of the repressor complexes we saw in Chapter 6, NoRC functions as a scaffold protein, coordinating the activities of macromolecular complexes that modify histones, methylate DNA, and establish a closed heterochromatic chromatin state. It performs this function through two domains known as TIP5 (TTF-I interacting protein 5) and SNF2h, which cooperate to assemble histone deacetylases, histone methyltransferases, and DNMTs at rDNA repeats selected for inactivation (**Figure 9.6**).

The binding of TTF-I to the promoter-proximal terminator element allows the recruitment of TIP5, which brings the other components of the NoRC complex into contact with the rDNA unit and its associated nucleosomes. This results in methylation of H3K9, H4K20, and H3K27, which may be a signal to the SNF2h remodeling protein to move the nucleosomes at the rDNA promoter on which these modifications have been established to a position 25 nucleotides downstream, a location that seems to interfere with formation of the pre-initiation complex. It is also thought that this action is responsible for exposing the critical cytosine at position −133 so that it can be methylated by DNMT1.

There is some evidence to suggest that the active rDNA repeats are enriched in histone N-terminal tail modifications such as H3K4 methylation, which is consistent with an active transcriptional status. It is known that histone demethylase enzymes such as JHDM1B can be localized to the nucleolus, which implies some sort of histone demethylation function, but at the time of writing there are no clues to how this enzyme is targeted to rDNA or how it might relate to NoRC-mediated silencing.

The repositioning of nucleosomes adjacent to critical control elements within the rDNA gene is an interesting mechanism. At potentially active genes, a nucleosome occupies sequences from −157 to the transcription start site, whereas at silent genes the nucleosome covers sequences from −132 to +22, indicating that specific nucleosome positions determine the transcriptional readout of rRNA genes. Positioned nucleosomes may either occlude or facilitate the binding of basal transcription factors to chromatin, thereby repressing or activating transcription. In the active rDNA genes, the nucleosome juxtaposes the core promoter element and the upstream control element (UCE), whereas both of these sequence elements are separated at silent genes, prohibiting the cooperative binding of UBF and SL1/TIF-IB, the factors that nucleate the assembly of the pre-initiation complex. The identification of NoRC as the major determinant of the silent nucleosome position suggests that remodeling complexes are the major determinants of chromatin dynamics at the rDNA loci, but it is also presumed that individual histone N-terminal tail modifications are the primary signals that direct the remodeling process.

TTF-I also functions in the designation of active rDNA genes through its ability to recruit Cockayne syndrome protein B (CSB), so essentially TTF-I forms the pivot on which the relative levels of rDNA transcription and repression are balanced. In view of this, it is highly likely that precise regulation of TTF-I activity or its concentration in the cell is the mechanism by which rDNA transcription is controlled; however, at present there is little information as to how such a control mechanism might function. We know that once a particular chromatin state has been established at an individual rDNA gene it has a tendency to spread toward adjacent parts of the rDNA repeat, but there is a lack of knowledge about how this spreading may be controlled.

Genes transcribed by RNA polymerase II show a different organization

Transcription by RNA Pol II is not associated with the large-scale organization that is observed for Pol I transcription in the nucleolus, but there is a growing body of evidence that Pol II transcription is organized into small structures termed **transcription factories** (see **Box 9.1**). This is based on observations that there are roughly one-eighth as many transcription processing regions in the nucleus as there are molecules of RNA polymerase II. However, visualizing small transcriptional units containing on average only eight RNA polymerase II molecules is difficult, so the evidence for structural organization of transcription "factories" is not as extensive as our characterization of the nucleolus. Essentially, much of the recent evidence supporting the existence of such factories is centered on observations of the intranuclear movement of green fluorescent protein (GFP)-tagged RNA polymerase II which suggest that the enzyme exchanges rapidly with a less mobile protein (or group of proteins) in the vicinity of the regions to which a potential transcription factory status has been assigned. This result (which needs to be confirmed by other methods) could be explained if the transcribing polymerase is constrained while the Pol II in the pre-initiation complex is in a state of dynamic exchange between factories.

Transcription factories have the potential to bring genes that are co-expressed into direct contact with one another, and two recent studies have suggested that tissue-specific genes that are activated in the same

Box 9.1 Structure of a transcription factory

Electron microscopy shows that transcription factories are 45–100 nm in diameter. However, apart from RNA polymerase II (RNAPII), it is not known what components are present in the factories, or what components are required for their formation and ultrastructure. The best we can do at present is to visualize the foci where transcription is taking place by using antibodies against RNA polymerase II or its phosphorylated form. Other evidence for the existence of transcription factories is the nuclear clustering of active genes. The fact that there are fewer observed sites of transcription in the nucleus than expected from estimates of the number of active RNAPII molecules, or of active genes, suggests that multiple (about eight) polymerases and actively transcribing genes are located at each transcription factory (**Figure 1**).

However, there may be fewer than expected sites of transcription not because of the existence of transcription factories but because of other factors. For example, we also need to consider that RNA polymerase II can be "preloaded" onto the transcription start sites and throughout the coding sequences of many inactive genes in a stalled or "poised" form that cannot enter the elongation phase of mRNA synthesis. The RNAPII at these poised genes is already phosphorylated at serine 5, but not at serine 2.

Detailed studies showing co-localization relative to Ser2P–RNAPII and Ser5P–RNAPII foci during induction, for genes known to be associated with poised polymerase, are required to determine whether this process is associated with movement of the gene into a transcription factory.

Are all transcription factories equal? The spatial separation of transcription by three different RNA polymerases (RNA pol I–III) is clear enough, but is there any further specialization of RNA polymerase II-mediated transcription? This question arises because artificially introduced plasmid-type gene expression constructs seem to cluster together during transcription, especially if the genes introduced into the cell in this manner have promoters of similar structure. However, for endogenous genes there is little evidence to suggest specialization of transcription factories. For example, β-globin in erythroid cells seems to have an equal probability of association with either active or inactive genes of the erythroid phenotype. If specialization does occur, it would most probably be mediated by the concentration of particular transcription factors to specific factories, but even here there is remarkably little relationship between transcription sites and the distribution of many specific transcription factors such as OCT1 or E2F.

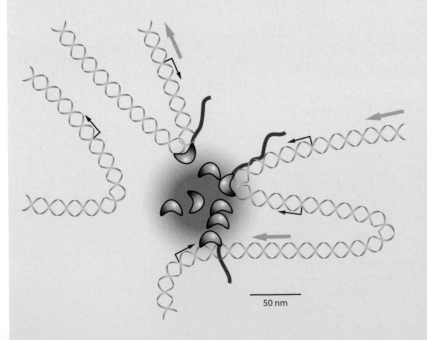

Figure 1 Model of transcription with RNA polymerase immobilized in transcription factories. The figure shows a transcription factory with a diameter of about 70 nm that contains eight RNA polymerase II enzymes (purple crescents). Genes are reeled through these polymerases (in the direction of the large arrows) as they are transcribed, and the nascent RNA (dark green) is extruded. Genes from the same chromosome or from different ones may associate with polymerases in the same factory. Small arrows indicate the direction of transcription at the transcription start site. [Adapted from Sutherland H & Bickmore WA (2009) *Nat. Rev. Genet.* 10, 457. With permission from Macmillan Publishers Ltd.]

50 nm

cell type co-localize in the nucleus at a higher frequency than would be expected through random association. An example of this type of interaction is seen in erythroid cells (those that produce erythrocytes in the bone marrow), in which α- and β-globin and *Eraf* loci (which are expressed in erythroid cells) are significantly closer together than in non-erythroid cells. Similarly, chromosome conformation capture technology has supported the idea of association between the β-globin and *Eraf* loci in erythroid cells but not in neurons, despite the considerable physical separation between these genes in the genome.

The idea of a transcription factory requires that the gene loci move to an immobilized polymerase already present in a factory, rather than the transcriptional machinery being recruited to and moving along the chromatin template to a gene. The DNA comprising the gene is then pulled through the transcriptional unit to make nascent RNA. But why should the cell have evolved this particular method? There is some logic to grouping the resources required for a particular synthetic operation into a defined area, much as the human species confines manufacturing to dedicated factories, and the method makes particular sense if more than one RNA polymerase molecule is required to work on the DNA at any one time. The possibility that the polymerase molecules might collide if allowed to move anywhere they like is removed by immobilizing them; this also eliminates the possibility of simultaneous bidirectional transcription, which would result in the DNA's being pulled in two directions at once and would most probably halt the formation of nascent RNA.

There is a significant problem with this model in that it actually might not allow multiple polymerases to work on the same gene simultaneously, because in theory the promoter region would have to be re-threaded through another polymerase unit to start the synthesis of another nascent RNA. The model is probably consistent with the number of RNA polymerase II molecules that are detectable in the cell, but it does not seem to allow for the very high transcriptional activity of some genes that have been estimated to need more than ten polymerases per transcription unit. Add to this the fact that multiple transcripts can actually be seen to form from the Balbiani ring genes of certain types of flies and it is clear that the current transcription factory models do not fully explain how the cell controls gene activity.

Transcription factories may be semi-permanent structures

It may also be true that the "factories" are not permanent structures within the nucleus in the way that other components of the cell such as the endoplasmic reticulum or mitochondria are. There is certainly evidence that some genes seem to become enriched in RNA polymerase II on activation without obvious movement of their positions in the nucleus, which is not consistent with the idea of a transcription factory. However, a further suggestion implies that it may also be possible to start the transcription of certain genes destined for high-level expression by an initial RNA polymerase II recruitment followed by the assembly of a transcription factory at that locus. More conclusive evidence of permanent or semi-permanent transcription factories would be the persistence of such structures in the absence of transcription, but unfortunately few studies have yet addressed this question. Mitosis is one instance in which all transcription stops, but during this time the isoforms of RNA polymerase II seem to be differentially distributed throughout the cell and not just in the nucleus. So far, no studies have followed the redistribution of these molecules to the positions where gene transcription activity begins anew in the G1 phase of either of the daughter cells. In view of this, it is impossible to say whether the transcription factories survive mitosis intact.

It is not impossible that the apparent localization of actively transcribing genes with RNA polymerase II does not actually indicate an association with a specialized subnuclear compartment dedicated to their transcription but instead indicates recruitment of the gene (or group of genes) to areas that are involved in the downstream processing of the nascent RNA transcript. We saw earlier that there was significant interaction between the α- and β-globin loci in human erythroid cells, but it turns out that the average distance between these genes (500–800 nm) is much larger than the predicted size of a transcription factory (70 nm). This leads us to ask whether two loci can be separated by such distances and still share the same factory. This discrepancy could be due to a limitation of the spatial resolution of current light microscopy (200 nm), but the lack of a skewed distribution of measurements toward distances smaller than the mean suggests that this is not the case. Surprisingly, many of the closely associated globin genes in human erythroblasts do not actually localize to the same RNA polymerase II foci; rather, they are organized around structures called nuclear speckles. These are relatively large subnuclear compartments enriched in factors involved in RNA splicing, suggesting that the function of transcription factories may have as much to do with the subsequent steps of RNA processing as with transcription itself.

9.3 THE EPIGENETIC CONTRIBUTION TO TRANSCRIPTION FACTORY ORGANIZATION

The exact molecular description of genomic features that promote association with a transcription factory has yet to be determined, but because there is a high degree of probability that the activities of clustered genes are controlled by epigenetic modifications of DNA methylation and histone modifications, the changes in chromatin conformation that reposition a gene targeted for transcription in the nucleus are very likely to be under similar systems of control. We will examine two well-characterized gene clusters that are subject to epigenetic expression control to understand how epigenetics relates to their abilities to form loops of chromatin that may be essential for entry into transcription factories.

The β-globin locus control region is subject to epigenetic control

The human β-globin locus provides an excellent example of gene expression that is regulated by a locus control region (LCR). The human β-globin locus contains five genes that are arranged from 5′ to 3′ in the order of their expression during development (**Figure 9.7**). Upstream of the gene cluster are five DNase I-hypersensitive sites within a 20 kb region that is referred to as the LCR. The LCR contributes to the developmentally regulated pattern of gene expression across the locus by modulating the chromatin structure and the ability of different regions of the extended genomic locus to enter into tertiary chromatin complexes. The tertiary complexes that form to regulate expression have been called an active chromatin hub (ACH).

Locus-wide acetylation of histones H3 and H4 has been observed that correlates with an open chromatin structure in the human β-globin locus in erythroid cells. Interestingly, on its own this is not sufficient to guarantee transcription of β-globin, but there is a striking increase in H3 acetylation at the LCR and the β-globin promoter when transcription begins. This preferential localized hyperacetylation of H3 during gene activation in the globin locus argues that HATs with H3 preference are recruited to the LCR and the globin gene. An open chromatin conformation of both of these regions of the locus is probably required so that they may interact

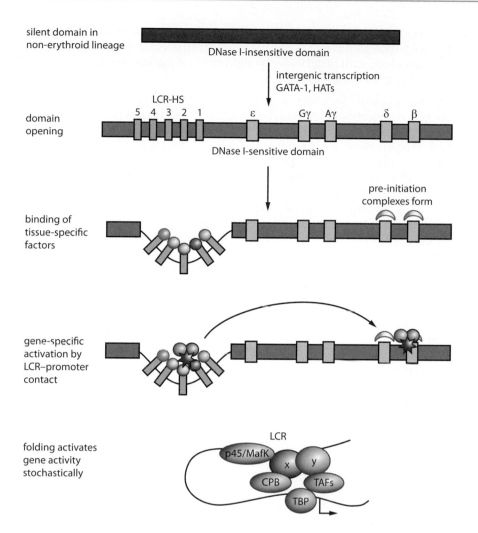

Figure 9.7 Organization of the β-globin locus. In mouse and human cells, globin genes are arranged sequentially (gray boxes on the brown bar), flanked by a locus control region (LCR) at the 5′ extremity. In non-erythroid lineages, the domain is inaccessible to nucleases such as DNAse, which is able to cleave the DNA sequence at five hypersensitive sites (HS). During sequential activation steps, the domain first becomes generally sensitive to DNase before the activation of specific genes. Tissue-specific transcription factors are thought to bind at or near the HSS (blue, green, orange, purple, and lilac shapes) and folding of the DNA brings the transcription factors into contact with pre-initiation complexes forming at multiple promoters. It is proposed that activation requires direct contact between as yet unspecified factors (x and y) bound to the LCR and the pre-initiation complex (CBP, TAFs, and TBP) bound at the promoter. [Adapted from Spector DL & Gasser SM (2003) *EMBO Rep.* 4, 18. With permission from Macmillan Publishers Ltd.]

to initiate transcription, but as we have seen from many of the other chromatin remodeling events discussed throughout this text, it is probably the initial binding of transcription factors that results in chromatin remodeling and the recruitment of protein complexes containing histone acteyltransferase activity. The localized H3 change is most simply explained by the specific recruitment of HAT activity to the LCR and to the active gene by an erythroid-specific transcriptional activator. In contrast, the domain-wide change in acetylation may result from the recruitment of HAT activity via protein(s) with binding sites throughout the locus, and because the active β-globin gene is positioned on a loop of chromatin protruding from the inactive chromatin territory, there are two possible scenarios through which this might be achieved.

It has been suggested that a locus-wide change in acetylation may be the consequence of the removal of the β-globin locus to a nuclear compartment enriched in factors able to bind specifically to the LCR and gene promoter. However, the alternative—that the acetylation of the gene results in its translocation to the same nuclear compartment—is equally possible. The LCR itself does not seem to be involved in this process, because its deletion from the genomic structure affects only globin transcription and localized H3 hyperacetylation but has no effect on either general acetylation or nuclear localization of the locus. In addition, there are mutant versions of the β-globin gene that, although incapable of transcription, nevertheless maintain an apparently open and acetylated chromatin structure that demonstrates the same nuclear localization as the wild-type locus; that is, away from centromeres. It therefore seems that broad areas of histone acetylation may be the key to loop formation and nuclear translocation.

The mechanisms by which a locus is recruited and/or sequestered into specific compartments remain to be determined. However, there is some evidence that certain regions of the β-globin locus may make important contributions to the process. One of the DNAse I-hypersensitive sites (5′-HS2, which contains an enhancer of the β-globin gene near the LCR) is sufficient for localization of a non-β-globin transgene away from centromeric heterochromatin. This results in transgene silencing. This finding led to the suggestion that enhancers or LCRs may maintain gene expression by preventing the gene's localization close to the repressive heterochromatic compartment. However, as we saw above, selective deletion of the LCR has no effect on β-globin position within the nucleus, so this mechanism is unlikely to apply in this case.

The large number of factor-binding sites, similar to those found in the LCR, that are scattered throughout the native locus alter subnuclear location and chromatin structure. Thus, specific cis-acting elements other than the LCR may maintain the β-globin locus in an open chromatin/acetylated configuration by disrupting or preventing its association with a nuclear compartment enriched in heterochromatin proteins and histone deacetylases. It is interesting to note that exposure of cells to histone deacetylase inhibitors results in more chromatin being located away from the nuclear periphery, and because this is associated with a global increase in histone acetylation there is a good possibility that acetylation levels are at least one of the determinants of subnuclear gene localization. This property may not be restricted solely to histone acetylation. It is possible that the presence of heterochromatic histone modifications such as H3K9me/me2 and/or H3K27me3 on the nucleosomes surrounding the β-globin locus may be involved in its localization into a region of heterochromatin and positioning near the nuclear periphery in a manner analogous to that suggested for the positioning of rDNA within the nucleolus (see Section 9.2).

The *HOX* clusters are also subject to epigenetic control of gene expression

We noted earlier that many of the RNA polymerase II-transcribed genes tend to occur in clusters, and some of the best examples of clustered genes are the *HOX* clusters. Please note that *HOX* refers to the human homologs of these genes, while *Hox* refers to the same genes of other species. Homeobox genes show a high degree of evolutionary conservation, being present in nearly all higher eukaryotic species examined so far as well as in the genomes of some simpler eukaryotes such as fungi. This highlights the primitive origin of these genes and underlines the importance of the Homeobox family as master regulators of development. The *Hox* genes are expressed throughout embryonic development in a highly coordinated manner and continue to be expressed in the vast majority of adult tissues in organ-specific patterns (**BOX 9.2**).

The molecular mechanisms controlling co-linear activation are poorly understood, although there is evidence to suggest that such control mostly affects the timing of gene activation because *Hox* genes inserted at ectopic locations maintain their spatial expression pattern but their temporal expression is misregulated. This also suggests that the clustered organization of *Hox* genes may be an essential part of the control mechanism, particularly because current evidence supports the theory

Box 9.2 *HOX* genes control formation of the body pattern during embryogenesis and are organized into epigenetically regulated clusters

In humans the *HOX* genes are organized into four clusters (*HOXA–HOXD*) located on different chromosomes (7p15, 17q21.2, 12q13, and 2q31). Each cluster contains 9–11 member genes encoding relatively small gene products. The proteins synthesized from these genes contain a highly conserved region of 60 amino acids (the homeobox) that is responsible for their ability to bind to specific DNA sequences and thus contributes to their action as transcription factors. One of the major functions of *HOX* genes seems to be to control the formation of the body plan during embryonic development, and the expression of members of all four clusters begins during early gastrulation, when the embryo establishes its principal body axis (**Figure 1**). To achieve this, undifferentiated cells in the embryo need information about their position in the structure, to ensure that they differentiate into the cell types required for the formation of tissue structures specific to that location. A remarkable feature of all four *HOX* clusters is co-linear gene activation. The members of each cluster are expressed in a precise spatio-temporal pattern along the anterior–posterior axis that corresponds to their position within the cluster. Expression of genes at the 3′ end begins at an early stage of development and is restricted to the anterior segments of the embryo, whereas expression of those genes at the 5′ end is up-regulated later in development and only in the posterior segments.

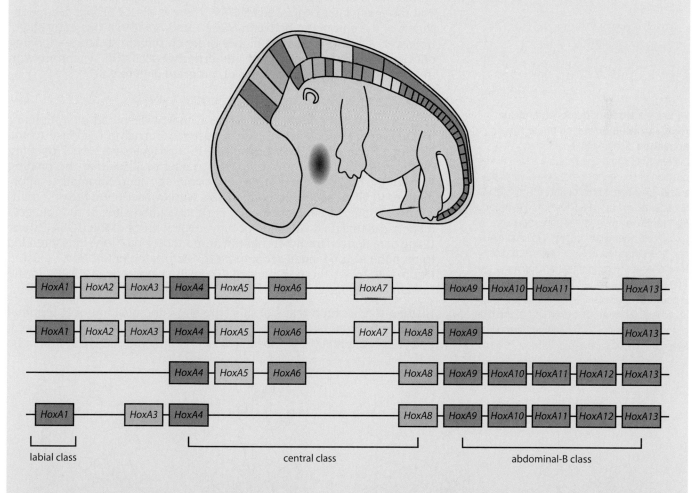

Figure 1 Expression of *Hox* genes during mouse (*Mus musculus*) development. The image shows a 12.5-day-old mouse embryo with approximate positions of the *Hox* gene expression domains. These are indicated by colored regions that correspond to the positions of the same *Hox* genes in the lower panel, which represents the relative positions of such genes in their respective *Hox* clusters. Development control of body pattern formation by *Hox* genes is a universal mechanism in vertebrates, so although there may be some interspecies differences between *M. musculus* and *Homo sapiens*, the former is still a reasonable model for processes that apply to human development. [Adapted from Pearson JC, Lemons D & McGinnis W (2005) *Nat. Rev. Genet.* 6, 893. With permission from Macmillan Publishers Ltd.]

that co-linear expression requires coordinated changes in the higher-order chromatin structure from one end of the cluster to the other. The chromosomal organization of the genes in each *Hox* cluster reflects its anterior–posterior expression in the body plan (spatial co-linearity). Unlike in *Drosophila*, vertebrate *Hox* genes are also temporally co-linear and are expressed in an anterior–posterior direction. In general, members of the same paralog group are expressed at the same time and have the same anterior boundary of expression.

This concept of co-linear expression has some parallels in the globin loci, which express their individual genes according to a developmentally specific timescale, as we saw earlier, but the *HOX* clusters take this mechanism to a new order of magnitude. The existence of co-linearity suggests a role for the structure of the *HOX* clusters in regulating gene function, and some elegant studies by Bickmore and her co-workers have demonstrated programmed changes in chromatin structure of the *HOXB* genes when their expression is induced in embryonic stem cells by exposure to retinoic acid. There is an early increase in the histone modifications that are marks of active chromatin at both the early expressed gene *HOXB1* and the much later expressed *HOXB9*. There is also a visible decondensation of the chromatin between *HOXB1* and *HOXB9* at this early stage. However, a further change in higher-order chromatin structure—looping out of genes from the chromosome territory—occurs in synchrony with the execution of the gene expression program (**Figure 9.8**).

RAREs occur in open-chromatin regions

These observations led to the proposal that widespread alterations of histone modifications at the *HOXB* cluster occurred in response to the binding of retinoic acid (RA), resulting in a progressive 3′ to 5′ opening of chromatin structure. However, this model requires that the binding sites for the retinoic acid be in regions of already open chromatin conformation, to allow access. In comparison with transcription factors, retinoic acid is quite a small molecule, but the binding sites of retinoic acid response elements (RAREs) do occupy defined areas of the *HOX* clusters. Using data derived from a chromatin immunoprecipitation study intended to map the histone modification patterns across the entire *HOXA* cluster, **Figure 9.9** shows the positions of two well-characterized RAREs in the human *HOXA* cluster.

Histone modification maps of this type show the position of N-terminal tail modifications such as H3 acetylation, H3K9me3, and H3K4me3, and are generated by labeling DNA fragments derived from the

Figure 9.8 Histone marks and nuclear reorganization during co-linear *Hox* activation. Schematic representation of histone marks and changes in the subnuclear position of *Hox* genes before (embryonic day 6.5; E6.5) or at the time of their first expression (E7.5 for *Hoxb1* (b1), E8.5 for *Hoxb4* (b4), and E9.5 for *Hoxb9* (b9)). The histone marks on histone H3, methylated lysine 4 (blue circle), and acetylated lysine 9 (green circle), poise the genes for transcription from the moment that the first *Hox* gene of the cluster (*Hoxb1*) is activated. Individual genes loop out of their chromatin territory (CT, brown line) at the time of their expression. [Adapted from Deschamps J & van Nes J (2005) *Development* 132, 2931.]

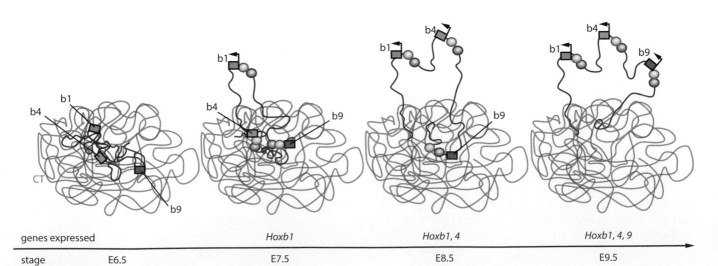

genes expressed | *Hoxb1* | *Hoxb1, 4* | *Hoxb1, 4, 9*

stage   E6.5   E7.5   E8.5   E9.5

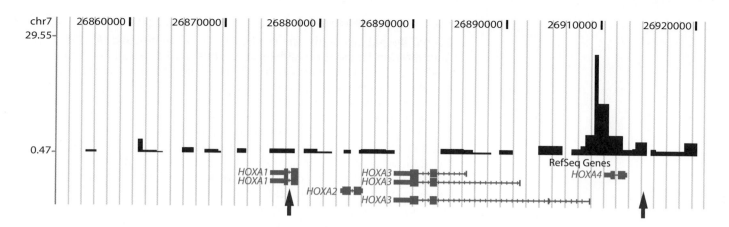

immunoprecipitations with a colored/fluorescent dye followed by hybridization to a microarray composed of tiled oligonucleotides designed to cover large regions of chromosomes such as the *HOXA* cluster. The height of individual blocks on the histograms is related to the intensity of fluorescence emitted by dye conjugated to the DNA hybridized to specific regions of oligonucleotides on the array surface. These intensities represent the enrichment of a particular modification relative to a similar array hybridized to non-immunoprecipitated DNA fragments. The peak heights and widths of the areas enriched for specific modifications strongly correlate with the number of nucleosomes at that position carrying the histone N-terminal tail modification.

In view of this, we can see from Figure 9.9 that the positions of the RAREs are in regions that either are not enriched for the specific modifications shown or have low nucleosome occupancy. As we saw in Chapter 4, segments of DNA that have been "released" from binding to the nucleosome surface are free to associate with transcription factors or other regulatory proteins, and the same is probably true of the RAREs. Unfortunately, using ChIP-on-chip (a technique that combines chromatin immunoprecipitation (ChIP) with microarray technology), we cannot easily distinguish between low nucleosome occupancy and low-level enrichment for a histone modification except by comparing data from a more general immunoprecipitation using antibodies directed against core histones such as H3.

In addition to keeping the RAREs in open-chromatin regions, it has been hypothesized that positions marked by activating histone modifications (such as H3K4me2 or H3K4me3) in developmentally specific genes function as recruitment sites for transcription factors during activation of those genes. ChIP-on-chip studies show that in undifferentiated embryonic stem cells or embryonal carcinoma cells (a type of aneuploid ES cells), there are small regions in the *HOXA* cluster enriched in H3K4me2/H3K4me3 and that these expand considerably when differentiation is induced, leading to the up-regulation of *HOX* genes. However, there is no evidence to suggest that these regions serve to recruit transcription factors to the genomic regions responsible for controlling *HOXA* cluster expression. In the few other examples in which the presence of such regions predisposes genes to activation, the regions serve also as assembly sites for pre-initiation complexes. However, RNA polymerase II is not highly enriched at these positions, so it is unlikely that these relatively small areas of enrichment function as recruitment centers. The up-regulation of *HOXA* genes may rely instead on the binding of transcription factors to areas of low nucleosome occupancy, in which the DNA is more easily accessible.

Figure 9.9 Positions of the retinoic acid response elements (RAREs) in the human *HOXA* cluster. Two well-characterized RAREs exist in close proximity to *HOXA1* and *HOXA4* in the human genome. The positions of these are indicated by the arrows. [Adapted from Atkinson SP, Koch CM, Clelland GK et al. (2008) *Stem Cells* 26, 1174. With permission from John Wiley and Sons.]

HOX gene expression levels

Even if the pattern of histone modifications at the inactive *HOXA* locus of undifferentiated pluripotent cells does not serve to recruit activating proteins, the effect of activating N-terminal tail modifications on the initiation of *HOX* expression is very impressive, indicating the widespread adoption of open or permissive chromatin. This does not necessarily mean that all the *HOX* genes will be expressed at the same level; it is simply a method to make them more accessible to the individual transcription factors or other control proteins that modulate their expression. **Figure 9.10** shows the relative expression levels of *HOXA* genes over 28 days of human ES cell differentiation. It is interesting to note from this diagram that the genes expressed at the highest levels are situated outside, or at least on the edges of, the area highly enriched by H3K4 trimethylation (shown as black vertical bars in Figure 9.9). This is also reflected in the levels of occupancy of the gene promoters by RNA polymerase II. However, why those *HOX* genes that lie outside the enriched region should be preferentially expressed is currently still under debate. It is possible that the pattern of histone modifications serves some other purpose, such as permitting the expression of noncoding RNAs that are involved in modulating gene expression (see Chapter 8, Section 8.4), or perhaps the precise expression level simply depends on the presence or absence of appropriate transcription factors.

In the context of our ongoing discussion of how epigenetic modification might assign genes destined for expression to transcription factories, it is possible that the enrichment of activating modifications at the *HOXA* cluster is part of the mechanism that ensures looping of the *HOXA* cluster locus out of the surrounding heterochromatin in a manner similar to that observed for the *HoxB* locus (see Figure 9.8). Perhaps the presence of H3K4 dimethylation or trimethylation may be required to sequester the *HOX* loci into a transcription factory. There is also a striking parallel with our description of the β-globin locus, in which we saw that enrichment with methylated H3K4 was only sufficient to cause translocation to the factory and did not guarantee transcription in its own right. The observation that *HOXA* genes seem to be enriched for RNA polymerase II even when they are not expressed at very high levels would seem to support this possibility.

In this chapter we have examined the contribution of epigenetic modifications such as the methylation of histone N-termini to the control of chromosome architecture, with specific focus on multiple genes present in groups such as the *HOX* clusters. The mechanisms described in this chapter are the basis of the epigenetic control of gene expression; having understood these concepts, we can proceed to examine how epigenetic control is exerted over diverse cellular functions such as mitosis.

Figure 9.10 Changes in expression levels of *HOXA* genes during the differentiation of human EC cells (hESCs). Expression levels of members of the human *HOXA* cluster in undifferentiated hESCs and their progeny resulting from 10, 19, and 28 days of differentiation. These were measured by quantitative real-time reverse transcription–polymerase chain reaction relative to the expression level of the glyceraldehyde-3-phosphate dehydrogenase (GAPDH) gene (blue). Also shown are the levels of RNA Pol II occupancy at the same chromosomal locations as those amplified by the primer sets used for expression analysis (red).

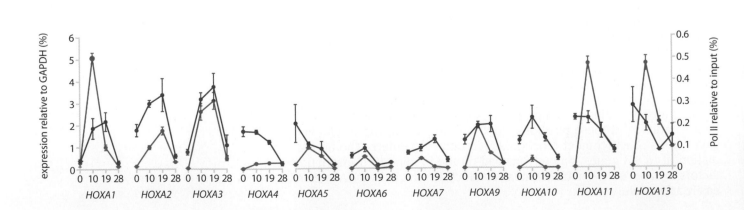

KEY CONCEPTS

- The nucleus is divided into several compartments into which chromatin is placed according to its state of activation. These compartments are known as chromosome territories.

- The chromosome territories are formed by the interaction of chromatin with the nuclear lamina. Heterochromatin is more likely to be held near the nuclear periphery through such interactions, and this tends to segregate inactive genes in such regions.

- Heterochromatin and euchromatin can be physically close to one another, and genes required for active transcription are often found on "loops" of DNA that protrude from the chromosomes held near the nuclear periphery into so-called "transcription factories" where RNA synthesis takes place.

- The β-globin and *HOX* gene clusters are examples of the epigenetic control mechanisms that influence looping of DNA from regions of heterochromatin so that transcription of specific regions can take place.

FURTHER READING

Barber BA & Rastegar M (2010) Epigenetic control of *Hox* genes during neurogenesis, development, and disease. *Ann Anat* 192:261–274 (doi:10.1016/j.aanat.2010.07.009).

Bartlett J, Blagojevic J, Carter D et al. (2006) Specialized transcription factories. *Biochem Soc Symp* 73:67–75.

Bottardi S, Aumont A, Grosveld F & Milot E (2003) Developmental stage-specific epigenetic control of human beta-globin gene expression is potentiated in hematopoietic progenitor cells prior to their transcriptional activation. *Blood* 102:3989–3997 (doi:10.1182/blood-2003-05-1540).

Chakalova L, Carter D, Debrand E et al. (2005) Developmental regulation of the beta-globin gene locus. *Prog Mol Subcell Biol* 38:183–206.

Chambeyron S & Bickmore WA (2004) Chromatin decondensation and nuclear reorganization of the *HoxB* locus upon induction of transcription. *Genes Dev* 18:1119–1130 (doi:10.1101/gad.292104).

Chan PK, Wai A, Philipsen S & Tan-Un KC (2008) 5′HS5 of the human beta-globin locus control region is dispensable for the formation of the beta-globin active chromatin hub. *PLoS One* 3:e2134 (doi:10.1371/journal.pone.0002134).

Cremer T & Cremer M (2010) Chromosome territories. *Cold Spring Harb Perspect Biol* 2:a003889 (doi:10.1101/cshperspect.a003889).

Cremer T, Kurz A, Zirbel R et al. (1993) Role of chromosome territories in the functional compartmentalization of the cell nucleus. *Cold Spring Harb Symp Quant Biol* 58:777–792.

Ferreira J, Paolella G, Ramos C & Lamond AI (1997) Spatial organization of large-scale chromatin domains in the nucleus: a magnified view of single chromosome territories. *J Cell Biol* 139:1597–1610 (doi:10.1083/jcb.139.7.1597).

Gavrilov AA & Razin SV (2008) Spatial configuration of the chicken alpha-globin gene domain: immature and active chromatin hubs. *Nucleic Acids Res* 36:4629–4640 (doi:10.1093/nar/gkn429).

Gavrilov AA, Zukher IS, Philonenko ES et al. (2010) Mapping of the nuclear matrix-bound chromatin hubs by a new M3C experimental procedure. *Nucleic Acids Res* 38:8051–8060 (doi:10.1093/nar/gkq712).

Goodfellow SJ & Zomerdijk JC (2012) Basic mechanisms in RNA polymerase I transcription of the ribosomal RNA genes. *Subcell Biochem* 61:211–236 (doi:10.1007/978-94-007-4525-4_10).

Hernandez-Verdun D, Roussel P, Thiry M et al. (2010) The nucleolus: structure/function relationship in RNA metabolism. *Wiley Interdiscip Rev RNA* 1:415–431 (doi:10.1002/wrna.39).

Kashyap V & Gudas LJ (2010) Epigenetic regulatory mechanisms distinguish retinoic acid-mediated transcriptional responses in stem cells and fibroblasts. *J Biol Chem* 285:14534–14548 (doi:10.1074/jbc.M110.115345).

Learned RM, Smale ST, Haltiner MM & Tjian R (1983) Regulation of human ribosomal RNA transcription. *Proc Natl Acad Sci USA* 80:3558–3562.

Liang Y & Hetzer MW (2011) Functional interactions between nucleoporins and chromatin. *Curr Opin Cell Biol* 23:65–70 (doi:10.1016/j.ceb.2010.09.008).

Meaburn KJ & Misteli T (2007) Cell biology: chromosome territories. *Nature* 445:379–381 (doi:10.1038/445379a).

Müller WG, Rieder D, Karpova TS et al. (2007) Organization of chromatin and histone modifications at a transcription site. *J Cell Biol* **177**:957–967 (doi:10.1083/jcb.200703157).

Noordermeer D & de Laat W (2008) Joining the loops: beta-globin gene regulation. *IUBMB Life* **60**:824–833 (doi:10.1002/iub.129).

Osborne CS, Chakalova L, Brown KE et al (2004) Active genes dynamically colocalize to shared sites of ongoing transcription. *Nat Genet* **36**:1065–1071 (doi:10.1038/ng1423).

Osborne CS, Chakalova L, Mitchell JA et al. (2007) *Myc* dynamically and preferentially relocates to a transcription factory occupied by *Igh*. *PLoS Biol* **5**:e192 (doi:10.1371/journal.pbio.0050192).

Soshnikova N & Duboule D (2009) Epigenetic regulation of vertebrate *Hox* genes: a dynamic equilibrium. *Epigenetics* **4**:537–540 (doi:10.4161/epi.4.8.10132).

Soshnikova N & Duboule D (2009) Epigenetic temporal control of mouse *Hox* genes *in vivo*. *Science* **324**:1320–1323 (doi:10.1126/science.1171468).

Sutherland H & Bickmore WA (2009) Transcription factories: gene expression in unions? *Nat Rev Genet* **10**:457–466 (doi:10.1038/nrg2592).

Ulianov SV, Gavrilov AA & Razin SV (2012) Spatial organization of the chicken beta-globin gene domain in erythroid cells of embryonic and adult lineages. *Epigenetics Chromatin* **5**:16 (doi:10.1186/1756-8935-5-16).

van de Corput MP, de Boer E, Knoch TA et al. (2012) Super-resolution imaging reveals three-dimensional folding dynamics of the β-globin locus upon gene activation. *J Cell Sci* **125**:4630–4639 (doi:10.1242/jcs.108522).

Verschure PJ, van Der Kraan I, Manders EM & van Driel R (1999) Spatial relationship between transcription sites and chromosome territories. *J Cell Biol* **147**:13–24 (doi:10.1083/jcb.147.1.13).

10

EPIGENETIC CONTROL OF THE MITOTIC CELL CYCLE

When the cell divides, it needs to replicate its DNA. Because histones carrying several post-translational modifications are an integral part of chromosome structure, it is essential to separate the DNA from histones temporarily to allow replication and the eventual partition of the copied DNA into the mother and daughter cells. Epigenetic control is central to this process.

The **cell cycle** is a collective term for the series of events taking place in a cell that culminates in its replication. The process comprises four distinct phases (G1, S, G2, and M), all of which are tightly regulated through the action of a large number of proteins, including cyclins and cyclin-dependent kinases (Cdks). Strictly speaking, there is a fifth phase called G0, but because this is populated only by cells that have temporarily stopped dividing, it is often neglected in discussions of cell cycle control mechanisms. A schematic representation of the mitotic cell cycle is shown in **Box 10.1**.

The entire purpose of the mitotic cell cycle is the production of new cells that for the most part contain the same DNA as the original mother cell, although progenitor and offspring DNA may not be subject to identical gene expression control because differentiation sometimes accompanies division, particularly in many types of adult or tissue-specific stem cells. Nevertheless, a key feature of cell replication of terminally differentiated cell types is that the phenotype of the daughter cells is generally identical (or at least highly similar) to that of the mother cell. This implies the existence of mechanisms to preserve the inheritance of epigenetic information from one generation to the next.

10.1 S PHASE INVOLVES DNA REPLICATION

From an epigenetic point of view, the most important phase of the cycle is S phase, during which the DNA temporarily dissociates from its nucleosomal structures to permit passage of the replication fork. Not unexpectedly, this is likely to inhibit transcriptional activity. In fact, it is likely that transcription is actively repressed when the cell enters the division cycle. It is important that the newly synthesized DNA reassociates rapidly with histones, to ensure its correct packaging within the nucleus. The physical manifestation of these processes is reflected in the decondensation and recondensation of chromatin at specific times during S phase. **Figure 10.1** is a reminder of the highly ordered three-dimensional structures adopted by large regions of chromatin outside mitosis.

10.1 S PHASE INVOLVES DNA REPLICATION

10.2 THE CELL DIVIDES IN M PHASE

Box 10.1 The mitotic cell cycle and protein interactions needed to control it

The complex regulatory and signaling pathways that govern cell cycle progression are highly conserved in eukaryotes. Two cell cycle checkpoints control the order and timing of cell cycle transitions (G1–S and G2–M) and ensure that critical events such as DNA replication and chromosome segregation are completed correctly before allowing the cell to progress further through the cycle. The first of these two checkpoints, the **restriction point (R)**, is located at the end of the G1 phase (**Figure 1**). Beyond this point, precursors of regulatory molecules will invariably complete the cell cycle.

The R point, which is the primary G1/S cell cycle checkpoint, controls the commitment of eukaryotic cells to transition through the "gap" phase (G1) and enter the DNA synthesis phase (S). This transition is accomplished by two cell cycle kinase complexes, CDK4/6–cyclin D and CDK2–cyclin E, which work in concert to relieve the inhibition of E2F by Rb (the retinoblastoma protein), both of which reside within the same dynamic transcription complex (**Figure 2**). More specifically, in G1 phase uncommitted cells, hypophosphorylated Rb binds to the E2F–DP1 transcription factors in a repressive complex containing HDACs and histone methyltransferases such as SuV39h1. This ensures that E2F target genes required for cell cycle progression are not expressed until phosphorylation of Rb by cyclin D–CDK4/6—and subsequently by cyclin E–CDK2—dissociates the repressor complex from Rb, permitting the transcription of key S phase-promoting genes,

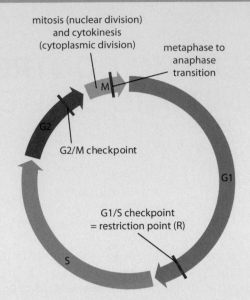

Figure 1 The eukaryotic cell cycle and its checkpoints. The checkpoints exist to ensure that DNA undergoes replication with minimal introduction of errors; they prevent the cell from progressing to mitosis if DNA damage is above threshold levels. [Adapted from Strachan T & Read A (2010) Human Molecular Genetics, 4th ed. Garland Science.]

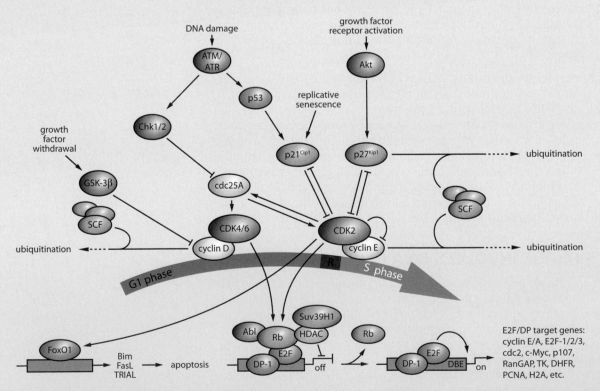

Figure 2 Signal transduction pathways that influence the G1–S transition. Growth factors promote entry into S phase; the detection of DNA damage transmits signals via the ATM/ATR mechanism to target cdc25a for degradation, thereby blocking cell cycle progress. [Adapted from Cell Signaling Technology, Inc. http://www.cellsignal.com/reference/pathway/Cell_Cycle_G1S.html]

including some that are required for DNA replication. CDK2 may also phosphorylate the proapoptotic FoxO1, which inhibits the transcriptional activity of FoxO1 by inducing its nuclear export and thereby allows the cell to survive and proliferate.

Several cellular systems are able to affect the G1–S transition. For example, growth factor signaling can influence CDK activity to effect Rb displacement, or DNA damage can block CDK activity, thereby ensuring that cell cycle progression genes are not expressed until appropriate damage repair has been completed and signal transduction by cell cycle arrest proteins such as ATM, ATR, and p53 is relieved. At a critical convergence point with the DNA-damage checkpoint, if DNA damage exists, cdc25A is ubiquitinated and targeted for degradation via the SCF ubiquitin ligase complex downstream of the ATM/ATR/Chk pathway (see Figure 2). Because cdc25A removes inhibitory phosphates from the active sites of CDKs, its ubiquitination prevents the activation of CDK complexes and stalls cell cycle progression. In contrast, after cdc25A has activated its CDK targets in M phase, its timely degradation—via the APC ubiquitin ligase complex—allows progression through mitosis. Furthermore, growth factor withdrawal activates GSK-3β, which in turn phosphorylates cyclin D, leading to its rapid ubiquitination and proteasomal degradation; this leaves E2F in the repressive grip of Rb, and the cell cycle does not progress past the R point. Collectively, ubiquitin/proteasome-dependent degradation and nuclear export are mechanisms commonly used to rapidly reduce the concentration of cell cycle control proteins.

Once S phase is complete, there is still the possibility that new errors will have been introduced into the DNA. There is therefore a second damage checkpoint at G2–M (**Figure 3**). The function of this checkpoint is the final elimination of errors before mitosis. The activity of the cdc2–cyclin B complex is pivotal in regulating this transition. During G2, cdc2 is maintained in an inactive state by the tyrosine kinases Wee1 and Myt1. It is thought that, as cells approach M phase, the phosphatase cdc25 is activated through phosphorylation by PLK1 (polo-like kinase 1) and the activated cdc25 in turn activates cdc2, which activates more cdc25, establishing a feedback amplification loop that efficiently drives the cell into mitosis. PLK1 specifically phosphorylates serine 198 of the cdc25 protein to achieve activation. Importantly, DNA damage activates the DNA-activated protein kinases ATM and ATR, which then initiate two parallel cascades that inactivate the cdc2–cyclin B complex. The first cascade rapidly inhibits progression into mitosis: the Chk kinases phosphorylate and inactivate cdc25, thereby preventing the activation of cdc2. Chk kinases achieve this by phosphorylating serine 216 on cdc25, which serves to inactivate its ability to activate cdc2 subsequently. The slower, second parallel cascade involves the phosphorylation of p53, which permits the dissociation of p53 from MDM2 and MDM4 and activates its DNA binding and transcriptional regulatory activity, the result being the transcription of genes encoding proteins that inhibit CDKs.

Figure 3 Entry into M phase. The newly synthesized DNA is checked for errors, and several signal transduction systems detect damage and can halt the G2–M transition. DNA damage such as double-strand breaks (red crosses) are caused by radiation or chemical damage by free radicals. [Adapted from Cell Signaling Technology, Inc. http://www.cellsignal.com/reference/pathway/Cell_Cycle_G2M_DNA.html]

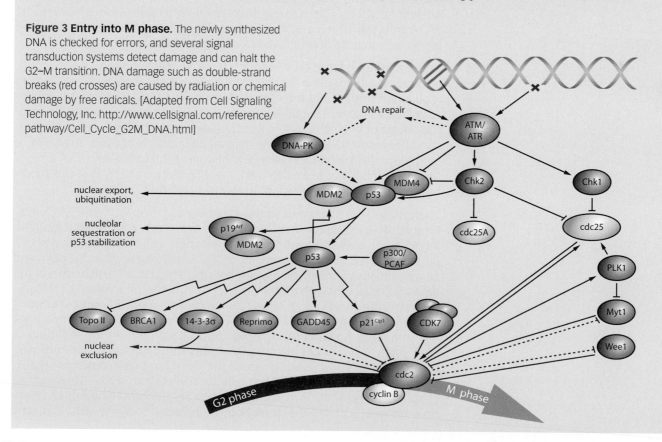

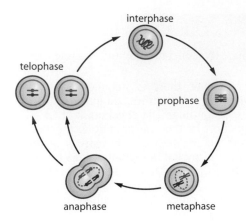

Figure 10.1 Chromatin condensation occurs during interphase in preparation for mitosis. The condensed structures that we recognize as chromosomes appear after chromatin condensation as the cell enters prophase and the chromosomes remain condensed throughout mitosis. [Adapted from Brown TA (2006) Genomes, 3rd ed. Garland Science.]

Replication of eukaryotic genomes involves the regulated and timely assembly of multiprotein complexes on chromatin templates *in vivo*. We know that the transcription complex modifies chromatin at promoters and within transcribed regions as transcription forks progress. It is therefore also likely that the DNA replication machinery is effective at chromatin remodeling, owing to the need for efficient access to the entire genetic complement.

We saw in Chapter 3 that an important component in higher-order chromatin packaging is histone H1, which binds to the linker DNA between nucleosome particles and stabilizes the higher-order structure of chromatin. Phosphorylation of H1 increases its ability to dissociate from the chromatin fiber. In the context of cell cycle progression, it is known that tethering of Cdc45—a protein associated with the transition from G1 to S phase—to a chromosomal locus promotes large-scale chromatin decondensation. Phosphorylated H1 becomes highly concentrated in the decondensed chromatin after Cdc45 targeting, possibly as a result of the recruitment of Cdk2 to the Cdc45-promoted open chromatin. Direct tethering of Cdk2 to the chromatin also induces chromatin unfolding coincident with the appearance of phosphorylated H1. The simplest interpretation of these data is that recruitment of Cdk2 to replication foci leads to H1 phosphorylation, chromatin decondensation, and facilitation of DNA replication. However, this mechanism is unlikely to be universally applicable: some loci are already in an open chromatin conformation because they have been subject to active transcription.

DNA replication does not occur at the same time for all loci in the genomes. In a given differentiated cell type, the differences between euchromatic and heterochromatic loci may provide a possible explanation for the timing of which genes are replicated during S phase. A correlation between replication timing and transcriptional activity of proximal genes has been observed in many organisms, suggesting a connection between these nuclear functions. For example, high-resolution replication profiles reveal an overall positive correlation between gene expression and timing of replication in both humans and *Drosophila melanogaster*.

In a similar manner to their regulatory role in transcription, histone modifications could regulate the access of replication factors such as Cdc45 to replication origins and therefore determine the time of origin activation. Exposure to HDAC inhibitors such as trichostatin A is known to advance the timing of normally late-replicating loci, and late-replicating chromosomal regions co-localize with hypoacetylated chromatin. However, a genome-wide correlation between histone acetylation and the time of origin firing has not been detected. It is therefore unlikely that this modification is the sole determinant of replication timing.

There is some evidence to suggest that methylation of lysine 36 on histone H3 defines the early-replicating loci and has a role in regulating the kinetics of Cdc45 association with replication origins. Its binding to origins is delayed in the absence of SET2 (which methylates H3K36) and cannot be advanced by increasing histone acetylation. Furthermore, the data are consistent with monomethylation and trimethylation of K36 having opposing functions in DNA replication initiation: K36me1 increases at replication origins on binding of Cdc45, suggesting a positive (that is, promoting) function for monomethylated K36 during the initiation of DNA replication; in contrast, early origins are depleted of K36me3 and this modification decreases at about the time that Cdc45 binds to origins, indicating a negative (inhibitory) role for trimethylated K36. However, to ensure rapid replication, it seems that H3K36 monomethylation and multiple histone acetylations must both be present. It is not yet clear why

H3K36 trimethylation should actively repress the binding of Cdc45 to the replication origin. However, because the timing of replication origin firing, and Cdc45 binding to replication origins, has been shown to be dictated by the chromosomal environment, it is possible that the effect of histone modifications on origin activation is purely context driven. How Cdc45 accesses the origins of late-replicating loci is not clear at present.

10.2 THE CELL DIVIDES IN M PHASE

After DNA replication, the chromatin must be reconstructed without delay to allow chromosome segregation immediately before cell division. When a cell divides, chromosomes need to reorganize into compact rod-shaped bodies to permit the segregation of their replicated sister chromatids to opposite spindle poles. The structural chromatin dynamics underlying the formation of mitotic chromosomes have been defined as mitotic chromosome condensation (see Figure 10.1).

The structural criteria for performing this process are quite exact. First, the chromatids require significant compaction to provide sufficient space for their individualized movements on the mitotic spindle. It is essential that their DNAs be completely disentangled from each other, and that their surfaces are non-adherent to neighboring chromosomes. The replicated sister chromatids need to be aligned in parallel compact rods to support bipolar attachment to the mitotic spindle. Mitotic chromosome arms need to be shorter than half of the spindle length to permit complete segregation to opposite spindle poles before cytokinesis. It is likely that dephosphorylation of histone H1 is involved in these structural alterations.

A fascinating aspect of the mitotic cell cycle is that the pattern of histone modifications present in the genome of specific cell types is identical (for all practical purposes) before and after mitosis. Such epigenetic "memory" of the cell's transcriptional profile is currently a very active area of investigation. One suggestion for how such a memory might be instilled in the histone structure is that certain DNA-binding factors withstand chromatin condensation and survive intact on the DNA during mitosis to facilitate the rapid reactivation of genes in the next cell cycle. This has become known as the **bookmarking theory** and is supported by the apparent retention of transcription factor IID (TFIID) on some gene promoters in mitotic chromosomes. The TFIID acts as a tag to recruit histone modification enzymes that can establish active marks, such as H3K3 methylation, on the new histones after reconstruction of the nucleosomes. However, this may not be enough to retain a complete memory of the previous transcriptional state of genes, because there are elements other than the promoter that control precise expression levels.

An example of the retention of epigenetic memory is provided by the globin genes. Studies on the α- and β-globin genes have shown that the tissue-specific factor NF-E2p45, which is crucial for activating globin genes, is preserved on mitotic chromosomes. This finding suggests that this transcription factor can serve as a molecular memory marker that maintains the locally hypersensitive state of the globin gene clusters. Chromatin remodeling complexes and histone modification enzymes, which are displaced from mitotic chromosomes, depend on other protein factors to reunite them with the chromatin. During globin gene activation, NF-E2p45 has a key role in recruiting other protein factors such as TAFII130 and CREB-binding protein (CBP), a co-activator of histone acetyltransferase activity. Interestingly, this phenomenon is not restricted to transcription factors: there is evidence of the retention of at least some modified histones at specific loci during mitosis.

In mitotic cells, some active modifications, especially H3 acetylation and H3K4 dimethylation, are well preserved on the mitotic chromosomes, thus providing markers to prompt the re-expansion of these marks when transcription is resumed. This could represent a type of epigenetic "memory," and there is some evidence that this might be reinforced by the association of histone variants with genes of different transcriptional activities. Histone variants are simply histone proteins with slightly different amino acid sequences, and the common variant histone H3.3 is often present in the same regions as those enriched by H3K4 methylation. The preservation of inactive marks such as H3K9 methylation during the cell cycle is less well documented. It is therefore less clear how the epigenetic memory of gene inactivity might function, although the presence (or absence) of histone variants may also serve as a marker of inactive genes, because regions enriched in H3K9 dimethylation tend to be associated with histone H3 rather than H3.3. These results strongly suggest that during mitosis the removal of histones and/or transcription factors from the DNA might not be as complete as originally envisaged, and that the retention of smaller amounts of these proteins might be the basis of epigenetic memory during the cell cycle.

The topics covered here and in Chapter 9 provide a basic understanding of the contributions made by epigenetic control of gene expression to the control of cellular functions such as differentiation and replication. Following on from this we can examine how perturbations to these processes can disrupt cellular functions and lead to the development of disease.

KEY CONCEPTS

- Eukaryotic cells replicate via mitosis, which occurs through a four-stage cell cycle.

- DNA replication is required to generate new daughter cells.

- A mechanism is needed to preserve the pattern of epigenetic modifications present in the mother cell to ensure that the daughter cell is the same type of cell.

- Phosphorylation of histone H1 loosens its association with the DNA, which allows the latter to adopt a more open conformation, allowing the synthesis of new DNA.

- The mechanism of epigenetic memory may involve the enrichment of variant histone types at the loci of transcriptionally active genes.

FURTHER READING

Aparicio OM (2013) Location, location, location: it's all in the timing for replication origins. *Genes Dev* **27**:117–128 (doi:10.1101/gad.209999.112).

Cimini D, Mattiuzzo M, Torosantucci L & Degrassi F (2003) Histone hyperacetylation in mitosis prevents sister chromatid separation and produces chromosome segregation defects. *Mol Biol Cell* **14**:3821–3833 (doi:10.1091/mbc.E03-01-0860).

Contreras A, Hale TK, Stenoien DL et al. (2003) The dynamic mobility of histone H1 is regulated by cyclin/CDK phosphorylation. *Mol Cell Biol* **23**:8626–8636 (doi:10.1128/MCB.23.23.8626-8636.2003).

Duan Q, Chen H, Costa M & Dai W (2008) Phosphorylation of H3S10 blocks the access of H3K9 by specific antibodies and histone methyltransferase. Implication in regulating chromatin dynamics and epigenetic inheritance during mitosis. *J Biol Chem* **283**:33585–33590 (doi:10.1074/jbc.M803312200).

Feng YQ, Desprat R, Fu H et al (2006) DNA methylation supports intrinsic epigenetic memory in mammalian cells. *PLoS Genet* **2**:e65 (doi:10.1371/journal.pgen.0020065).

Gréen A, Sarg B, Gréen H et al. (2011) Histone H1 interphase phosphorylation becomes largely established in G1 or early S phase and differs in G1 between T-lymphoblastoid cells and normal T cells. *Epigenetics Chromatin* **4**:15 (doi:10.1186/1756-8935-4-15).

Hake SB & Allis CD (2006) Histone H3 variants and their potential role in indexing mammalian genomes: the "H3 barcode hypothesis". *Proc Natl Acad Sci USA* **103**:6428–6435 (doi:10.1073/pnas.0600803103).

Halmer L & Gruss C (1996) Effects of cell cycle dependent histone H1 phosphorylation on chromatin structure and chromatin replication. *Nucleic Acids Res* **24**:1420–1427 (doi:10.1093/nar/24.8.1420).

Hans F & Dimitrov S (2001) Histone H3 phosphorylation and cell division. *Oncogene* **20**:3021–3027

Kelly TK & Jones PA (2011) Role of nucleosomes in mitotic bookmarking. *Cell Cycle* **10**:370–371 (doi:10.4161/cc.10.3.14734).

Kruhlak MJ, Hendzel MJ, Fischle W et al (2001) Regulation of global acetylation in mitosis through loss of histone acetyltransferases and deacetylases from chromatin. *J Biol Chem* **276**:38307–38319 (doi:10.1074/jbc.M100290200).

Li Y, Kao GD, Garcia BA et al. (2006) A novel histone deacetylase pathway regulates mitosis by modulating Aurora B kinase activity. *Genes Dev* **20**:2566–2579 (doi:10.1101/gad.1455006).

Medina R, Ghule PN, Cruzat et al. (2012) Epigenetic control of cell cycle-dependent histone gene expression is a principal component of the abbreviated pluripotent cell cycle. *Mol Cell Biol* **32**:3860–3871 (doi:10.1128/MCB.00736-12).

Miles J, Mitchell JA, Chakalova L et al. (2007) Intergenic transcription, cell-cycle and the developmentally regulated epigenetic profile of the human beta-globin locus. *PLoS One* **2**:e630 (doi:10.1371/journal.pone.0000630).

Ng RK & Gurdon JB (2008) Epigenetic memory of an active gene state depends on histone H3.3 incorporation into chromatin in the absence of transcription. *Nat Cell Biol* **10**:102–109 (doi:10.1038/ncb1674).

Park JA, Kim AJ, Kang Y et al. (2011) Deacetylation and methylation at histone H3 lysine 9 (H3K9) coordinate chromosome condensation during cell cycle progression. *Mol Cells* **31**:343–349 (doi:10.1007/s10059-011-0044-4).

Rice JC, Nishioka K, Sarma K et al. (2002) Mitotic-specific methylation of histone H4 Lys 20 follows increased PR-Set7 expression and its localization to mitotic chromosomes. *Genes Dev* **16**:2225–2230 (doi:10.1101/gad.1014902).

Sarge KD & Park-Sarge OK (2005) Gene bookmarking: keeping the pages open. *Trends Biochem Sci* **30**:605–610 (doi:10.1016/j.tibs.2005.09.004).

Sarge KD & Park-Sarge OK (2009) Mitotic bookmarking of formerly active genes: keeping epigenetic memories from fading. *Cell Cycle* **8**:818–823 (doi:10.4161/cc.8.6.7849).

Sharif J, Endoh M & Koseki H (2011) Epigenetic memory meets G2/M: to remember or to forget? *Dev Cell* **20**:5–6 (doi:10.1016/j.devcel.2010.12.012).

Swarnalatha M, Singh AK & Kumar V (2012) The epigenetic control of E-box and Myc-dependent chromatin modifications regulate the licensing of lamin B2 origin during cell cycle. *Nucleic Acids Res* **40**:9021–9035 (doi:10.1093/nar/gks617).

Wako T, Murakami Y & Fukui K (2005) Comprehensive analysis of dynamics of histone H4 acetylation in mitotic barley cells. *Genes Genet Syst* **80**:269–276 (doi:10.1266/ggs.80.269).

Wu S, Wang W, Kong X et al. (2010) Dynamic regulation of the PR-Set7 histone methyltransferase is required for normal cell cycle progression. *Genes Dev* **24**:2531–2542 (doi:10.1101/gad.1984210).

Xu M, Wang W, Chen S & Zhu B (2011) A model for mitotic inheritance of histone lysine methylation. *EMBO Rep* **13**:60–67 (doi:10.1038/embor.2011.206).

Zaidi SK, Young DW, Montecino M et al. (2011) Bookmarking the genome: maintenance of epigenetic information. *J Biol Chem* **286**:18355–18361 (doi:10.1074/jbc.R110.197061).

CHAPTER ELEVEN

THE EPIGENETIC BASIS OF GENE IMPRINTING

Another cellular function that seems to be subject to epigenetic control is genomic imprinting. This phenomenon is observed in mammals and many species of plants, and it has been suggested that this may serve to promote the survival of genes from one parent at the expense of the other. How this might relate to the way in which organisms continue to evolve is a matter of debate, but the imprinting phenomenon is still a good example of the epigenetic control of gene expression.

Genomic imprinting refers to monoallelic gene expression that occurs in a manner that is specific to the parent of origin. For the vast majority of genes present in the human genome, expression occurs from both alleles simultaneously (provided, of course, that expression is permitted in the cell type of interest), however, a small percentage (less than 1%) of genes are expressed from only one allele. Because each of these alleles is derived from a different parent, the imprinted genes are sometimes described as being maternally or paternally imprinted. For example, the gene encoding insulin-like growth factor 2 (*IGF2/Igf2*) is expressed only from the allele inherited from the father and thus is paternally imprinted. Shortly before completion of this book, the total number of imprinted genes in humans was 90, although the consensus of opinion is that more remain to be discovered.

The basis of imprinting seems to be epigenetic. Although the precise nature of the initial epigenetic imprint is currently a matter of intense investigation, it is assumed that the parental imprint is set in the germ line, because this is the time when the genomes are in distinct compartments and can be differentially modified. It is worth noting that although a number of imprinted genes remain imprinted throughout the life of the organism, many genes are imprinted in a tissue-specific or temporal-specific way.

11.1 CONTROLLING MONOALLELIC EXPRESSION OF IMPRINTED GENES

Imprinted genes share few characteristics in common

One of the hallmarks of imprinted genes is that many are found in clusters throughout the genome, although this characteristic is by no means common to all imprinted genes. Moreover, it is probably true that no single characteristic distinguishes an imprinted gene from a non-imprinted gene. Rather, imprinted and non-imprinted genes are distinguished by a list of parameters that often vary. However, there are a few common genomic features that are shared by imprinted genes, in addition to their presence in clusters. For example, imprinted genes often contain fewer and smaller introns than do non-imprinted genes. In addition, they tend to exhibit some degree of repetitive sequences; in corroboration of this, they seem to contain unusually high numbers of retrotransposable elements, with 23 of the known imprinted genes sharing this characteristic. It has been suggested that the repetitive elements facilitate the initiation and spreading of a heterochromatic state. However, although this process may occur in the repetitive sequences of the X chromosome (which, as we saw in Chapter 3, undergoes extensive epigenetic silencing), there is less evidence to support the existence of a similar mechanism that is common to all imprinted genes. For example, the *Impact* gene is imprinted in mouse, rat, and rabbit. Tandem repeats within the mouse and rat *Impact* gene are methylated in a manner that is specific to the parent of origin. The rabbit gene lacks the tandem repeats but is apparently still imprinted.

The clustering feature of imprinted genes is probably their most common property. In mammals, approximately 80% of imprinted genes exist in clusters, or at the very least in close proximity to each other. These clusters contain two or more imprinted genes over a region that can span in excess of 1 Mb, and in most cases these genes, which can be either maternally or paternally expressed, are jointly regulated through an **imprinting control region (ICR)**.

Imprinting control regions (ICRs) regulate the imprinted expression of genes

The power of the ICR to direct imprinting has been demonstrated by experiments in which the ICRs have been removed from known imprinted genes, leading to the loss of the imprinting—that is, to expression from both alleles. Perhaps more importantly, the parallel approach of inserting the same ICR into a gene that is normally not subject to imprinting can induce the phenomenon at that locus, the result being the loss of expression from one of the alleles. Examples of this type of experiment include the creation of a non-imprinted derivative of the mouse gene *RSVIgmyc* by mapping and deleting its ICR. Imprinting of the non-imprinted derivative was restored by the substitution of an ICR from the *Igf2/H19* locus. In a similar experiment, the *H19* ICR was able to impart imprinting status to the normally non-imprinted β-globin locus.

The ICRs exhibit parental-specific epigenetic modifications, such as DNA methylation, that govern their activity. The regions affected by such DNA methylation events are called differentially methylated regions (DMRs) and there may be several such regions situated within an imprint control center. The term arises because the DMR of the nonexpressed allele is subjected to higher levels of DNA methylation than that of the expressed allele; this is consistent with the known functions of DNA methylation (described in Chapter 5). However, the mechanisms that direct methylation to either the paternal or maternal copy of an imprinted gene are not entirely clear. Because much of the epigenetic control of gene imprinting is focused on the DMRs, we will refer to these regions rather than the imprint control regions.

As their apparent function might suggest, DMRs are enriched in CpG dinucleotides, but overall there are fewer of these than would normally be found in CpG islands. Interestingly, the repetitive elements mentioned above are usually located within the DMR, although once more it is unclear why this should be so. Depending on the methylation status, the DMRs (which effectively function as *cis*-acting sequences) may recruit different regulatory factors, such as methyl-CpG-binding domain (MBD) proteins, which contain a domain that recognizes and binds methylated DNA. This is not unexpected, because it is part of the mechanism by which methylated DNA is recognized and leads to repression of gene function.

Differentially methylated regions contain imprinting signals

The DMR-swapping experiments mentioned above suggest the presence of an imprinting signal in the DMRs of different imprinted loci. Because DMRs are defined by divergent sequences, primary sequence is not likely to be the only signal. It is possible that the unusually high incidence of repetitive elements within the DMR may contribute to the imprinting signal or even that the precise order or number of CpG dinucleotides may direct this phenomenon. However, these factors alone may not be sufficient, given that they exist in other parts of the genome that are not subjected to imprinting. Conversely, imprinting is invariably associated with the presence of 5-methylcytosine, so it is possible that differences in regional methylation and/or the degree of CpG island methylation may provide additional marks to distinguish between the maternal and paternal alleles. For example, the DMRs of some imprinted genes are often completely devoid of methylation on the expressed allele.

Despite the ubiquity of DNA methylation in imprinting, many investigators concede that it is not known whether cytosine methylation is the cause or merely the manifestation of the imprinting process. In some cases, DNA methylation seems to be essential for the proper expression of imprinted genes. For example, aberrant DNA methylation has been linked to several imprinting-related disorders in humans, apparently owing to improper spatial and/or temporal allele-specific expression. In *Arabidopsis*, maintenance of DNA methylation is required for the proper imprinted expression of the genes *FWA* and *FIS2*. However, DNA methylation seems to have only a partial role in regulating imprinting of the *Arabidopsis Medea* gene. Examples of a nonessential role for DNA methylation in genomic imprinting come from *Drosophila* and *Caenorhabditis elegans*. Although cytosine methylation has been detected in *Drosophila*, its role in the ability of *Drosophila* to imprint endogenous genes has not been established. DNA methylation has not been reported for *C. elegans*, yet this species can apparently imprint a subset of transgenes during early embryogenesis. And in contrast with mammals and plants, in which DNA methylation is generally associated with gene inactivation, mealybugs show the opposite relationship between DNA methylation and gene activity. The inactive heterochromatinized paternal genome is hypomethylated relative to the active, hypermethylated maternal genome. Hence, although DNA methylation seems to be widely associated with the imprinting process, its true mechanistic role remains largely unknown.

Chromatin modifications at DMR sites affect gene imprinting

Chromatin modification by covalent post-translational modification of histones also seems to have a role in imprinting. DMRs are sites of differential chromatin structure (which thereby regulates the accessibility of regulatory factors) between the two alleles of an imprinted gene. For example, the DMRs of *IGF2/H19*, *IC2/KCNQ1*, and *PWS/IC* are sensitive to digestion by DNase I in a manner that depends on the parent of origin, suggesting that the DNA is more accessible to the enzyme in one of the alleles. Both activating (that is, H3 and H4 acetylation) and deactivating histone modifications (H3 lysine 9 and lysine 27 methylation) are found at these DMRs, and the pattern of histone modification seems to be correlated with the pattern of DNA methylation. For example, activating histone marks are found at hypomethylated DMRs, whereas deactivating marks are generally associated with highly methylated DMRs. As discussed in Chapter 6, the presence of activating marks tends to promote the formation of euchromatin, which adopts a more "open" conformation that may allow access of DNAse I more readily, whereas the opposite is true of the heterochromatin-promoting repressive marks. The association between DNA methylation and histone modification does not always hold true for imprinted genes: there are examples of activating and repressing chromatin conformations in DMRs with low and high levels of DNA methylation, respectively. In addition to this, differential histone modification can also occur outside DMRs to modulate imprinted gene expression. For example, at the *IGF2r* locus, it is allele-specific histone modification at the promoter, possibly in conjunction with DNA methylation at the ICR/DMR, that is thought to regulate imprinting. Background information about *IGF2* and *IGF2r* gene products is given in **Box 11.1**.

11.2 EXAMPLES OF IMPRINTING

The imprinting of *IGF2/H19* is well documented

How is *IGF2* imprinted? As noted above, many imprinted genes are found clustered in small genomic regions, and *IGF2* is no exception because it is found in close proximity to the *H19* locus in both mouse and human chromosomes. The DMRs for these genes in both of these species have been described in considerable detail (**Figure 11.1**). Mouse *Igf2* has three DMRs: there is a single maternally methylated, placenta-specific DMR located at exon U1 (DMR0), and a paternally methylated DMR1 and DMR2 located upstream of promoter 1 and in exon 6. *H19* has only one paternally methylated DMR, and it is located 2–4 kb upstream of its transcription start site. The *IGF2* gene is normally expressed from the paternal allele, while *H19* is expressed from the maternal allele.

The DMRs seem to have rather different effects on the global gene expression from the *IGF2/H19* pair. Deletion of the maternal *H19* DMR results in loss of imprinting of *IGF2*, with biallelic expression occurring

---

Box 11.1 An example of gene imprinting—insulin-like growth factor II

Insulin-like growth factor-II (IGF-II or IGF2) is an important fetal growth factor, acting as a mitogen for many different cell types and in particular modulating the growth and differentiation of muscle cells. The protein is encoded by the *IGF2* gene located on chromosome 11 in humans and chromosome 7 in mice. *IGF2* is interesting because the human and mouse homologues were among the first genes to be identified as genomically imprinted.

The *IGF2* gene product belongs to the class of insulin-like growth factors that are polypeptides with a high sequence similarity to insulin. These factors are part of the system, often referred to as the insulin-like growth factor axis, that is used by cells to communicate with their surroundings in the body. This system comprises IGF2 and its counterpart IGF1 with their cognate receptors (IGF1R and IGF2R), a family of six high-affinity IGF-binding proteins (IGFBP1–IGFBP6), and several proteases that degrade IGFBPs.

The IGF axis in the adult body is controlled mainly by the liver in response to growth hormone signaling and relies mainly on IGF1 for the regulation of normal physiology. Almost every cell in the body is targeted by IGF1, especially those in the muscles, cartilage, bone, liver, kidneys, skin, and lungs. IGF2 is thought to be a primary growth factor required during fetal development and thus seems to be more specific in its range of cellular targets.

IGF2 is thought to function through binding to the type 1 IGF receptor on the cell surface, which then starts a signal transduction cascade. It also binds to the type 2 IGF receptor (IGF2R), which instead of initiating a signaling response, causes IGF2 to be transported to lysosomes and ultimately degraded. In this respect, the IGF2 receptor acts only as a clearance receptor, as it does not activate any intracellular signaling pathways. However, the IGF2R protein is also thought to have a critical role in tumor suppression and immunity, particularly through its ability to activate T cells by internalization of their CD26 antigen. This latter function arises from the involvement of IGF2R in lysosomal trafficking.

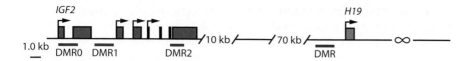

**Figure 11.1 Structure of the *IGF2/H19*
locus.** The *IGF2* and *H19* genes are separated
by 90 kb of intervening sequences. *IGF2* DMR1
and DMR2 and the *H19* DMR are shown.
[Adapted from Murrell A, Heeson S & Reik W
(2004) *Nat. Genet.* 36, 889. With permission from
Macmillan Publishers Ltd.]

in most tissues studied, and deletion of the maternal *IGF2* DMR1 results
in biallelic expression of *IGF2* in mesodermal tissues. Deletion of *IGF2*
DMR2 has no effect on imprinting but reduces the transcriptional activation of *IGF2*. These data suggest that DMR1 has a silencer function and
that DMR2 has an activator function. Methylation studies have shown
that on the maternal chromosome, the unmethylated *H19* DMR protects
IGF2 DMR1 and DMR2 from methylation, and DMR1 protects the DMR2
from methylation in a hierarchical manner. These results suggest that the
DMRs interact directly in such a way that the *H19* DMR protects the *IGF2*
DMRs from becoming methylated in somatic tissues. The interaction
between the DMRs requires considerable looping of the 90 kb of DNA
sequence between the two genes. A model for this is shown in **Figure
11.2**, which illustrates the association between either DMR1 or DMR2 of
IGF2 and the single DMR of *H19*, depending on the methylation status of
the *IGF2* DMRs.

Binding of CTCF at the *IGF2/H19* imprint control region to an insulator mechanism to control imprinted gene expression

The *H19* DMR contains a consensus binding site for the CTCF protein,
a binding factor named for its affinity for CCCTC sequences. CTCF is a
transcriptional repressor normally associated with blocking the access of
transcriptional activators to enhancer sites in a wide range of genes. Not
surprisingly, this protein is involved in different aspects of gene regulation, including promoter activation or repression, hormone-responsive

(a)

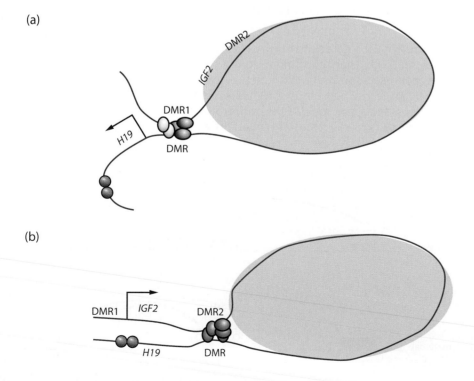

(b)

**Figure 11.2 Chromatin looping controls
allele-specific expression.** The differentially
methylated regions of the *IGF2/H19* locus
interact to determine which allele is expressed.
(a) On the maternal allele, the unmethylated
H19 DMR (which is bound by CTCF and possibly
other proteins; yellow and green ovals) and
IGF2 DMR1 interact, resulting in two chromatin
domains, with *H19* in an active domain with its
enhancers (pink circles) and *IGF2* in an inactive
domain away from the enhancers (shaded
area). (b) On the paternal allele, the methylated
H19 DMR associates with the methylated *IGF2*
DMR2 through putative protein factors (blue
ovals), moving *IGF2* into the active chromatin
domain. [Adapted from Murrell A, Heeson
S & Reik W (2004) *Nat. Genet.* 36, 889. With
permission from Macmillan Publishers Ltd.]

gene silencing, methylation-dependent chromatin insulation, and genomic imprinting. One model of CTCF function at the *IGF2/H19* locus is that binding of CTCF to the unmethylated maternal DMR protects the DMR from *de novo* methylation and prevents downstream enhancers from activating *IGF2*, leaving them available to activate transcription at *H19*. CTCF is unable to bind the methylated paternal DMR, resulting in expression of *IGF2* while *H19* is silenced.

Another possibility is that binding of CTCF to the DMR alters its ability to function as a genomic insulator. These are regions of DNA that can prevent the activation of promoters when they are present between an enhancer sequence and the promoter sequence. Some DMRs may function as insulators, depending on their methylation status (**Figure 11.3**), and the binding of CTCF seems to be implicated in this method of control although there is still some debate about the mechanism by which this operates.

The mechanism by which insulation occurs is uncertain

It is well established that CTCF binding occurs at the *H19* DMR, but our knowledge of the mechanism of insulation remains incomplete. It may simply be a modification of the chromatin looping model (see Figure 11.2), but although there is fairly solid evidence for the interactions between the shared enhancer elements and the promoter of the *IGF2* gene, the evidence for the formation of broader loops by the interaction of the separate DMRs of the *IGF2* and *H19* genes has been questioned. This uncertainty arises from the "chromatin conformation capture" methodology that was used to gather the data that led to the looping model. Isolated nuclei were treated with formaldehyde, and physically interacting loci and the proteins mediating the interaction were cross-linked. The cross-linked chromatin was digested with a restriction enzyme, followed by ligation of DNA fragments present in the same complex (intramolecular ligation). After reversal of cross-links, the interaction was detected by PCR using primers designed to amplify the ligation products. Because different research groups made use of different restriction enzymes, the suggestion has arisen that the method did not distinguish between the DMRs and other regulatory regions of the *IGF2/H19* locus with sufficient accuracy to determine their degree of proximity.

Fortunately, the interaction between the DMRs has been determined using other methods than chromatin conformation capture. One alternative approach used chromatin immunoprecipitation. Three copies of a GAL4-binding motif (which binds a regulatory sequence known as upstream activation sequence, or UAS) were introduced into the mouse *H19* DMR by a targeted knock-in strategy. In addition, a transgenic mouse line was

Figure 11.3 The insulator model of imprinted gene regulation. DMRs that lie upstream and in the promoter region directly regulate the expression of *H19*, and the specific binding of CTCF to the upstream primary DMR indirectly regulates the expression of *IGF2* by inhibiting the effect of downstream enhancer(s). Because CTCF binding is sensitive to DNA methylation, reciprocal expression of *H19* and *IGF2* occurs. DNA methylation of the paternal allele is indicated by the blue circles. [Adapted from Kaneko-Ishino T, Kohda T & Ishino F (2003) *J Biochem* 133, 699. With permission from Oxford University Press.]

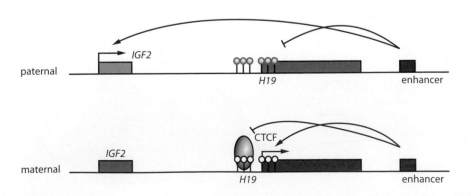

produced with a construct carrying the GAL4 DNA-binding domain fused to a human MYC epitope tag. Liver cells from mice carrying both the *H19* DMR-UAS knock-in allele and the *GAL4-Myc* transgene were then subjected to chromatin immunoprecipitation (ChIP) with an antibody against Myc. If there was an interaction between the *H19* DMR and DMR1 or DMR2, DNA fragments from DMR1 or DMR2 should be enriched in the immunoprecipitated chromatin. The results showed that, when the *H19* DMR-UAS was inherited maternally, DMR1 was immunoprecipitated together with the *H19* DMR-UAS. This suggests that DMR1 is in physical proximity to the maternal *H19* DMR *in vivo*. By contrast, neither DMR2 nor an intergenic region was enriched, indicating that the interaction between the maternal *H19* DMR and DMR1 is specific. When the *H19* DMR-UAS was inherited paternally, DMR2, but not DMR1 or an intergenic region, was significantly enriched, suggesting an interaction between the *H19* DMR and DMR2 on this chromosome. In view of these data, it appears that chromatin looping induced by the interaction of DMRs can prevent activation of either the *IGF2* or *H19* loci by placing them in an inactive chromatin domain. This fulfills one of the defining characteristics of an insulator because although there is no inactivating segment of DNA present in the sequence of nucleotides between the two loci, inactivation is still achieved by conformational changes in the DNA.

In short, although we know the location of several DMRs, there is still some uncertainty as to their exact modes of action. What is clear from the above discussion is that some DMRs have more importance for controlling groups of imprinted genes than others. For example, we have seen that the DMR situated between *H19* and *IGF2* is the critical determinant of *IGF2* gene activity because it can bind or repel DMR1 or DMR2, depending on its methylation status. There is also dependence on the methylation status of DMR1 and DMR2, of course, but as we shall see in the next section, this is secondary to the methylation of the *H19* DMR, which justifies the latter's position as an imprinting control region.

There are other examples of imprinting on the same stretch of DNA

Consistent with the observation that imprinted genes often occur in clusters, the *IGF2/H19* locus is part of a much larger imprinted domain comprising a total of 1 Mb of DNA on the distal region of mouse chromosome 7. This shares a high degree of homology with the corresponding region on human chromosome 11 (Chr11p15.5). In fact, in both human and mouse, the region lies very close (less than 3 Mb) to the chromosome end (the telomere) and contains at least 10 imprinted protein-coding genes that are known to be under the control of two ICRs. One of these ICRs (ICR1) is the DMR found between *IGF2* and *H19* that we mentioned above (*H19* DMR), but there is a further imprint control region (ICR2) situated between the genes *KCNQ1* and *CDKN1C* (**Figure 11.4**). This *CDKN1C–KCNQ1* DMR is situated between exons 10 and 11 of the *KCNQ1* gene and is responsible for a variety of allele-specific functions such as promoter, enhancer, and CTCF-binding activities. Transgenic and knockout studies suggest that the functions of the ICR1 and ICR2 are independent.

There is also a highly enriched region of repeats between the *Mash2* and *Ins2* loci that correlates well with the known characteristics of imprinted gene clusters. The function of this particular group of repeats is unknown, but it seems to be an area that forms heterochromatic structure quite readily and therefore may, for reasons that are not clear, act as some form of boundary between the *Mash2* and *Ins2* areas of the cluster. It has also been suggested that the repeat region may contain nuclear matrix attachment sites, because consensus DNA sequences of such sites in

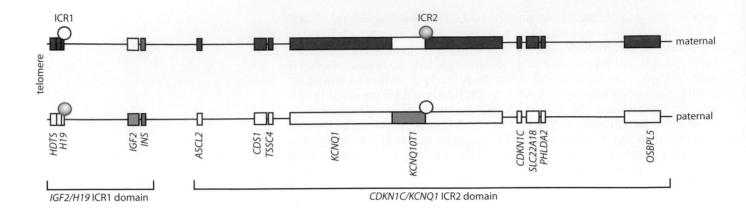

Figure 11.4 Relative positions of the *IGF2*/*H19* and *CDKN1C*/*KCNQ1* imprinted domains. The human 11p15 region is organized in two imprinted domains, the *IGF2*/*H19* and *CDKN1C*/*KCNQ1* domains, each of them under the control of its own imprinting center, imprinting control region 1 (ICR1) and ICR2, respectively. ICR1 and ICR2 are methylated on the paternal and maternal allele, respectively (blue circles). [Adapted from Shmela ME & Gicquel CF (2013) *J. Med. Genet.* 50,11. With permission from BMJ Publishing Group Ltd.]

other genomic loci have been detected. These speculations are interesting because nuclear attachment sites, if present, could be part of a mechanism to form chromatin loops that position one part of the cluster in an inactive nuclear compartment near the nuclear membrane, in a manner similar to that described in Chapter 9. So far, this loop-formation model has only been invoked to explain the effects of the *H19* DMR on *IGF2* expression, but it may also apply to the selection of one group of genes on the mouse distal chromosome 7 complex for expression in preference to another. In this respect, it is interesting that the maternally and paternally expressed genes occupy different regions.

It should by now be apparent that there are many examples of imprinted genes across a range of species and that these have several control systems in common. Unanswered by the preceding discussion is the question of how and why imprinting was established in the first place.

11.3 ESTABLISHING DIFFERENTIALLY METHYLATED REGIONS

To answer the questions of how and why imprinting was established, we need to consider the cycle of imprinting changes that occurs during embryogenesis. Gametes produced within the adult body do not share the imprinting pattern of the parents' genomes; instead they have their own unique profile, generated during the process of primordial germ cell specification and the subsequent development of oocytes and spermatozoa. This seems to be important, especially because the differential expression of imprinted genes is apparently linked to development of the embryo and the placenta that supports it *in utero*, and for the most part this is where imprinting has its effects.

Most genes undergo demethylation after fertilization

Imprinted genes hold a special status in the pre-implantation embryo in that they seem to resist the genome-wide demethylation events that take place shortly after fertilization. Upon fertilization, there is a series of events that involve the incoming sperm as it encounters the egg cytoplasm. The initial event after fertilization is the decondensation of the sperm nucleus, resulting in the unwinding of the tightly packaged sperm DNA held in a unique, almost toroidal, conformation by the sperm-specific protamines. So highly ordered is this chromatin organization in sperm that it is effectively dehydrated, and hence rehydration is an essential, very early post-fertilization event. Upon decondensation, the protamines are replaced rapidly by nucleohistones derived from the oocyte cytoplasm, usually in the first hour after fertilization, and the DNA is wound onto the histone octamers in an ATP-dependent process. It is during this same time period that rapid and paternal-specific demethylation of the genome takes place in the absence of transcription or DNA synthesis. The maternal genome also loses DNA methylation, but this takes place more slowly than for the paternal genome. The membrane-bound vesicles known as pronuclei are most probably the structures that permit this differential demethylation activity, because the gamete genomes are processed in separate pronuclei for several hours after sperm entry into the oocyte (**Figure 11.5**).

The progress of these demethylation events is shown in **Figure 11.6**, where we can see that both the maternal and paternal genomes reach their minimum levels of 5-methylcytosine by the morula (eight-cell) stage of development. Thereafter, the genomes (now combined, of course; they were only distinguishable for these experiments by genetic polymorphisms) undergo remethylation until they have basically the same levels of 5-methylcytosine that were observed in the separate gametes.

Imprinted genes retain their DNA methylation patterns at their DMRs during fertilization

As was mentioned above, imprinted genes seem to be largely resistant to this process of genome-wide demethylation upon fertilization, although why this should be so is not completely clear. One possibility is that an oocyte-specific form of DNA cytosine-5-methyltransferase 1 (DNMT1oo) contributes to maintenance methylation of the imprints, but, again, why it is only able to act on the methylated sequences is not known.

It is possible that non-histone proteins could be responsible for maintaining imprinting during embryonic development or for preserving the differentiated state (the cellular memory) of somatic cells. An additional possibility is that histone modifications on the chromatin associated with the methylated allele provide a specific mark that ensures maintenance of methylation. The imprint control regions of the inactive (methylated) alleles are known to have methylation of lysine 9 on histone H3 (H3K9me) and hypoacetylation of H3 and H4, whereas the active (unmethylated) allele is characterized by H3/H4 hyperacetylation and H3K4 methylation. In addition, recent studies have shown that trimethylation of H3K36 overwhelmingly corresponds to maternally imprinted alleles of some genes in mouse. As yet, no underlying biochemical mechanism has been described to link these histone modifications to any specific means for maintenance of chromatin states in imprinted and non-imprinted regions alike.

Genes of progenitor germ cells undergo two rounds of demethylation

The situation is rather different in the cell types that contribute to germline development. Despite having gone through the genome-wide remethylation as all the other cells during pre-implantation development, these cells undergo a second round of DNA demethylation shortly after

Figure 11.5 The formation of pronuclei allows the maternal and paternal gamete genomes to be kept separate during the first few hours after fertilization. After entry of the spermatozoon during fertilization, its DNA content is maintained in a separate membrane-enclosed structure distinct from the female nuclear material. These structures are referred to as pronuclei, and DNA demethylation proceeds at different rates in these regions. The paternal genome is shown in blue and the maternal in red. [Adapted from Armstrong L, Lako M, Dean W & Stojkovic M (2006) *Stem Cells* 24, 805. With permission from John Wiley and Sons.]

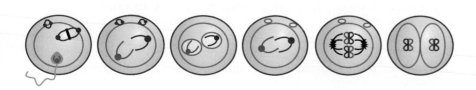

Figure 11.6 Reprogramming of DNA methylation during pre-implantation embryonic development. Methylation levels throughout pre-implantation development of embryos. The paternal genome (blue) of normally derived embryos undergoes rapid active demethylation, whereas the maternal genome (red) undergoes passive demethylation until the morula stage of pre-implantation development, when *de novo* methylation commences. [Adapted from Armstrong L, Lako M, Dean W & Stojkovic M (2006) *Stem Cells* 24, 805. With permission from John Wiley and Sons.]

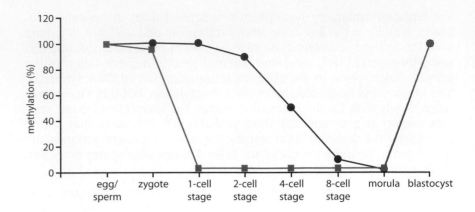

the embryo implants in the uterine wall. There is always a finite possibility that the germ cells arise from cells that have simply escaped the remethylation program, but this is admittedly unlikely. This second round of demethylation is needed to eliminate parental gene imprints derived from the gametes so as to permit the developing germ cells to establish their own sex-specific set of gene imprints, but the demethylation process is not restricted to the imprinted genes.

The germ line of most mammalian species is specified quite early in development. Primordial germ cells (PGCs) are the earliest recognizable precursors of the gametes and are detectable at 7.2 days after conception in the developing mouse embryo as a cluster of approximately 50 alkaline phosphatase-positive cells located in the extraembryonic mesoderm at the base of the allantois (**Figure 11.7**). PGCs are present in the hindgut epithelium of the 4-week human embryo and thereafter escape into the neighboring mesenchymal tissues to begin their migration via the dorsal mesentery to the developing gonads, where they arrive at approximately week 6 of gestation. In both mouse and human, the early gonads are referred to as the genital ridges (or, sometimes, the gonadal anlagen); in these structures the PGCs can multiply and complete the process of gamete development.

The removal of DNA methylation from the imprinted genes begins during the PGCs' migratory phase and is generally complete before the cells undergo sexual differentiation. One of the odd features of PGCs is that although they may be genetically male or female before arrival in the genital ridge, they can adopt the morphology of either male or female gametes, depending on the sex of the genital ridge they end up in (whether or not male oocytes and female sperm would work is another issue!).

Establishment of new imprints begins soon after the onset of sexual differentiation, but this is a longer process than simply removing imprints, with the exact duration depending on the gene in question and the gamete identity. For example, some imprinted sequences in oocytes undergo *de novo* methylation only just before ovulation. Compared with the mechanism of demethylation, about which we know very little, our knowledge of the remethylation mechanism used to establish *de novo* imprints is virtually nonexistent. Some data suggest that most *de novo* DNA methylation is directed to transposons and their remnants and to clustered repeats (primarily pericentric satellite DNA), whereas the DMRs of imprinted loci represent a much smaller subset of targeted sites.

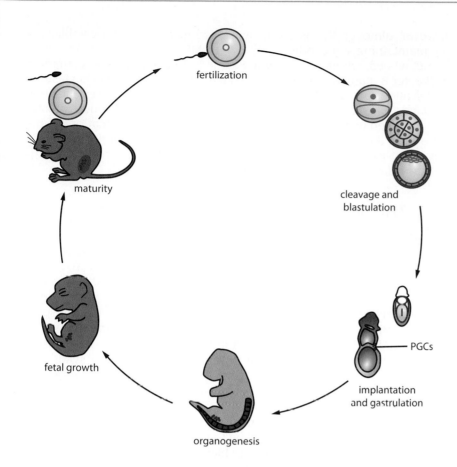

fertilization

cleavage and
blastulation

PGCs

implantation
and gastrulation

organogenesis

fetal growth

maturity

Figure 11.7 Development of the mammalian germ-line, using the mouse as an example. The germ-cell lineage originates from the proximal epiblast, in which the primordial germ cells (PGCs) are detectable at embryonic day 7.25 (E7.25) during the gastrulation stage. Thereafter, PGCs migrate rapidly to the hindgut and dorsal mesentery during organogenesis at E8.5. During fetal growth, they also proliferate and sexually differentiate until E13.5. At puberty in both sexes, gametes mature and lead to the production of haploid oocytes and sperms. [Adapted from Sabour D & Schöler HR (2012) *Curr. Opin. Cell. Biol.* 24, 716. With permission from Elsevier.]

As we saw in Chapter 5, the *DNMT3A/DNMTB* and *DNMT1* families of DNA cytosine-5-methyltransferases are responsible for the establishment and maintenance of methylation patterns, respectively. *DNMT3B* is expressed only in germ cells and only at the stages at which *de novo* methylation occurs. Thus it is possible that this enzyme has some role in establishing imprinting. Most *DNMT* genes contain sex-specific germ-line promoters that are activated at specific stages of gametogenesis. These promoters give rise to germ-cell-specific transcripts that lead to the production of truncated forms of the proteins, or to untranslated mRNAs that may have regulatory properties. The functional consequences of the alternative mRNAs and proteins are largely unknown, although, not surprisingly, *DNMT3L*-null embryos have significant defects in *de novo* methylation. Loss of *DNMT3L* results in very different phenotypes, depending on the sex examined. Deletion of *Dnmt3L* in female mice prevents the establishment of maternal methylation imprints in oocytes without having marked effects on retrotransposon methylation. The result is that any embryos derived from such oocytes die before mid-gestation. The *DNMT3L*-null phenotype of male germ cells is rather different. These cells show extensive expression of retrotransposons and long interspersed repeat elements and are incapable of entering into meiosis; this causes the immature gametes to enter apoptosis. Surprisingly, the small number of paternally methylated DMRs are almost normally methylated.

The DNMT3L protein seems identical in both male and female germ cells, with the only difference being the timing of its expression in the two sexes. This could imply that DNMT3L is merely acting in response to a mark or signal that is already established on the DNA sequences that are destined for imprinting. One possibility is that specific histone modifications recruit the DNMT3L enzyme through some as yet unknown mechanism.

However, although this is probably a good mechanism for establishing and maintaining some indicator that the DMRs of the target gene should be methylated, we have exactly the same problem that we encountered in Chapter 8, namely the lack of an explanation for how specific histone modifications are directed to particular genomic loci. In most genes that we have looked at so far, specificity has been achieved through interactions between gene-specific transcription factors (or combinations of the same) and the epigenetic modification machinery. Genomic imprinting may be no exception to this, because RNA-directed DNA methylation is one candidate mechanism for its regulation.

11.4 THE NEED FOR IMPRINTING

The process of methylating one DMR as opposed to another (and even the necessity to evolve DMRs in the first place) seems an extraordinary activity to engage in if there is no useful purpose for doing so. To set the whole necessity question in context, it is useful to remind ourselves that the consequences of disrupting imprint establishment during embryogenesis can be serious or even fatal. The cluster on human chromosome 15q11–13 that contains the *IGF2/H19* locus and other imprinted genes was originally discovered because of its involvement in the progression of two disease states: Prader–Willi syndrome and Angelman syndrome. These conditions produce neurological and behavioral disorders (characterized by excessive appetite, poor muscle tone, immature physical development, and learning disabilities) that correlate closely with the occurrence of imprint dysfunctions in the above cluster. These physiological responses are very likely to be the result of inappropriate expression levels of the imprinted genes (which, as we have seen, are central to metabolic control and neurogenesis); we should therefore not be surprised at the range of observed effects.

The nature and function of the imprinted genes give us some clues as to why the phenomenon may have evolved, and one current theory has gained a degree of acceptance, although hard evidence to support it is somewhat lacking. The theory is known as the **parental conflict hypothesis**, and it states that the inequality between parental genomes due to imprinting is a result of the differing interests of each parent in terms of the evolutionary fitness of their genes. The father is more "interested" (as interested as bits of DNA get) in ensuring that his offspring grow, survive, and go on to pass their (and therefore his) genes on to subsequent generations. The mother is more "interested" in ensuring that her own body can continue to provide support for generating future offspring, and this means that she places more emphasis on resource conservation. In consequence, this hypothesis would demand that paternally imprinted genes should promote growth while maternally expressed genes should be growth-limiting. In this respect, it is interesting that the paternally expressed *IGF2* gene seems to promote embryonic development but is partly controlled by the DMR adjacent to the maternally expressed *H19* gene. Perhaps this represents the parental struggle encapsulated in a single genomic locus.

Because imprinting is involved in more processes than simply growth, some researchers have attempted to extend the conflict hypothesis to other aspects of animal behavior. The adult mouse brain shows spatial expression of maternally and paternally imprinted genes. These distributions may give the mother's genes a key developmental role in the cerebral cortex, striatum, and hippocampus, with the father's genes being more significantly expressed in the mid-brain emotional centers—the hypothalamus, amygdala, and pre-optic area. The suggestion that the "balance" between patterns of maternally and paternally imprinted genes can account for behavior and personality must remain a speculation until more solid data are produced to support it.

In conclusion, we have discussed some of the basic mechanisms through which epigenetic modifications may contribute to the phenomenon of gene imprinting. The biological reasons why this phenomenon has arisen during evolution are still unclear, but this does not detract from the utility of genomic imprinting in underlining the impact of epigenetic modification to control gene dosage levels.

KEY CONCEPTS

- Genomic imprinting is the term used to describe allele-specific expression of some genes. Expression is dependent on the parent of origin of that gene.

- The phenomenon is found mostly in mammals and some types of flowering plants.

- The primary epigenctic modification that controls such allele-specific expression seems to be DNA methylation, but the effects of this can be reinforced by specific histone modification patterns.

- Genomic imprinting is established early in embryonic development, when the primordial germ cells are specified.

FURTHER READING

Cattanach BM & Jones J (1994) Genetic imprinting in the mouse: implications for gene regulation. *J Inherit Metab Dis* **17**:403–420.

Driscoll DJ (1994) Genomic imprinting in humans. *Mol Genet Med* **4**:37–77.

Feinberg AP, Kalikin LM, Johnson LA & Thompson JS (1994) Loss of imprinting in human cancer. *Cold Spring Harb Symp Quant Biol* **59**:357–364.

Giannoukakis N, Deal C, Paquette J et al. (1993) Parental genomic imprinting of the human IGF2 gene. *Nat Genet* **4**:98–101 (doi:10.1038/ng0593-98).

Gold JD & Pedersen RA (1994) Mechanisms of genomic imprinting in mammals. *Curr Top Dev Biol* **29**:227–280.

Goshen R, Ben-Rafael Z, Gonik B et al. (1994) The role of genomic imprinting in implantation. *Fertil Steril* **62**:903–910.

Hall JG (1992) Genomic imprinting and its clinical implications. *N Engl J Med* **326**:827–829.

Henckel A, Chebli K, Kota SK et al. (2011) Transcription and histone methylation changes correlate with imprint acquisition in male germ cells. *EMBO J* **31**:606–615 (doi:10.1038/emboj.2011.425).

Kato Y & Sasaki H (2005) Imprinting and looping: epigenetic marks control interactions between regulatory elements. *BioEssays* **27**:1–4 (doi:10.1002/bies.20171).

Kim MJ, Choi HW, Jang HJ et al. (2013) Conversion of genomic imprinting by reprogramming and redifferentiation. *J Cell Sci* **126**:2516-2542 (doi:10.1242/jcs.122754).

Li E, Beard C & Jaenisch R (1993) Role for DNA methylation in genomic imprinting. *Nature* **366**:362–365 (doi:10.1038/366362a0).

Li E, Beard C, Forster AC et al. (1993) DNA methylation, genomic imprinting, and mammalian development. *Cold Spring Harb Symp Quant Biol* **58**:297–305.

Li X (2013) Genomic imprinting is a parental effect established in mammalian germ cells. *Curr Top Dev Biol* **102**:35–59 (doi:10.1016/B978-0-12-416024-8.00002-7).

Ling JQ, Li T, Hu JF et al. (2006) CTCF mediates interchromosomal colocalization between Igf2/H19 and Wsb1/Nf1. *Science* **312**:269–272 (doi:10.1126/science.1123191).

Lyon MF (1994) The X inactivation centre and X chromosome imprinting. *Eur J Hum Genet* **2**:255–261.

Murrell A (2011) Setting up and maintaining differential insulators and boundaries for genomic imprinting. *Biochem Cell Biol* **89**:469–478 (doi:10.1139/o11-043).

Nicholls RD (1994) Imprinting: the embryo and adult point of view. *Trends Genet* **10**:389.

Razin A & Cedar H (1994) DNA methylation and genomic imprinting. *Cell* **77**:473–476.

Reik W (1992) Genomic imprinting in mammals. *Results Probl Cell Differ* **18**:203–229.

Sasaki H, Allen ND & Surani MA (1993) DNA methylation and genomic imprinting in mammals. *EXS* **64**:469–486.

Singh P, Cho J, Tsai SY et al. (2010) Coordinated allele-specific histone acetylation at the differentially methylated regions of imprinted genes. *Nucleic Acids Res* **38**:7974–7990 (doi:10.1093/nar/gkq680).

Surani MA (1991) Genomic imprinting: developmental significance and molecular mechanism. *Curr Opin Genet Dev* **1**:241–246.

Surani MA (1994) Genomic imprinting: control of gene expression by epigenetic inheritance. *Curr Opin Cell Biol* **6**:390–395.

Tycko B (1994) Genomic imprinting: mechanism and role in human pathology. *Am J Pathol* **144**:431–443.

Wolf JB (2013) Evolution of genomic imprinting as a coordinator of coadapted gene expression. *Proc Natl Acad Sci USA* **110**:5085–5090 (doi:10.1073/pnas.1205686110).

Zhang H, Niu B, Hu JF et al. (2011) Interruption of intrachromosomal looping by CCCTC binding factor decoy proteins abrogates genomic imprinting of human insulin-like growth factor II. *J Cell Biol* **193**:475–487 (doi:10.1083/jcb.201101021).

CHAPTER TWELVE

EPIGENETIC CONTROL OF CELLULAR DIFFERENTIATION

All the information needed to construct an adult organism can be found in the genomes of any number of its nucleated cells, but because the body is composed of a wide variety of different cell types capable of performing very different functions, there must be mechanisms to control the way in which information encoded in the genome can be selectively controlled. Different types of cells arise early in embryonic development, and epigenetic control of gene expression is used to establish specific gene expression patterns that distinguish individual types of cells. The process by which different cell types arise is known as differentiation.

12.1 FROM CELLULAR TOTIPOTENCY TO PLURIPOTENCY

The early embryos of most multicellular eukaryotes begin as groups of cells that are termed **totipotent** because each cell seems to have an equal ability to become any of the cell types generated as the embryo develops. **Figure 12.1** shows some of the stages of pre-implantation development that apply to mammals. The blue spheroids are referred to as blastomeres and are totipotent. The first differentiation of these cells takes place between the morula and early blastocyst stages, when those blastomeres that happen to be on the outside of the morula become trophoblast cells, which are destined to become parts of the placenta. Only the cells referred to as **the inner cell mass (ICM)** of the blastocyst will give rise to the tissues of the developing embryo.

Once formed, the trophoblast cannot produce any of the cells that normally derive from the inner cell mass. Moreover, trophoblast cells can never revert to totipotent blastomeres under normal circumstances. This brings up an important point that must be borne in mind in any consideration of the process of cellular differentiation: it is almost always unidirectional. Similarly, the ICM cannot produce trophoblast cells (in a normal embryo); its cells are therefore referred to a **pluripotent** (that is, able to give rise to *many* cell types but not all).

The ICM is of great interest to science because removing this structure and culturing it under special growth conditions gives rise to embryonic stem cells, which may be thought of as a "snapshot" of a very brief period of embryonic development. It is brief indeed, because a few days after the blastocyst has implanted in the wall of the uterus, the ICM differentiates into three primordial layers that will produce the organs and structures of the body (**Figure 12.2**) and pluripotent cells no longer exist in the body.

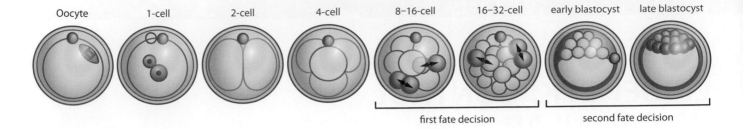

| Oocyte | 1-cell | 2-cell | 4-cell | 8–16-cell | 16–32-cell | early blastocyst | late blastocyst |

first fate decision   second fate decision

Figure 12.1 The stages of pre-implantation development. The totipotent fertilized oocyte undergoes a series of divisions to generate an 8–16-cell stage referred to as the morula. After this stage, the outer cells differentiate to become the trophectoderm (the first fate decision) while the inner cells are destined to become the inner cell mass, which subsequently differentiates into the primitive endoderm and epiblast by the late blastocyst stage of development. [Adapted from Zernicka-Goetz M, Morris SA & Bruce AW (2009) *Nat. Rev. Genet.* 10, 467. With permission from Macmillan Publishers Ltd.]

There are some possible exceptions to this rule, because recent studies suggest that the bone marrow and reproductive organs may contain small populations of apparently pluripotent cells. However, the evidence for this is inconclusive so far.

With the exception of erythrocytes and post-meiotic gametes, most of the cells of the developing organism contain an identical copy of the original genome that was generated by the combination of the two parental genomes shortly after fertilization. The basis of cell differentiation is that specific patterns of gene expression apply to different types of cells, but the fact that they all probably possess the same genome implies that epigenetic regulation must control the cell-type-specific expression pattern. It is important to understand how this works, but the subject is worthy of an entire volume in its own right and we can only give a few select examples. A useful first step is to examine how the pluripotent state is maintained, because this will help us to understand how the cells of the ICM may be able to make fate decisions that allow them to produce multiple cell types. We will follow this with some examples of the pathways leading to the differentiation of specific cell types, although it must be said

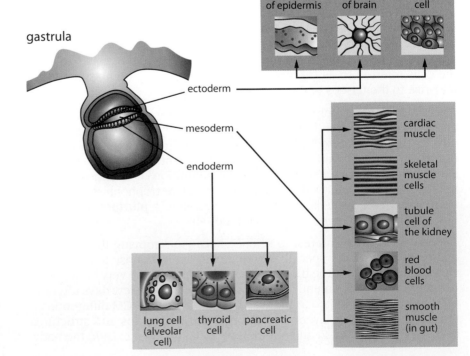

Figure 12.2 Three primordial germ layers arise during embryonic development. These originate from the ICM and are known as ectoderm, mesoderm, and endoderm. These produce different parts of the body, as shown. For example, the cells of the ectoderm give rise to the skin, peripheral nerves, and CNS, whereas the mesoderm produces bones, blood, kidneys, muscles, and heart. The endoderm produces organs such as the lungs, liver, and digestive tract.

that the information provided is incomplete because the science simply has not progressed to the level where step-by-step descriptions of all the required epigenetic controls are possible at this time.

12.2 MAINTENANCE OF PLURIPOTENCY IN EMBRYONIC STEM CELLS

Embryonic stem cells (ESCs) are derived from the ICM of day 5–8 blastocyst-stage or morula-stage embryos (see Figure 12.1). ESCs are pluripotent and are capable of karyotypically stable, prolonged self-renewal. They are characterized by their potential to differentiate into cells of the three germ layers both *in vitro* and *in vivo*. In contrast to the specific gene expression programs observed in differentiated cells, ESCs are defined by their potential to activate all of the gene expression programs that are found in embryonic and adult cell lineages.

Various studies have endeavored to understand the molecular mechanisms that give ESCs their properties. These attempts to identify a molecular "signature" of ESCs have borne some fruit, including the identification of the transcription factors OCT4 and NANOG as markers for ESCs. However, these experiments also show that there is very little other overlapping of gene expression among different ESC lines and also among the same ESC lines studied in different laboratories. This lack of overlap raises the question of whether a common molecular identity for ESCs can be uncovered. However, recent studies have indicated that epigenetic mechanisms may regulate self-renewal, pluripotency, and lineage-specific differentiation, giving ESCs their unique characteristics. Recall that epigenetic mechanisms include modification of histone proteins, DNA methylation, ATP-dependent remodeling, incorporation of variant histones, changes in local and higher-order conformation of DNA, and RNAi. Through the combined efforts of these epigenetic mechanisms, gene expression patterns can be tightly and dynamically regulated.

As an example of this regulation, the chromatin structure of mouse embryonic stem cells (mESCs) has now been demonstrated to be hyperdynamic, in that major architectural proteins, for example H1, H2B, H3, and HP1, are loosely bound to chromatin with very short residency times. Differentiation has been found to lead to a decrease in the dynamic nature of these proteins, as demonstrated by an increase in their residency time. These findings suggest that this dynamic nature of chromatin is specific to pluripotent cells and that differentiation leads to a restructuring of the genome. Additionally, it was demonstrated that heterochromatic markers change from a dispersed localization in mESCs to more concentrated, distinct foci in differentiated cells, with increased global levels of trimethylated lysine 9 H3 (H3K9me3)—a heterochromatic histone modification linked to gene repression (see Table 6.2)—and decreased levels of acetylated histones H3 and H4 (H3ac and H4ac). Recall that acetylation of histones is linked to euchromatin and permissivity of gene expression, and so the noted decrease in this modification is in keeping with the dampening of gene expression. Analyses of global histone modification patterns in ESCs have previously suggested that the ESC genome is subject to generalized histone acetylation and lysine 4 H3 methylation (H3K4me). As these are both transcription-activating modifications, these changes in global genomic architecture and global histone modifications suggest that the chromatin environment in ESCs is highly euchromatic, and the genome is therefore highly permissive for gene expression. This would account for the pluripotent nature of ESCs, with the genome becoming more structured, condensed, and heterochromatic during differentiation, leading to loss of pluripotency.

Another study of mESCs sought to understand regulatory mechanisms involved in development by examining highly conserved noncoding elements (HCNEs). HCNEs were studied because they are present in regions where genes encoding developmentally important transcription factors are concentrated. Large-scale chromatin immunoprecipitation (ChIP) assays found large areas of a repressive histone modification, dimethylation and trimethylation of lysine 27 histone H3 (H3K27me2/me3), alongside smaller regions of a permissive modification, H3K4me, at HCNEs. This is unusual inasmuch as these modifications are usually mutually exclusive, and so these regions were termed **"bivalent" domains** (**Figure 12.3**). Interestingly, the bivalent domains coincided with differentiation-associated transcription factor genes expressed at very low levels in the ESCs. It was therefore proposed that the bivalent domains act to silence such genes so as to maintain pluripotency while also allowing the genes to remain poised for transcription, so they can be rapidly activated on differentiation. This echoes the presence of activating histone modification such as H3K4me2 and H3K4me3 in the *HOXA* cluster, as described in Chapter 9. On differentiation, these bivalent domains "resolve," so that silent genes become concentrated for H3K27me2/me3 only, and active genes become enriched for euchromatic modifications.

12.3 DIFFERENTIATION OF EMBRYONIC STEM CELLS

The earliest observable differentiation events for ESCs *in vitro* are the decrease in pluripotent character and the appearance of characteristics of the three primordial germ layers. It is probable that several changes in gene expression precede these events, but either they are too transient to be detected with our current techniques or they make no observable changes to the morphology or behavior of the cells. Another problem is that differentiation of ESCs seems to be partly random in the laboratory. One would not expect this to be so in the developing embryo, in which

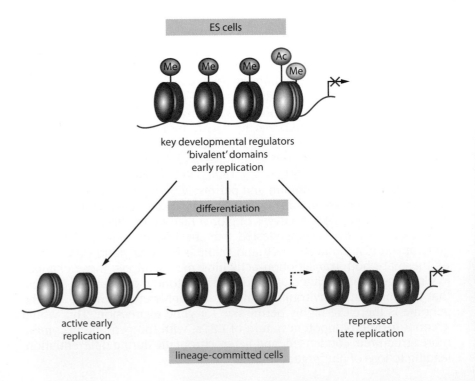

Figure 12.3 Bivalent chromatin domains. In pluripotent stem cells, a large number of genes encoding key developmental regulators are marked by a combination of permissive (H3K9me2 and H3K9ac, shown in blue and green, respectively) and repressive (H3K27me3, shown in red) epigenetic marks. Such domains are usually resolved during cell differentiation when specific genes must either be expressed or permanently repressed. (Courtesy of Véronique Azuara, Imperial College London.)

individual cells should respond to precisely regulated developmental cues telling them what to do and where to go; however, in a culture dish in the laboratory, such information is scrambled or nonexistent, and the resulting mixture of differentiated cells contains small percentages of nearly all the cell types found in the adult body. This makes the process leading to differentiation very difficult to study because we have to separate out individual cells for analysis, a task more easily done for some types of cells than for others. For example, early hematopoietic cells can be isolated in abundance because some of the proteins they express on their surfaces can be used to separate them from the cellular "soup" by using flow cytometry, but we do not have this luxury for many other cell types.

Because epigenetic control systems are highly dependent on cell identity, we must restrict our description of such systems to those that can be studied in well-defined and easily accessible cells. That said, it is nevertheless worth describing some of the more global changes that take place during ESC differentiation. These changes tell us little about the mechanisms that control individual genes, but they are useful for understanding the differences in genome organization between pluripotent and differentiated cells.

We saw earlier that ESCs have characteristic blocks of euchromatin throughout much of their genome with relatively little heterochromatin. Heterochromatin is detected by the presence of H3K9 dimethylation and trimethylation, and although these histone modifications do exist in the ESC genome, they do not accumulate into large blocks of H3K9 dimethylated chromatin as seems to be the case for differentiated cells. These regions known as **large organized chromatin K9 modifications (LOCKs)** can be up to 4.9 Mb in length and are highly conserved between human and mouse. Although they cover only 4% of the genome in mouse ESCs, they cover 31% of the genome in differentiated ESCs, 46% in mouse liver cells and 10% in brain. The positions of these LOCKs vary between cell types; some of these are shown in **Figure 12.4**, along with the relative positions of the genes in these regions.

The level of acetylation at H3K9 is also decreased. This modification is normally associated with euchromatin, and its progressive loss is needed to make way for the H3K9 dimethylation and trimethylation associated with heterochromatinization.

Figure 12.4 Summary of data from a 10 Mb region of mouse chromosome 8. Locations of LOCKs in undifferentiated mouse ES cells (green bars), differentiated ES cells (red bars), liver cells (orange bars), and brain cells (blue bars) are shown. [From Wen B, Wu H, Shinkai Y et al. (2009) *Nat. Genet.* 41, 246. Macmillan Publishers Ltd.]

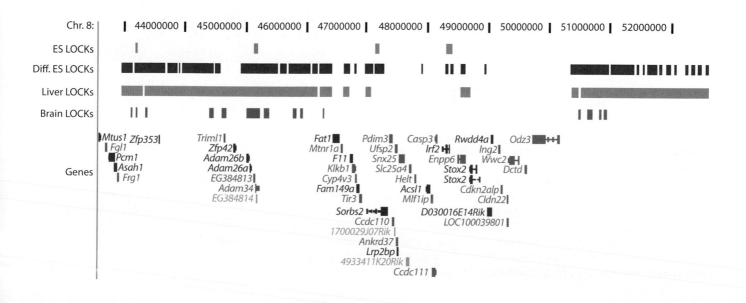

12.4 BIVALENT CHROMATIN DOMAINS IN NEURAL STEM CELLS

Generating specific differentiated cell types from ESCs can help us to describe and perhaps better understand the mechanism by which epigenetic modification controls the gene expression pattern of that cell. A commonly occurring developmental pathway from ESCs and embryonal carcinoma cells is neurogenesis; given the prevalence of defined surface-marker proteins for the neural progenitor stage and the neural stem cell stage, separation and analysis of these cells is relatively straightforward. During human embryonic stem cell (hESC) differentiation into the neural lineage, two stages of specification can be purified by the use of flow cytometry for the CD133 surface marker and β-III tubulin (**Figure 12.5**). (CD133 is a marker of multipotent stem cells of many tissue types (including multipotent neural progenitor cells), and β-III tubulin is neuron-specific and indicative of more terminally differentiated cells.)

The CD133-expressing neural progenitor cells show substantial down-regulation of pluripotency-associated genes compared with their ESC expression levels, and up-regulation of neural marker genes such as *SOX1*, *SOX3*, *PAX7*, and *Nestin*. Even though these latter genes are differentiation-specific, they are still expressed at low levels in ESCs; however, these levels are so much lower than in the neural progenitor cells that the expression is probably of no functional significance. In line with

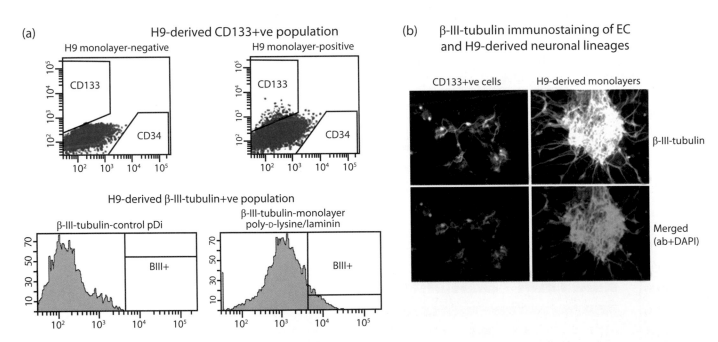

Figure 12.5 Flow cytometric analysis and sorting of CD133-expressing neural progenitor cells and terminally differentiated neurons from differentiating hESCs. (a) Multipotent neural progenitors express the surface antigen CD133, whereas terminally differentiated neurons arising from the progenitors express β-III-tubulin. This allows the use of flow cytometry to separate neural progenitors from other cell types produced during hESC differentiation and to isolate relatively pure populations of mature neurons. The diagrams in (a) are the outputs from a flow cytometer in which each of the dots on the plotted data represents a single cell expressing the CD133 protein on its surface. This expression is detected by a fluorescent antibody that binds specifically to CD133. CD133-expressing cells are represented by blue dots. The flow cytometer can recognize the cells and direct them into a different container to the other non-CD133-expressing cells. The histogram shows how the cytometer can detect β-III-tubulin-expressing neurons, which may be sorted in the same way as the CD133-expressing cells. (b) Immunostaining of hESC-derived neurons to confirm the presence of β-III-tubulin. [From Golebiewska A, Atkinson SP, Lako M, Armstrong L (2009) *Stem Cells* 27, 1298. With permission from John Wiley and Sons.]

expectations, the neural progenitors did not display H3K4me3 at the pluripotency-associated gene promoters, which correlates with the much lower expression levels of genes associated with pluripotency in these cells; however, significant levels of H3K4me2 were still detected at the pluripotency-associated gene promoters in cells expressing CD133 and β-III-tubulin, in keeping with the possible need for rapid up-regulation of these genes.

Most of the genes specific to the neural lineage have bivalent chromatin domains at their promoters in the undifferentiated ESCs, and these show the expected resolution on differentiation into the neural progenitor cells. This causes some loss of the repressive H3K27me3 modification, with retention of the activating H3K4me3 mark. Markers of other lineages, such as *GATA4* (which is expressed by cells found in the mesoderm and endoderm), lost H3K4me3 and displayed increased levels of H3K27me3 (which correlates with their lower expression during neurogenesis), but surprisingly the resolution does not seem to be absolute because at least some level of both types of histone modifications seemed to be retained for some of the bivalent genes after differentiation. This was unexpected for genes that are not likely to require expression in cells that differentiate from the neural progenitors, so it is currently unknown why bivalent domains are needed. It is possible that the functionality of a bivalent domain depends on relative levels of the two histone modifications, but this has not yet been determined with any certainty. However, it is interesting that H3K9 methylation levels also increase at the apparently retained bivalent promoters. Perhaps this is related to the suppression of gene expression at these loci, which would make the nature of bivalent domains rather different in pluripotent cells than in the multipotent progenitors (such as neural stem cells) that derive from them.

12.5 CHROMATIN PROFILE OF HEMATOPOIETIC PROGENITORS

Neural stem cells are not unique in having lineage-specific genes that are required for differentiating into multiple cell types further down the differentiation pathway. Hematopoietic stem cells (HSCs) are multipotent cells that at the single-cell level have the potential to differentiate into all cells of the erythromyeloid and lymphoid lineages, as well as to maintain their numbers by means of controlled self-renewal. In short, they are responsible for the lifelong regeneration of the blood and immune systems.

Progression from HSCs to their differentiated progeny involves the coordinated regulation of multiple gene expression programs that lead to the activation or repression of lineage-specific genes. At the single-cell level in HSCs, low-level transcription of lineage-affiliated genes has been observed, a phenomenon known as **lineage priming**. It is possible that a specific chromatin structure exists at lineage-affiliated genes in HSCs that mediates low-level expression for the propagation of transcriptional memory during the differentiation process. As in the neural progenitors, it is apparent that H3K4 dimethylation marks the genes that need to be kept in a "poised" state for rapid up-regulation when HSC differentiation takes place. Because multiple fate decisions are possible, many different genes must be maintained in this state. An excellent example of this is the mouse γ5–*VpreB1* locus.

Transcription of the *VpreB1* and γ5 genes in mice is activated during the pre-B-cell stage, before the heavy-chain rearrangement that is central to the immune function of the B cells that are eventually produced. The γ5-*VpreB1* domain is already marked by histone H3 acetylation and histone H3K4 methylation at a discrete site in ES cells; these modifications are not present in the rest of the locus. The marked region expands in early B-cell progenitors and becomes a localized center for the recruitment of transcription factors and RNA polymerase II, but it disappears in mature B cells and is not present in cells of other lineages such as liver.

This is a rather extreme example because other elements of this type have not been discovered, but it demonstrates the general principle that lineage-specific genes may have regions of H3K4 methylation that prevent them from being incorporated into heterochromatin while not actually having sufficient activation capacity to increase expression probability beyond a threshold level. In any case, gene up-regulation probably needs the activity of a specific transcription factor, but at least the promoter is more likely to be accessible when required.

There is also evidence that differential levels of H3K4 dimethylation and trimethylation mark developmentally poised hematopoietic genes. Multipotent hematopoietic cells have a subset of genes that are differentially methylated (H3K4me2+/me3–). These genes are transcriptionally silent, lineage-specific hematopoietic genes that are uniquely susceptible to differentiation-induced H3K4 demethylation in developmental pathways other than that of the hematopoietic system. Many of the genes are not formally recognized as having bivalent chromatin domains in ESCs. These examples suggest that formal recognition of the presence of bivalent chromatin domains within a gene locus does not mean that gene is not able to undergo up-regulation of expression at some point during differentiation of a progenitor cell. It merely suggests that bivalency is one mechanism by which some genes can be held in a "poised" state that permits rapid expression when such genes are needed. The alternative mechanisms controlling the spatio-temporal expression of genes without bivalent chromatin domains remain to be fully elucidated.

We have attempted to introduce the basic concepts behind the cellular differentiation that permits development of the embryo, but this discussion is far from exhaustive. More detailed information can be obtained from the references given in the Further Reading section below.

KEY CONCEPTS

- Embryonic development begins with a totipotent cell produced by the fusion of sperm and oocyte during fertilization.

- The resulting "zygote" is called totipotent because it can differentiate into any of the cell types found in the developing embryo.

- Changes in gene expression patterns both cause and result from the differentiation of totipotent cells, and this is mostly controlled by epigenetic modifications.

- Embryonic stem cell differentiation can be used to model the processes of cellular differentiation occurring in the embryo.

FURTHER READING

Atkinson S & Armstrong L (2008) Epigenetics in embryonic stem cells: regulation of pluripotency and differentiation. *Cell Tissue Res* 331:23–29.

Barrero MJ & Izpisua Belmonte JC (2012) Epigenetic mechanisms controlling mesodermal specification. In StemBook [Internet]. Cambridge (MA): Harvard Stem Cell Institute. http://www.ncbi.nlm.nih.gov/books/NBK133266/

Bibikova M, Laurent LC, Ren B et al. (2008) Unraveling epigenetic regulation in embryonic stem cells. *Cell Stem Cell* 2:123–134 (doi:10.1016/j.stem.2008.01.005).

Bogliotti YS & Ross PJ (2012) Mechanisms of histone H3 lysine 27 trimethylation remodeling during early mammalian development. *Epigenetics* 7:976–981.

Calvanese V & Fraga MF (2012) Epigenetics of embryonic stem cells. *Adv Exp Med Biol* 741:231–253 (doi:10.1007/978-1-4614-2098-9_16).

Han YM, Kim SH & Kang YK (2006) Analysis of DNA methylation profiles in preimplantation embryos using bisulfite mutagenesis. *Methods Mol Biol* 325:251–260.

Hattori N & Shiota K (2008) Epigenetics: the study of embryonic stem cells by restriction landmark genomic scanning. *FEBS J* 275:1624–1630 (doi:10.1111/j.1742-4658.2008.06331.x).

Khavari DA, Sen GL & Rinn JL (2010) DNA methylation and epigenetic control of cellular differentiation. *Cell Cycle* 9:3880–3883 (doi:10.4161/cc.9.19.13385).

Li M, Liu GH & Izpisua Belmonte JC (2012) Navigating the epigenetic landscape of pluripotent stem cells. *Nat Rev Mol Cell Biol* 13:524–535 (doi:10.1038/nrm3393).

Liu H, Kim JM & Aoki F (2004) Regulation of histone H3 lysine 9 methylation in oocytes and early pre-implantation embryos. *Development* 131:2269–2280.

Lund RJ, Närvä E & Lahesmaa R (2012) Genetic and epigenetic stability of human pluripotent stem cells. *Nat Rev Genet* 13:732–744 (doi:10.1038/nrg3271)

Maroof AM, Keros S, Tyson JA et al. (2013) Directed differentiation and functional maturation of cortical interneurons from human embryonic stem cells. *Cell Stem Cell* 12:559–572 (doi:10.1016/j.stem.2013.04.008).

Melcer S & Meshorer E (2010) Chromatin plasticity in pluripotent cells. *Essays Biochem* 48:245–62 (doi:10.1042/bse0480245).

Menendez P, Wang L & Bhatia M (2005) Genetic manipulation of human embryonic stem cells: a system to study early human development and potential therapeutic applications. *Curr Gene Ther* 5:375–385.

Mitalipov S & Wolf D (2009) Totipotency, pluripotency and nuclear reprogramming. *Adv Biochem Eng Biotechnol* 114:185–199 (doi:10.1007/10_2008_45).

Oswald J, Engemann S, Lane N et al. (2000) Active demethylation of the paternal genome in the mouse zygote. *Curr Biol* 10:475–478.

Ruzov A, Tsenkina Y, Serio A et al. (2011) Lineage-specific distribution of high levels of genomic 5-hydroxymethylcytosine in mammalian development. *Cell Res* 21:1332–1342 (doi:10.1038/cr.2011.113).

Senner CE, Krueger F, Oxley D, et al. (2012) DNA methylation profiles define stem cell identity and reveal a tight embryonic-extraembryonic lineage boundary. *Stem Cells* 30:2732–2745 (doi:10.1002/stem.1249).

Smith A (2005) The battlefield of pluripotency. *Cell* 123:757–760 (doi:10.1016/j.cell.2005.11.012).

Song J, Saha S, Gokulrangan G et al. (2012) DNA and chromatin modification networks distinguish stem cell pluripotent ground states. *Mol Cell Proteomics* 11:1036–1047.

Torres-Padilla ME (2008) Cell identity in the preimplantation mammalian embryo: an epigenetic perspective from the mouse. *Hum Reprod* 23:1246–1252 (doi:10.1093/humrep/dem434).

van Heesbeen HJ, Mesman S, Veenvliet JV & Smidt MP (2013) Epigenetic mechanisms in the development and maintenance of dopaminergic neurons. *Development* 140:1159-1169 (doi:10.1242/dev.089359).

Wang J, Hevi S, Kurash JK et al. (2009) The lysine demethylase LSD1 (KDM1) is required for maintenance of global DNA methylation. *Nat Genet* 41:125–129 (doi:10.1038/ng.268).

Wen B, Wu H, Shinkai Y et al. (2009) Large histone H3 lysine 9 dimethylated chromatin blocks distinguish differentiated from embryonic stem cells. *Nat Genet* 41:246–250 (doi: 10.1038/ng.297).

Yamagata K (2008) Capturing epigenetic dynamics during pre-implantation development using live cell imaging. *J Biochem* 143:279–286 (doi:10.1093/jb/mvn001).

13

REVERSIBILITY OF EPIGENETIC MODIFICATION PATTERNS

Epigenetic control of gene expression may often seem to have evolved as a semi-reversible mechanism that responds to the changing requirements of the cell. For example, we saw in Chapter 10 how certain elements of the mitotic cell cycle are under epigenetic control, but of course any changes in histone modification patterns that influence the expression of cell-cycle-related genes cannot be permanent because the cell might have to repeat the entire cell cycle a few hours after the current mitosis. Control of gene expression needs to be highly adaptable and capable of rapid alteration in response to environmental triggers. Despite this, the basis of cell identity is the stable maintenance of control over most of the genome. Cells do not normally change their phenotype unless this is a normal part of their behavior—or unless something goes very wrong.

Maintaining the patterns of epigenetic modifications present on the genomes of individual cells is central to the maintenance of cell function and identity. For example, it is essential that a specific type of cell, such as a fibroblast, be able to express only the set of genes that are needed to perform the functions of fibroblasts. Expression of genes associated with neuron or hepatocyte functions would diminish the ability of the fibroblast to perform fibroblast-like functions. Similarly, if neurons were suddenly to behave as fibroblasts, the consequences for the operation of our brains would be unfortunate. We will see in Chapter 14 that dysregulation of the epigenetic control of gene expression is a contributing factor to a number of diseases, so it will be clear that epigenetic stability is of great importance to the survival of the organism. The key message of this chapter is that once established during cell differentiation, epigenetic modification patterns are highly stable, and only major changes in the mechanisms that maintain the "epigenome" can alter them to the point where cell identity is changed.

Patterns of epigenetic modification are established mostly during embryonic development (as we saw in Chapter 12), and their imposition on the genome begins very soon after fertilization. In the zygote and up to about the 32-cell stage of the embryo, all the cells (blastomeres) are more or less equal in their ability to produce all of the body's cell types and may be thought of as totipotent, a fact underlined by their ability to replace blastomeres that have been mechanically removed from the embryo. By the blastocyst stage, differentiation of the outermost blastomeres has taken place to produce the trophectoderm, which is an outer layer of large flattish cells distinct from the protruding inner cell mass of smaller cells at the base of the embryo (see Figure 12.1). Only these smaller cells will give rise to the developing embryo (the trophectoderm will only produce certain components of the placenta and can never become part of the inner cell mass), but even these can no longer be described as totipotent because, in some mammalian species, they lack the ability to produce trophectodermal cells. They are, however, pluripotent. This first cell differentiation event highlights an important lesson about cell differentiation in that it is almost always unidirectional. Once differentiation has taken place, cells cannot find a way to reverse this process unless something goes wrong with them (such as might occur in cancer cells), and this underlying rule is the basis of all embryogenesis.

In this chapter we look at circumstances that are capable of altering the cell-specific patterns of epigenetic modifications other than those resulting from disease. We begin by looking at the processes by which animals have been cloned, because this requires extensive reprogramming of the epigenotype (the pattern of epigenetic modifications present in a specific type of cell), and extend our discussion to examine more recently developed methods of creating apparently pluripotent cells from differentiated somatic cell types.

13.1 REPROGRAMMING THE EPIGENOME BY SOMATIC CELL NUCLEAR TRANSFER

At the time of writing, it is more than 60 years since the experiments of Briggs and King showed that the genome taken from a single animal cell could be used to create a whole new organism. Even earlier experiments had shown that some invertebrate species could be "cloned" simply by dividing them into smaller pieces, each of which was capable of regenerating itself into a complete organism. However, the first really groundbreaking study was Briggs and King's demonstration that somatic cell nuclear transfer (SCNT) could be used to clone frogs. By removing the nuclei from oocytes of the frog *Rana pipiens* and replacing them with nuclei from the relatively undifferentiated cells of early embryos of the same species, they found that the "reconstructed" embryos were sometimes capable of undergoing development to the tadpole stage. The process seemed to work only with cells from early embryos, and this led Briggs and King to assume that once the cells of the embryo had differentiated or developed to a certain stage, the cells were no longer capable of contributing to the formation of a new organism. They did not know quite why this should be so, because although the concept of epigenetics had been established by Conrad Waddington in the 1940s, the molecular mechanisms through which epigenetics could control cell development were unknown.

By the time of John Gurdon's contribution to the cloning of frogs in the early 1960s, the science of epigenetics had progressed only slightly, but Gurdon's use of intestinal cells from tadpoles demonstrated that differentiated somatic cells were capable of producing viable embryos. These observations suggested that, in principle, the genome could be "reset" to a totipotent state. Early attempts to clone higher animals such as mammals met with failure. It was therefore assumed that the cells of adult vertebrates were simply too specialized to revert to a totipotent state. This opinion was changed in a very dramatic fashion by the cloning of Dolly in 1996 by the fusion of a mammary gland epithelial cell from a Finn Dorset ewe with the enucleated oocyte from a separate donor (**Figure 13.1**). Many studies since then have confirmed the feasibility of SCNT-based cloning of cows, mice, pigs, and several other species. In addition, the cytoplasm of oocytes from sheep, cows, and rabbits seems to be able to reprogram the somatic cells from other species and support the growth of such interspecies-cloned embryos to blastocysts.

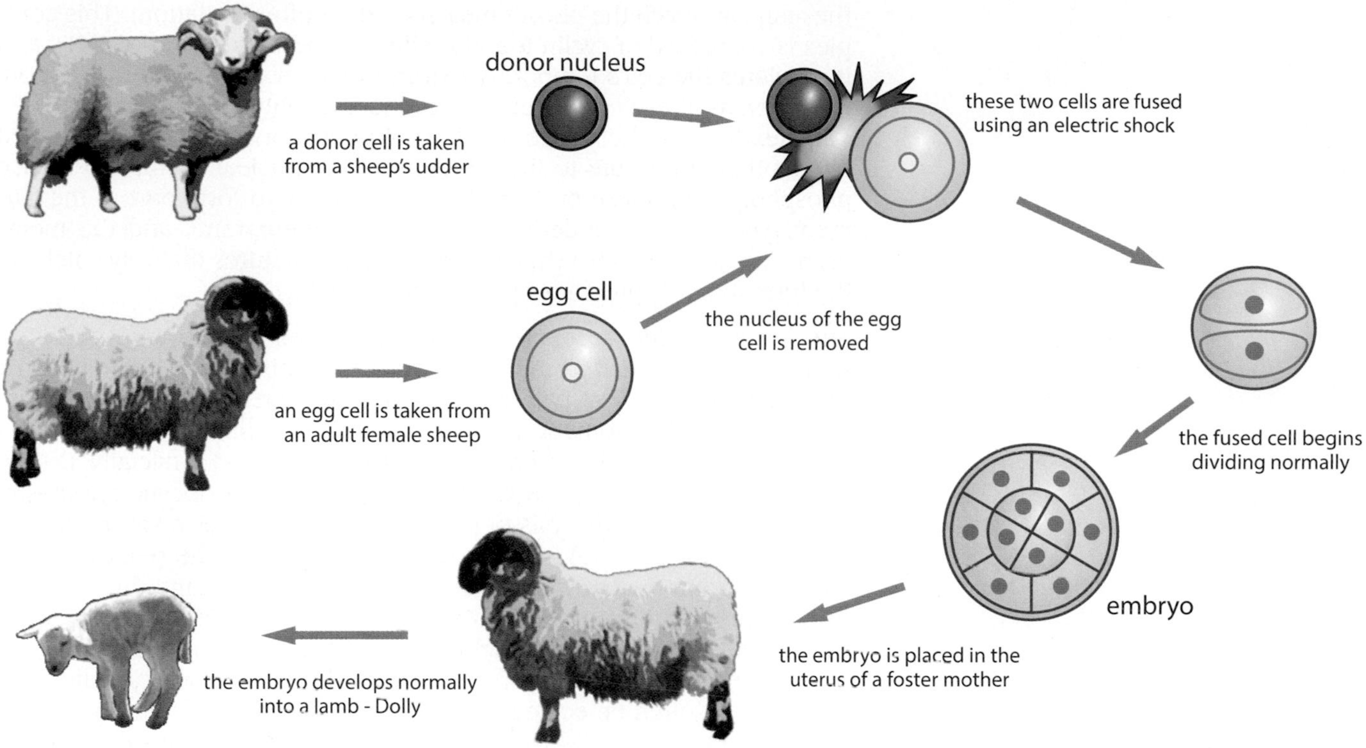

Figure 13.1 **The method used to create "Dolly" the sheep.** The nuclear material was removed from an oocyte taken from an adult female and replaced with that of a somatic cell from another sheep. Fusion and activation of this reconstructed zygote gave rise to an embryo that was surgically transferred to a surrogate mother, wherein development to term was completed. [Adapted from Armstrong L, Lako M, Dean W & Stojkovic M [2006] *Stem Cells* 24, 805. With permission from John Wiley and Sons.]

One of the most significant developments in recent years has been the production of a cloned human embryo, although there can be no doubt that this is also a highly controversial area of science. Indeed, there seem to be few topics that initiate such intense debates about the rights and wrongs of science as cloning research, and the greatly polarized opinions about the possible benefits and drawbacks of the technique have produced some data that were perhaps subjected to less rigorous initial scrutiny than is typical for published results. The first attempts at human cloning were not performed with the aim of reproducing a live human being, because this is strictly prohibited under international law. Rather, the objective was to derive embryonic stem cells (ESCs) from cloned blastocyst-stage embryos, because in theory these would be genetically identical to the somatic cell used as the nucleus donor. This process was termed "therapeutic cloning" because the objective was to produce an isogenic or "patient-specific" ESC whose differentiated progeny would be unlikely to be attacked by the immune system when transplanted back into the patient from whom the nucleus donor cell was derived. That, at least, is the theory behind therapeutic cloning. So far, it has not been shown to work in practice, because although cloned human blastocysts have been obtained in small numbers, they have never given rise to ESC colonies after placement of their inner cell masses onto feeder cell layers.

What happens to the somatic genome during SCNT?

Regardless of the method used to transfer the donor nucleus, the nucleus undergoes disassembly soon after transfer into the oocyte cytoplasm, even if the steps needed to activate the embryo into its division cycle are omitted. The disassembly involves the breakdown of the nuclear membrane, with an associated condensation of the chromosomes that exposes the chromatin to the epigenetic reprogramming systems of the oocyte. Breakdown of the membrane is thought to be a response to the high levels of a protein complex called maturation promotion factor (MPF), which is present in the cytoplasm of MII-stage oocytes (MetaphaseII stage, is the stage at which the oocyte meiosis arrests after ovulation). This complex is composed of cyclin B and cyclin-dependent kinase 1 (CDK1), and it regulates the cell's temporary exit from meiosis just before ovulation. However, as a result of the instrinsic kinase activity it possesses, the MPF complex has an alternative function of phosphorylating proteins called lamins that contribute to the structure of the nuclear membrane. Once phosphorylated, these proteins lose their ability to form part of the filament-type network underlying the nuclear membrane, and the membrane breaks down into smaller vesicle-like structures that rely solely on the formation of spheroids by phospholipid bilayers.

The levels of MPF decrease in zygotes derived from normal fertilizations to allow the exit from metaphase arrest; this process is induced by the spermatozoon that performed the fertilization. Perhaps not surprisingly, the injected somatic nucleus cannot exit from metaphase arrest in this way, and the SCNT embryo has to be activated artificially. During fertilization, phospholipase C-ζ is released into the oocyte cytoplasm, which causes the hydrolysis of phosphatidylinositol 4,5-bisphosphate to phosphatidylinositol 1,4,5-trisphosphate, resulting in the release of calcium ions from the oocyte's intracellular stores of calcium. The increase in calcium ion concentration is ultimately responsible for the chain of events that we recognize as fertilization. Thus, strategies to activate the SCNT-reconstructed embryo rely on methods to introduce calcium ions, and this can often be achieved through the use of a calcium ionophore This works by increasing the transport rate for calcium ions through the oolemma (the lipid bilayer surrounding the oocyte cytoplasm), which the ionophore molecule can do because of its hydrophobic, lipid-soluble domain and its ion-binding region that delocalizes the charge of the ion, thereby shielding it from the hydrophobic regions on the membrane. The MPF level decreases in response to the calcium influx until the MPF can no longer maintain high levels of phosphorylated lamins, whereupon the lamins begin to reassemble the nuclear membrane. This leads to the formation of pseudo-pronuclei; these are reminiscent of the true pronuclei that form after normal fertilization but they do not (and cannot) separate the genetic material from the male and female gametes. It is not even certain that such pronuclei always contain DNA, especially because multiple pronuclei can form. It is therefore more likely that the formation of these vesicle-like structures is simply a normal response of oocyte activation, and the fact that the injected diploid genome is not what the oocyte expects to find when it starts the process that would lead to encapsulation of the separate male and female genomes does not seem to stop it from trying.

It has been argued that a lengthy duration of exposure to high levels of MPF in the oocyte is beneficial for epigenetic reprogramming of the transferred genome. There are several reports describing how long the somatic nucleus should be exposed to high levels of MPF to complete reprogramming, and recommendations range from as little as 15 minutes to 6 hours between the nuclear transfer step and artificial activation. However, it is probable that the differentiation state of the donor cell has a significant impact on this timing. Reprogramming also seems to be partly dependent on the cell cycle stage of the nuclear donor cells. Donor cells in S phase also undergo premature chromosome condensation, but this may cause damage to the DNA duplexes of the donor nuclei and result in a "pulverized" chromatin appearance. Cytoplasts derived from pre-activated oocytes (generated by exposure of the oocyte to a calcium ionophore before enucleation) do not induce nuclear envelope breakdown and premature chromosome condensation, and they can thus accommodate donor cells from any stage of the cell cycle; however, the rate of progression of embryos to blastocysts using this method is low for differentiated somatic cells.

Pre-activated cytoplasts work better with blastomeres than with somatic cells. Blastomeres are less developmentally committed to the somatic cell lineages of later-stage embryos and are therefore theoretically easier to reprogram. This suggests that pre-activation removes some of the oocyte-derived factors that are capable of reprogramming the somatic genome. The inference from this is that the chromatin modifications induced in the somatic genome by premature chromosome condensation in the MII cytoplast may facilitate its reprogramming to a totipotent state, but it is not clear whether a longer exposure of the somatic genome to these factors (that is longer exposure than would normally occur during oocyte activation) ensures that epigenetic reprogramming is complete. As noted above, the differentiation state of the donor cell probably has a significant impact on the optimal exposure time.

The nature of the oocyte derived factors responsible for reprogramming is largely unknown, although because the reprogramming ability is lost after oocyte activation, it seems that their existence is transitory. From the point of view of the normally fertilized oocyte, their limited persistence is undoubtedly sufficient for the task of rapidly demethylating the incoming paternal DNA, but the highly differentiated state of a transplanted somatic donor karyoplast may be more problematic. Donor cells from early pre-implantation stage embryos may be more easily reprogrammed because they are pluripotent and thus may have lower levels of genomic DNA methylation and repressive histone modifications per blastomere. These nuclei may require less reprogramming of the genes required for early embryo development than nuclei from the types of donor cells that are most accessible for the purpose of nuclear transfer. This would explain Briggs and King's observation that frogs were more easily cloned from very early embryonic cells; it would also explain the success of serial nuclear transfer, which uses the blastomeres from four-cell SCNT mouse embryos as the nuclear donors for SCNT. The embryos from the transfer of blastomeres from four-cell SCNT embryos into oocytes progress to blastocysts at a much higher rate (83%) than original SCNT-derived embryos, and 57% of these can develop to term when transferred into a foster mother. This is much better than the survival rate obtained for embryos generated by a single nuclear transfer, which is usually around 5%.

Epigenetic modification is the basis of SCNT reprogramming

The greater success rate of serial nuclear transfer suggests that multiple exposure to some stage of the fertilization process or the very early embryonic development process increases the ability of the somatic genome to contribute to the formation of a new organism. This is most likely to be an epigenetic reprogramming step, but it seems that the factors present in the oocyte responsible for this activity are not always able to reprogram a somatic genome effectively. On transfer of a somatic nucleus to an oocyte during the cloning process, several essential changes must ensue. First, the somatic nucleus must cease to express its unique repertoire of gene products. Second, that nucleus must become responsive to the instructions provided by the oocyte cytoplasm to unfold a new pattern of development-specific gene transcripts. Third, the heritable memory endowed by the chromatin that defined the molecular characteristics of the donor cell must be erased. This is an enormous group of tasks that must be achieved rapidly, and above all precisely, within the limited timeframe available during early embryonic development. It is therefore perhaps not surprising that SCNT-derived embryos express genes at different levels from those of their counterparts derived from normal fertilizations.

Microarray studies of the placentas of cloned mice show that approximately 4% of their genome is misexpressed compared with genomes from placentas of normal animals, and other organs (such as liver) also show considerable gene dysregulation. The imprinted genes seem to be particularly affected, even though they are normally exempt from the genome-wide demethylation of DNA that occurs during pre-implantation development. As we have already seen in Chapter 11, imprinted genes are a unique group that is important for fetal growth and development, especially in the placenta, as well as for postnatal behavior and cognition. The expression of imprinted genes does not follow a Mendelian pattern of inheritance; instead, it depends on the parent of origin to dictate its expression. Regulatory regions of such genes are typically methylated in the silent allele, but they are also particularly sensitive to environmental changes. For example, embryos generated by using assisted reproduction technologies show a higher rate of imprinting defects, a problem that has been attributed to exposure of the embryo to sub-optimal culture conditions. Thus it is probably not surprising that SCNT embryos show widespread DNA methylation defects in this sensitive gene group.

The fact that the placentas of cloned mice are less affected than the tissues of the embryo supports the likelihood that epigenetic reprogramming is not complete by the time that the cells of the trophectoderm differentiate from those that will form the inner cell mass. Similarly, this implies that the genomes of the inner cell mass may be more highly reprogrammed, but the fact that there is gene dysregulation in the embryo at all suggests that the reprogramming cannot be completely effective. Further evidence to support this emerges from SCNT studies with ESCs as nuclear donors, in which a smaller subset of genes showed inappropriate expression compared with genes of surviving clones derived from cumulus cells (specialized cells that surround the oocyte in the ovary). That there were still apparent errors in the ESC-derived clones is evidence that the reprogramming effected by the oocyte is imperfect.

Epigenetic reprogramming is a normal feature of fertilization that is hijacked by SCNT

Why the oocyte should be able to reprogram a somatic genome at all is an important question. The appearance of a terminally differentiated nucleus in the MII oocyte is a wholly unnatural event. It is therefore surprising that the molecular systems present in the oocyte are able to recognize the types of epigenetic modifications on the somatic chromatin and try to alter them to be more representative of pluripotent cells. Probably the only reason that this works at all is because epigenetic reprogramming of the gamete genomes is a normal feature of the fertilization process. In effect, SCNT hijacks one of the mechanisms active during fertilization for its own purposes.

After fertilization, there is a series of events that involve the incoming sperm as it encounters the egg cytoplasm. The head of a spermatozoon is not very large, so the haploid genome has to be packaged very tightly into it. This means that the spermatid (the haploid precursor of the spermatozoon) cannot rely on the compaction produced by wrapping the DNA around histone octamers, as is the normal method for other cell types, because this simply does not compress the DNA into a small enough space. To get around this, a special DNA packing protein called protamine has evolved that is present only in the male germ cells. This protein packs the DNA into an almost toroidal conformation (**Figure 13.2**), assisted by the fact that the DNA is very heavily methylated, presumably to increase the protamine–DNA interaction. The packing is so tight as to be quite similar to the effects of dehydrating the DNA. Naturally, this

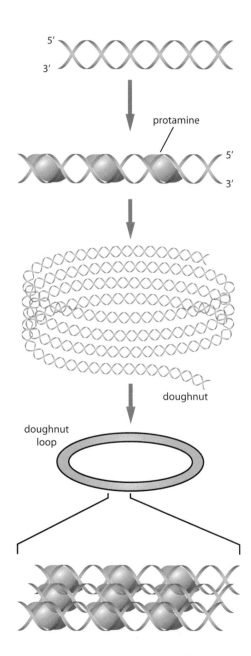

Figure 13.2 Packaging of DNA into the spermatozoon requires protamine. In the sperm nucleus, protamines replace the histones and the protamine–DNA complex is coiled into a doughnut shape. [Adapted from Braun RE (2001) *Nat. Genet.* 28, 10. With permission from Macmillan Publishers Ltd.]

compacted genome is not very useful for transcription, so as soon as it is released from the sperm head into the oocyte cytoplasm it has to be unwrapped and combined with histones to form nucleosomes that can interact with the transcriptional machinery. This usually occurs within 1 hour of fertilization, and the remodeling of the sperm nucleus into an accessible, transcriptionally competent chromatin configuration is coincident with the formation of the pronuclear membrane and demethylation of the paternal genome.

As the end of telophase approaches, centromeric proteins A and B, which function as part of the kinetochore complexes, are assembled onto the DNA. On completion of active demethylation and with the initiation of S phase, transcription factors (such as TATA box protein and Sp1) bind to prime the genome for transcription late in the first cell cycle. There can be little doubt that DNA demethylation at this stage is an active process, for it occurs far too rapidly to be simply a passive process reliant on a lack of maintenance DNA methyltransferase activity, but the nature of such an active demethylation is not well understood. As yet, the origin of this activity has not been unequivocally assigned to either the oocyte cytoplasm or the sperm itself; however, indirect evidence of partial demethylation on SCNT points to the activity's residing in the oocyte cytoplasm. Demethylation of up to five supernumerary male pronuclei obtained by polyspermic fertilization of zona-free mouse oocytes suggests a high abundance of this activity.

There are several possible mechanisms by which the somatic genome might be remodeled in SCNT

We discussed some possible mechanisms for DNA demethylation in Chapter 11. The candidate proteins that have been described as possible demethylaters in SCNT cover the three basic mechanisms for methyl group removal. The first of these removes methyl group directly from the major groove of DNA. Although the mechanism by which this is achieved is uncertain, methyl-CpG-binding domain protein 2 (MBD2) has been shown to possess demethylase activity, with methanol as the stable leaving group. The purported demethylase activity of MBD2 has not been without some controversy, especially because crosses between MBD2-null mice showed no apparent differences in DNA methylation at the one-cell zygote stage. The DNA methylation levels in these embryos were measured by using an immunocytochemical analysis of 5-methylcytosine, which is not absolutely quantitative. Nevertheless, the results led to the proposal of alternative demethylation mechanisms.

A second possible mechanism envisages the replacement of 5-methylcytosine by cytosine or the removal of the CpG dinucleotide by either base or nucleotide excision repair. For this reason, the uridine deglycosylase MBD4 (methyl-CpG-binding domain protein 4), which may have a role in DNA repair, was proposed as a potential demethylase. The caveat here is that paternal-specific demethylation seemed to occur normally in MDB4-null fertilized oocytes.

A third possibility proposes the hydrolytic deamination of 5-methylcytosine, resulting in the conversion of 5-methylcytosine to thymidine. However, this process would require a considerable input of energy and is therefore the least likely mechanism. Mechanistically, the loss of a methyl group poses a considerable enzymatic challenge, and these activities and their epigenetic regulation are a research "hotspot" at present.

Why the embryo needs to actively demethylate the paternal genome so rapidly after fertilization remains a mystery. It has been suggested that de-repression of a number of paternal alleles is required to accommodate the burst of transcriptional activity that occurs at the end of the first cell cycle. An alternative hypothesis has been suggested that largely focuses on imprinted alleles. It proposes that active demethylation arose as a protective response of the maternal genome to reduce the influence of the paternal genome, which may have alleles optimized for the production of large, highly competitive offspring. Thus, the rapid demethylation of the paternal genome may serve the maternal interest by ensuring the survival of a larger number of offspring overall.

The rapid genome-wide loss of 5-methylcytosine from the paternal genome, with the exception of some elite sequences (for example imprinted genes, centromeric satellites, and some endogenous retroviruses), is followed by a slower, passive demethylation arising from the exclusion of DNA methyltransferase 1 (DNMT1) from the nuclei of the blastomeres during pre-implantation development. There is an oocyte-specific form of DNMT1—designated DNMT1oo—which is thought to be involved in the maintenance of DNA methylation at imprinted genes, but the general absence of maintenance methylation at other sequences results in a stepwise decline in the level of 5-methylcytosine, which reaches its lowest point at the morula stage. Thereafter, *de novo* DNA methylation begins simultaneously with the first embryonic differentiation event, which establishes the cell lineages that will give rise to the ICM and trophectoderm of the blastocyst. The maternal genome seems to be subject more to this type of DNA demethylation rather than any active form, because its demethylation takes place over a much longer period than the demethylation in the paternal genome. For a comparison of the rates of 5-methylcytosine loss from the respective genomes, see Figure 11.6.

DNA methylation does not influence transcription in isolation; rather, it exerts its influence through interactions with other epigenetic features of chromatin. It is therefore unsurprising that histone modification patterns undergo similar changes immediately after fertilization. Detailed examinations of methylation on histone H3 lysine 9 (H3K9) and histone H3 lysine 27 (H3K27) by immunohistochemistry indicate very precise and distinctive patterns up to and during the first cell cycle of mouse oocytes fertilized *in vitro*. Immediately after entry into the oocyte cytoplasm, the decondensed sperm nucleus acquires H3K9 acetylated histones in addition to monomethylated H3K9 and H3K27, but dimethylated and trimethylated H3K9 are undetectable, suggesting a lack of any heterochromatin regions in the genome. The slower DNA demethylation of the maternal genome may imply a slower reprogramming rate (or even, perhaps, less of a need for reprogramming). This is reflected in the persistence of dimethylated and trimethylated H3K9 and H3K27 signals in the female pronucleus, although this seems to be associated with pericentromeric and centromeric satellite sequences.

During the phase when paternal-specific active DNA demethylation occurs (up to 6 hours after fertilization), the male pronucleus has not yet acquired dimethylated or trimethylated H3K9 modifications. Several hours later, after full maturation of the pronuclei, dimethylated, but not trimethylated, H3K9 is first detectable in the male. In contrast, double labeling for dimethylated or trimethylated H3K9 and 5-methylcytosine indicated that these marks are abundant in the female pronucleus. There is apparent co-localization between dimethyl H3K9 and DNA methylation and no apparent association with trimethylated H3K9, even in pericentromeric regions. The timing of the disappearance of DNA methylation suggests that demethylation proceeds in, and is perhaps permitted by, the absence of genome-wide H3K9 methylation in the male. In turn, the dimethyl H3K9 modification may confer full protection from demethylation in the female.

During the earliest stages of decondensation and the completion of meiosis, only the maternal chromatin is positive for trimethyl H3K27, reinforcing an early asymmetry between the two pronuclei. However, trimethyl H3K27 is first detected in the male by late pronuclear stages. By the time that mature pronuclei have formed (approximately 8–10 hours after fertilization), trimethylated H3K27 is also detectable in the male, although rather less intensely than in the female. This time point is significant because it corresponds to the beginning of DNA replication. Interestingly, the enzymatic activities necessary to establish the H3K27 trimethylation mark are already present in the pronuclei, as are other histone methyltransferases such as SuV39h (which establishes the trimethyl H3K9 mark) and histone deacetylases. Also present is heterochromatin protein 1-β (HP1-β), although it does not seem to be able to take part in its usual interaction with the methylated forms of H3K9. Because of the apparent absence of trimethylated H3K9, it is no surprise that HP1-β is unable to form heterochromatin, but it does seem to be able to bind to monomethylated H3K9; it has been suggested that this interaction may be able to block the trimethylation normally carried out by SuV39h. Similarly, EED/EZH2 (see Section 7.2) does not modify monomethylated H3K27 to trimethylated H3K27 in the same compartment, so possibly another type of inhibition system is acting on this enzyme. This mechanism of delayed histone methylation may be critical to ensure that rapid and efficient DNA demethylation of the male pronucleus proceeds unhindered. The detection of trimethylated H3K27 in advance of trimethylated H3K9 suggests that H3K9 may have a more prominent role in protecting from demethylation. Indeed, in the male pronucleus, monomethylated H3K9 is localized in the pericentromeric regions, which are the only visible structures that do not lose 5-methylcytosine.

The epigenetic remodeling that occurs in SCNT differs from the remodeling that occurs after fertilization

The preceding discussion of DNA methylation and histone modification changes that occur immediately after fertilization applies only partly to the SCNT-derived embryo. The somatic genome does not seem to respond to the DNA demethylating activity of the MII oocyte, and in most cases the level of methylated DNA remains much higher than in embryos from normal fertilizations; this higher level of methylation is a state reminiscent of somatic rather than pluripotent cells. In addition to reduced passive loss of DNA methylation, the onset of *de novo* methylation frequently begins much earlier (at the four-cell stage) than in normal embryos. This suggests that incomplete remodeling of the donor nucleus impairs the normal temporal progression of epigenetic reprogramming, leading to transcriptional misregulation. One might expect that, given the apparent abundance of demethylating activity in the oocyte, the somatic genome would be rapidly demethylated, but this is not the case.

A number of explanations may account for inadequate epigenetic remodeling of the donor nucleus. For instance, the enucleation process may remove essential components that are both intimately associated with the MII chromosomes and required for demethylation. Precedence for an essential component of the mitotic apparatus has been reported in nonhuman primates. Removal of a critical cytoplasm factor may also account for impaired demethylation, but because SCNT has a low efficiency irrespective of the cloning protocol, this seems unlikely. Continued expression of demethylating proteins can be discounted because very little transcription occurs at this time. This leaves the possibility that the endogenous DNMTs continue to methylate target sites, coupled with the likelihood that the chromatin of a differentiated cell differs from that of the diploid zygote and hence is resistant to the demethylating activity of the oocyte.

The last possibility fits well with the earlier discussion about the resistance to DNA demethylation imposed by the presence of H3K9 methylation at specific loci, because a key observation of SCNT embryos is that they consistently show levels of H3K9 (monomethylation, dimethylation, and trimethylation) in the very early stage of development after nuclear transfer. Even at the earliest stage, the SCNT embryos have substantially higher levels of both of these histone modifications, and although the acetylation of H3K9 may assist the reestablishment of pluripotency (or perhaps not, if it happens to be associated with a somatic-specific gene that should be repressed!), it is unlikely that dimethyl H3K9 will be of any help at all because it probably inhibits the removal of DNA methylation from genes that are repressed in the somatic cells but need to be up-regulated for pluripotency. Similar problems seem to exist with the trimethyl H3K4 modification, because it is present in excess from the 2-cell to the 32-cell stage. Trimethylated H3K27 is even worse. It starts out looking fairly normal up to the 4-cell stage, but then despite a slightly abnormal increase at the 8-cell stage, it disappears and by the blastocyst stage it is undetectable.

Trimethylation of H3K27 is an essential epigenetic modification present on a wide range of so-called "bivalent" chromatin domains in pluripotent cells. These include many of the genes needed for embryonic development, so the absence of trimethylated H3K27 from the inner cell mass in SCNT embryos is a potentially serious matter. The consequences of this loss of control for the inner cell mass can be serious indeed. SCNT-derived mouse embryos often express much lower levels of the pluripotency gene *Oct4* than their counterparts derived from normal fertilizations. Because *Oct4* is one of the key genes required to maintain pluripotency, this could mean that the inner cell mass of the SCNT embryo is less able to differentiate into all the cell lineages of the developing embryo. Conversely, the SCNT embryo does express some *Oct4*, so it is possible that this level may be sufficient. However, the clear difference in *Oct4* expression in normally fertilized embryos and SCNT embryos means that the developmental potential of SCNT embryos will always be viewed with suspicion.

Some aspects of reprogramming of the somatic epigenome are outside the oocyte's capacity

So why is the oocyte unable to reprogram a somatic genome completely? The fact that the failure of this process is not total suggests that the mechanism used to remodel the incoming paternal genome is able to perform most of the operations needed to alter the somatic epigenome sufficiently to allow a more or less pluripotent cell population to form. However, the inability to do so fully indicates that there must be some aspects of reprogramming that are outside the oocyte's capacity. Several studies have shown the expression of chromatin-modifying proteins in mature MII oocytes to have a similar profile to that observed for ESC. The Jumonji domain containing JARID histone demethylases (particularly JARID 2A) and DNMT3B are present at MII, in addition to histone acetyl-transferases and ATP-dependent chromatin remodeling proteins such as SMARCA5. The MII oocyte should therefore possess the capacity to alter epigenetic modification patterns. However, these enzymes are only components in the multiprotein machines that process specific genomic loci, depending on the cellular context. If appropriate transcription factors or microRNAs do not exist in the oocyte to direct the assembly of a chromatin-remodeling complex and then recruit and control another complex that will alter the epigenetic modification pattern of that locus in a manner precisely related to the reacquisition of pluripotency, then the locus will not undergo appropriate reprogramming.

In contrast, some data show that components of the polycomb repressive complexes (such as EED, Ezh2, and Suz12) can be underexpressed at the pseudo-pronuclear stages of SCNT mouse embryos, and this has been highlighted as one of the possible reasons why establishment of the H3K27 trimethyl mark could fail. Even if these three polycomb components were present at significant levels in the MII oocyte, there is some evidence to suggest that some of the protein components of the polycomb complexes dissociate from the chromatin during meiotic transcriptional arrest, and thus a completely functional polycomb may not be available to interact with the incoming somatic genome immediately, which could hinder its reprogramming. However, PCR amplification indicates that the mouse SCNT embryo possesses much lower levels of mRNA for Eed, Ezh2, and Suz12, so on balance of probability it is more likely that the trimethyl H3K27 defect arises from lack of an appropriate enzyme complex to establish this histone mark than that it arises from dissociation of its components.

Interestingly, the low mRNA levels for these enzymes in SCNT embryos would seem to imply that the expression of the polycomb subunits might come from the somatic genome rather than from maternal mRNA stored within the oocyte. If it were the latter, we might expect the amounts of mRNA for these enzymes to be the same whether the embryo was derived by SCNT or by normal fertilization, but if the mRNAs have to be transcribed *de novo* from a somatic genome, then lower levels in SCNT embryos would be expected because the somatic genome may have adapted to low-level production of these mRNAs during differentiation. Reduced production of these mRNAs would probably depend on the type of cell used as the nuclear donor, because down-regulation of EED, EZH2, and SUZ12 is not observed during the early stages of hESC differentiation. Another possibility is that the somatic nucleus brings with it the ability to degrade mRNAs via the siRNA mechanism. In this case, we would expect to observe levels of the RNAi proteins Dicer, Drosha, and so on, to be higher in the somatic cells, as a consequence of hESC differentiation. An alternative scenario is therefore that failure to immediately eliminate these RNAi proteins after nuclear transfer may enhance the breakdown of maternal mRNAs that were supposed to allow the synthesis of epigenetic reprogramming proteins (such as EED, EZH2, and SUZ12), which may in turn affect the ability of the SCNT embryo to establish appropriate patterns of epigenetic marks.

Somatic gene expression must be turned off for epigenetic reprogramming to occur in SCNT embryos

An essential step during SCNT-based reprogramming is to switch off the somatic gene expression program. Genes actively transcribing are typically acetylated on lysine residues by histone acetyltransferases (HATs). To turn off gene expression, histones are deacetylated by histone deacetylases. These deacetylated residues are substrates for HMTs, which leave methyl groups on key lysine residues. Mammalian oocytes typically express very low levels of histone deacetylases, and this seems to be the case during much of pre-implantation development. Hence, this may favor the sustained transcription of SCNT donor genes. Conversely, oocytes do express higher levels of the histone acetyltransferases HAT1 and GCN5, so one might imagine that remodeling of chromatin at inactive genes needed for reprogramming could be possible. However, this would potentially require the presence of histone demethylases. A potential problem for successful and specific histone demethylation is that the somatic genome might be expected to retain a high level of expression of the histone methyltransferases that normally methylate H3K9 (such as G9a, which generates dimethyl H3K9). If the activity of this enzyme cannot be easily repressed, then it could contribute to the continued presence of methylated H3K9, which, as we have seen, disrupts the ability of the oocyte to effect demethylation of the DNA at the loci covered by this modified histone. We know from studies of hESC differentiation that the G9a enzyme is expressed at much higher levels in somatic cells. It is therefore one of the key epigenetic components that would need to be shut down immediately after SCNT. If some of the G9a protein is transferred along with the nucleus, the oocyte can try to repress its encoding gene as much as it likes, but this will still not stop H3K9 methylation unless there is some other mechanism by which the protein itself can be inhibited.

At present, it is not clear whether the oocyte is uniquely competent to remodel and reprogram the wide variety of chromatin modifications, both nucleosomal and otherwise, in a more efficient manner. Perhaps the focus of attention should be on presenting inherently more compatible donors during SCNT. Irrespective of the limitations to reprogramming, a small number of SCNT embryos do survive, suggesting that in rare cases the embryo is capable of at least partial resetting of the genome. It may be that the reprogramming activity is simply overwhelmed by the enormous task of having to modify or replace somatic histones, remove polycomb complex proteins, and demethylate areas of the genome that may be a lot less accessible than the corresponding areas in gamete-derived genomes. Alternatively, it may be that such reprogramming is actually "forbidden" for the genomes of somatic cells and that it is only when the mechanism controlling this malfunctions that successful clones arise.

The MII oocyte is not the only repository of epigenetic reprogramming activity. Pluripotent cells themselves also seem to have this capacity, which has led to the suggestion that it may be the pluripotency-associated genes that are the ultimate regulators of the epigenome of hESC. This concept has been greatly reinforced by the development of techniques to selectively overexpress a small range of these genes in somatic cells, which seems to turn a few of the cells back to a pluripotent state, albeit at very low efficiency. The MII oocyte also has abundant mRNA encoding the pluripotency gene *OCT4*. It is therefore possible that the OCT4 protein is responsible for the ability to reprogram the paternal genome (and to a smaller extent the maternal genome). However, because OCT4 is a transcription factor with no inherent chromatin or histone modifying ability, this effect must be indirect in nature.

13.2 REPROGRAMMING THE EPIGENOME BY CELL FUSION

Fusion of somatic cells with pluripotent cells can reprogram the somatic genome

It is possible to fuse pluripotent and somatic cells by exposing mixtures of them to polyethylene glycol (**Figure 13.3** shows a schematic representation of this process). Only a few cells will undergo the initial fusion, and fewer still will survive to form stable heterokaryons. Those that do will appear to be pluripotent, even though they contain the chromosomes of the somatic cell.

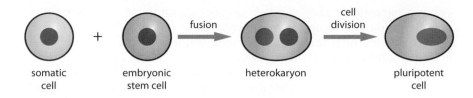

Figure 13.3 Fusion of somatic cells and embryonic stem cells can reprogram the somatic cell's genome. [Adapted from Narbonne P, Miyamoto K & Gurdon JB (2012) *Curr. Opin. Genet. Dev.* 22, 450. With permission from Elsevier.]

The interesting thing about these heterokaryons is that the former somatic chromosomes behave in the same way as pluripotent ones and express their own copies of *OCT4*, *SOX2*, and so on. This implies that epigenetic reprogramming has taken place. In addition, after cell fusion, ESC–somatic cell hybrids have the growth characteristics of ESCs and express many pluripotency markers while differentiation genes from the somatic genome are down-regulated. Furthermore, ESC–somatic cell hybrids have the potential to differentiate into derivatives of the three germ layers both *in vitro* and *in vivo*. At the chromosomal level, cell fusion of two diploid cells leads to the hybridization and union of their respective nuclei, producing tetraploid hybrids. The heterokaryons can be cultured just like normal ESCs; however, they are not stable, and after multiple passages they begin to lose the formerly somatic chromosomes as a result of an apparently unequal partitioning of these chromosomes between the daughter cells at each mitosis. Eventually, ESC-like cells are recovered that possess very little of the DNA from the former somatic cells. This unequal segregation of parental chromosomes and preferential loss of somatic chromosomes observed in some interspecies and intraspecies mouse ESC–somatic cell hybrids suggest that pluripotency is maintained by the ESC chromosomes in a *cis*-acting rather than a *trans*-acting manner.

That reprogramming can occur at all is fascinating, because it implies that ESCs have their own mechanisms to recognize and alter the epigenotype of a somatic nucleus. The reprogramming is also quite rapid. In the particular case of fusion induced between mouse ESCs and human B cells, expression of human *OCT4*, *NANOG*, *CRIPTO*, *DNMT3B*, and *TLE1* was detected in cells as early as 1 day after fusion and human *REX1* after 2 days, although the levels were low in comparison with those normally detected in human ESCs. These expression levels increased over time, but, crucially, they were undetectable in non-fused (or self-fused) human B cells or in control mouse ESCs. Mouse lymphocyte-specific gene transcripts (mCD19, mCD37, and mCD45) were not detected throughout the span of the 8-day analysis, confirming the dominance of ESCs in conversion. Increased expression of human pluripotency-associated genes over this 8-day period was mirrored by a reduction in expression of several human lymphocyte-associated genes within the second day (*hCD45*, *hCD37*, and *hCD19*) or third day (*hCD20* and *hPAX5*) of heterokaryon formation.

A key issue about cell fusion is that to be truly recognized as a hybrid cell, the nuclei of the individual cells must fuse to give a cell with a single nucleus containing the chromosomes of both original cell types. The process of nuclear fusion takes much longer than fusion of the individual cells, but it is interesting that the expression of human pluripotency genes begins at the heterokaryon stage when the nuclei are still separate. Collectively, these data show that after dominant reprogramming, the activation and silencing of tissue-specific gene programs begin ahead of, and therefore do not require, nuclear fusion and cell division. However, this also implies that the reprogramming activity is present in the ESC cytoplasm and is capable of entering the somatic nucleus.

The reprogramming step involves large-scale changes in DNA methylation, as one might expect, especially because, before fusion, the human B cells showed high levels of DNA methylation throughout the *hOCT4* promoter and across a single *IGF2/H19* allele. After cell fusion, DNA methylation of *hOCT4* in reprogrammed B cells declined, consistent with a trend toward a hypomethylated state similar to that seen in human ESC lines. Demethylation of the *hOCT4* promoter is seen before nuclear fusion and cell division, a result that is consistent with active chromatin remodeling of the locus before expression. No changes in DNA methylation at the *IGF2/H19* ICR (imprint control region) locus were detected over this period, suggesting that imprinted sequences may be exempt from this reprogramming mechanism, as we see in fertilization.

OCT4 is involved in genome reprogramming in heterokaryons

So what is the mechanism of reprogramming in the heterokaryons? Experiments in which the mouse pluripotency-associated transcription factor Oct4 is tagged with a fluorescent marker (Flag tagging; **Figure 13.4**) show that it enters the human B-cell nucleus 3–6 hours after cell fusion, which precedes the onset of expression from the human *OCT4* locus (which happens around 24 hours after fusion).

This possible involvement of mouse *Oct4* was investigated further by making mouse ESCs in which the repression of *Oct4* (and a few other pluripotency genes such as *Sox2*) was dependent on the addition of tetracycline to the culture medium, and then fusing these ESCs with human B cells. Successful reprogramming, as judged by the induction of several human genes (*OCT4*, *NANOG*, *CRIPTO*, *DNMT3B*, *SOX2*, *TLE1*, *TERT*, and *REX1*), was reduced (+6 hours) or eliminated (+12 hours) by pretreatment of the heterokaryons with tetracycline. These results confirm that *mOct4* expression is critically important for initiating successful reprogramming. However, it was also observed that the extinction of human B-cell-specific genes was not impaired by *Oct4* removal, which may suggest that the activation and silencing of gene expression programs in heterokaryons are mechanistically distinct processes.

Elimination of other pluripotency genes such as *Sox2* from the mouse ESCs by this method seems to have much more modest effects on their ability to reprogram the human B cells, although there is undoubtedly some degree of redundancy as the human *SOX2* transcript begins to be synthesized. Another valuable observation is that the renewed expression of human genes seems to be stably maintained, because withdrawal of the mouse *Oct4* has no effect on the human *OCT4* expression 24–48 hours after cell fusion.

The big question is this: How does a transcription factor such as Oct4 induce an enormously complex process such as epigenetic reprogramming? It does not seem to have any intrinsic chromatin-modifying ability and so can only function by up-regulating or repressing other genes, or through hitherto unknown interactions with other proteins. This presents us with something of a problem. The number of genes involved in epigenetic reprogramming either as direct participants (such as those proteins that mediate DNA demethylation) or as bystanders (such as silenced somatic-specific genes) is most probably enormous. Is it possible that Oct4 is able to interact with the promoter sequences of all these genes, or does it initiate a cascade of other transcription factors that can do this? Moreover, is Oct4 able to recruit chromatin-modifying or chromatin-remodeling complexes?

The latter two points are matters of intense investigation, but there is evidence supporting the interaction of Oct4 with a large number of

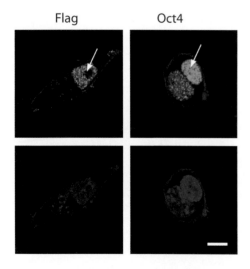

Flag   Oct4

Figure 13.4 Immunofluorescence analysis of cultured heterokaryons 6 hours after cell fusion, showing the presence of mouse ESC-derived Oct4 (Flag–Oct4, green) in a human nucleus (white arrows). The human nuclei were distinguished from mouse nuclei on the basis of diffuse versus punctuate staining with DAPI (blue); 6 hours after fusion, the human nuclei are showing expression of their own OCT4, indicating that this gene has been reprogrammed by the presence of the ESC genome from mouse. [From Pereira CF, Terranova R, Ryan NK et al. (2008) *PLoS Genet.* 4:e1000170.]

gene promoters from the data describing bivalent chromatin domains in ESCs. Studies using chromatin immunoprecipitation to extract DNA sequences associated with the *OCT4* promoter in human ESCs, followed by hybridization to whole-genome microarrays, demonstrated that OCT4 was associated with 623 (3%) of the promoter regions for known protein-coding genes and 5 (3%) of the promoters for known microRNA genes. Similar analyses showed occupancy of many gene promoters by SOX2 and NANOG; crucially, however, many of the binding sites for these transcription factors were identical to the OCT4-binding sites. These data suggest that OCT4, SOX2, and NANOG function together to regulate a significant proportion of their target genes in human ESCs.

There are several possible mechanisms by which the OCT4/SOX2/NANOG trinity of pluripotency factors may work to reprogram genomes

The targets of the three pluripotency factors (OCT4, SOX2, and NANOG) can be active or inactive genes. Transcripts were consistently detected in ESCs for approximately half of the genes that were bound simultaneously by OCT4, SOX2, and NANOG. Among these active genes, several that encode transcription factors (for example OCT4, SOX2, NANOG, STAT3, and ZIC3) and components of the transforming growth factor-β (TGF-β; for example TDGF1 and LEFTY2/EBAF) and Wnt (for example DKK1 and FRAT2) signaling pathways were notable targets. Recent studies have shown that signaling by TGF-β and by Wnt has a role in pluripotency and self-renewal in both mouse and human ESCs. These observations suggest that OCT4, SOX2, and NANOG promote pluripotency and self-renewal through positive regulation of their own genes and genes encoding components of these key signaling pathways. Among the transcriptionally active genes bound by OCT4, SOX2, and NANOG there are many transcription factor genes that have been implicated in developmental processes. These include genes that specify those transcription factors that control differentiation into extra-embryonic, mesodermal, and endodermal cell lineages. In addition, many of the homeobox genes that establish the initial plan for constructing the embryonic body are targets of this trinity of pluripotency transcription factors.

At present it is unclear how the three key regulators can activate some genes and repress others. It is likely that the activity of these key transcription factors is further controlled by additional cofactors or by the precise numbers of OCT4, SOX2, and NANOG that bind to specific genomic loci. Protein–protein interactions between the three transcription factors and other cellular proteins have been demonstrated by immunoprecipitation of NANOG, tagged with biotin and followed by proteomic analysis using liquid chromatography–mass spectrometry. This identified a group of proteins that once again are mostly transcription factors or OCT4 and SOX2. However, there are also associations with proteins that would normally be part of chromatin-remodeling complexes such as HDACs and REST.

The interactions of OCT4 with Arid3b and HDAC are suggestive of an ability to recruit at least some components of chromatin-modifying complexes, but this is far from being conclusive proof that pluripotency genes possess this ability. Genome-wide chromatin immunoprecipitation experiments using antibodies directed against RNA polymerase II and the PRC2 complex subunit SUZ12 showed that the vast majority of SUZ12-bound sites were found at gene promoters of the key developmental genes identified as targets of the OCT4, SOX2, and NANOG transcription factors. Each of the three DNA-binding transcription factors occupied approximately one-third of the PRC2-occupied genes that encode developmental transcription factors; remarkably, however, almost all of the PRC2 targets showed simultaneous occupation by OCT4, SOX2, and NANOG. In view of this, it is possible that OCT4 and its cognate pluripotency factors are able to reprogram not only by activating genes necessary for the reestablishment of pluripotency but also by repressing genes that are central to the phenotype of the differentiated cell used in the fusion experiments mentioned above.

Reprogramming may not be the sole purview of ESCs

The tetraploid hybrid cells resulting from polyethylene glycol-induced cell fusion of ESCs with somatic cells show a level of pluripotency similar to that of the original ESCs, including the ability to contribute to live chimeras developed from mouse blastocysts injected with these cells. There is, however, a significant problem with this reprogramming approach in that the tetraploid cells are unstable because they frequently segregate chromosomes unequally on cell division. Indeed, after a while in culture, many of the hybrids may revert to an almost diploid state. This might give us some clinically useful cells were it not for the fact that the segregation is almost always unilateral owing to the elimination of the chromosomes from the original somatic partner. This observation is even more disturbing because it implies that there must be some characteristic that distinguishes the cellular origins of the chromosomes. Such a characteristic is probably epigenetic in nature, which means that the somatic genome may not have undergone complete reprogramming to a pluripotent state.

The reasons for such unequal chromosome segregation are far from clear, but the marked differences of chromosome segregation patterns in ESC–ESC and ESC–fibroblast hybrid cells suggest that the differentiation status of the parental cells has a significant role in subsequent chromosome segregation. In contrast, similarity in developmental status of the parental cells results in a stable tetraploid cell hybrid. Moreover, a difference in developmental status of the parental cells seems to determine the direction of segregation of the parental chromosomes. This is demonstrated by unilateral segregation of chromosomes derived from the somatic partner in most ESC–splenocyte hybrid cells and in all ESC–fibroblast clones, which suggests that ESCs do not have "all-powerful" reprogramming capacity. Instead, it seems that developmentally determined epigenetic modifications can restrict the capacity of ESCs to reset chromosomes into a pluripotent state. Although the nature of these modifications is unknown, this is reminiscent of the epigenetic "memory" we encountered in our discussion of SCNT.

Enhanced unilateral segregation appears most frequently for fusion hybrids made from cells that would be expected to have very different epigenotypes, whereas cells with higher degrees of similarity are able to form stable tetraploids that can be cultured in the long term. This has some interesting consequences, in that reprogramming activity may not be the sole province of pluripotent cells. Fusions between myoblasts (a broad population of muscle cells that includes the skeletal muscle stem cells) and keratinocytes (relatively undifferentiated cells from the dermis) result in reprogramming of the keratinocyte genome toward muscle-specific gene expression at the heterokaryon stage. How the epigenotype of one cell type can dominate another was not immediately clear because, at least in theory, myoblasts and keratinocytes should not be far apart in terms of differentiation (both are progenitor cell type populations). Help in resolving this issue came from experiments in which myoblasts from humans were fused with keratinocytes from mice, which allowed selective PCR amplification of the genomes from the two species. These studies showed that significant numbers of myoblast-specific genes were up-regulated in the keratinocyte genome but equal numbers of keratinocyte-specific genes were up-regulated in the myoblast genome. The balance of reprogramming activity could be altered by changing the ratio of the two cell types. For example, more myoblast-like heterokary-ons could be obtained if the ratio of myoblasts to keratinocytes was higher, and vice versa.

This does not seem to apply to ESC fusions to somatic cells, which could mean that the ESCs still have some special mechanism that allows them to effect a more extensive reprogramming activity. However, it does suggest that the presence of appropriate transcription factors or micro-RNAs is able to direct not only the phenotype but also the epigenotype of the hybrid cells. This is interesting because it suggests the possibility of reprogramming by taking transcription factors/microRNAs, signal trans-duction-related proteins, or other gene expression control factors out of one cell and introducing them into another type of cell. In essence, this is a bit like nuclear transfer in reverse: instead of taking the nucleus out and putting it into contact with the cytoplasm of a cell that has the capacity to reprogram, we are taking the cytoplasm (or factors derived from it) and putting it into contact with the genome we want to reprogram, albeit in a mixture with the cytoplasm that the nucleus was already associated with.

13.3 REPROGRAMMING THE EPIGENOME BY CELL EXTRACTS

Cell extracts can effect epigenetic reprogramming by providing the needed regulatory factors

As an alternative to fusion, somatic nuclear function may also be altered by using nuclear and cytoplasmic extracts, the underlying rationale for this being that extracts provide the necessary regulatory compo-nents. This was demonstrated by the work of Philippe Collas's group in Denmark. They caused 293T epithelial cells to express T-cell genes by treating these somatic cells with the membrane-damaging extracel-lular toxin streptolysin O (SLO, produced by hemolytic streptococci) to make them transiently permeable and then exposing them to a whole-cell extract of Jurkat T cells. Similarly, lysates of cardiomyocytes or insu-linoma cells elicit the expression of cardiomyocyte or beta-cell markers in adipose stem cells, and fibroblasts and extracts from pneumocytes (a cell type lining the respiratory tract) have recently been shown to induce the differentiation of ESCs into a pneumocyte phenotype. However, the most exciting aspect of Collas's work was the demonstration that 293T and NIH 3T3 cells can be programmed by extracts of undifferentiated embryonal carcinoma cells or mouse ESCs to acquire characteristics of pluripotency.

293T cells were made permeable with SLO and exposed for 1 hour to an extract of undifferentiated embryonal carcinoma cells. After sealing the holes made in the membrane by SLO (by simply adding calcium chloride to the medium), *in vitro* culture of extract-treated cells was accompa-nied by morphological, immunological, gene-expression, and functional analyses over an 8–12-week period. Within 2 weeks, for example, cells treated with embryonal carcinoma cell extracts, developed colonies that were reminiscent of embryonal carcinoma cells; however, most impor-tantly, 1 week after extract treatment, more than 60% of the 293T cells displayed intranuclear Oct4, whereas 293T cells treated only with an extract made from other 293T cells showed no Oct4 expression at all. Other reprogramming-related events were apparent in the cells treated with embryonal carcinoma cell extract. A-type nuclear lamins were grad-ually replaced by B-type lamins, which are another marker of pluripotent cells, and demethylation of the *Oct4* promoter from the 293T genome was almost complete. Significantly, there was also widespread down-regulation of the 293T-specific gene expression profile.

Similar results were obtained with a whole-cell extract from mouse ESCs, although in this case the level of Oct4 present in the treated cells showed a biphasic response. First, Oct4 was detected as early as 1 hour after recovery of the cells from the extract, and the level peaked at 24 hours. This peak was followed by a marked decrease in Oct4 by 36 hours, to a level barely detectable by 48 hours. By 72 hours, however, a second wave of Oct4 was detected that had an amplitude similar to or higher than the first wave, and it persisted for at least 5 days.

A transient induction of *Oct4* transcription and translation—independent of uptake of residual Oct4 protein from the extract—was triggered within the first hours after extract exposure. This early *Oct4* up-regulation may be explained by nuclear uptake of extract-derived transcription factors and chromatin remodelers that target the *Oct4* promoter. The transient (24–48-hour) nature of this first wave of *Oct4* activation presumably results from the depletion of regulatory factors (most transcription factors have a half-life of hours). This suggests that transcription factor synthesis and targeting are not optimally sustained during the first hours after extract treatment. The second wave of *Oct4* up-regulation, however, was sustained for several days and weeks. Long-term *Oct4* expression is consistent with DNA demethylation taking place on the promoter of the *Oct4* gene in the extract-treated cell, and the timing of this event is broadly consistent with DNA demethylation events that occur during SCNT or after the injection of nuclei into *Xenopus* oocytes. The exact timing of demethylation in this system remains to be determined. Nevertheless, if *Oct4* demethylation is required for expression of the gene, then the rapid induction of *Oct4* transcription soon after exposure to the ESC extract suggests that demethylation is triggered very rapidly.

Cell extract reprogramming has the potential to be clinically useful

One of the potential benefits of extract-based reprogramming is its apparent ability to direct the differentiation of ESCs into clinically useful cell types. For the most part, ESC-differentiation protocols can at best give the ESCs a push onto the desired differentiation pathway, but large numbers of cells still seem to ignore the signals and go off down some completely unrelated route. Attempts to replicate the signals experienced by specific cells during embryonic development have had some successes, but any protocols arising from these studies still have to rely on adding soluble factors to the culture media of ESCs, and they tend not to give very high yields of the desired cell types.

We mentioned above that extracts from pneumocytes were able to induce ESCs to differentiate into pneumocytes. These are crucial to the natural regenerative process of the alveoli after injury, and murine ESCs respond to the extracts of alveolar epithelial cells by increasing their expression of genes specific to the pneumocyte phenotype. The drawback of this technique is that the new gene-expression profile of the treated cells does not seem to be permanent. Fourteen days after treatment, the putative pneumocytes showed a decline in the expression of certain marker genes. It is possible that this is simply due to differentiation of the newly derived pneumocytes to another cell type (for example, differentiation of type II pneumocytes to type I) rather than reversion to the ESC phenotype (which would have been unlikely because they were no longer in culture conditions that would maintain their pluripotency). That notwithstanding, the use of extracts from different types of differentiated cells could be useful in defining which genes have to be up-regulated or down-regulated at specific time points during ESC differentiation so as to follow defined developmental pathways. In addition, there is little knowledge of which cell types are capable of inducing this effect. It is possible that the effect simply relies on excess concentrations of the transcription factors that define a specific cell type, as we saw in the fusion of developmentally dissimilar cell types.

13.4 REPROGRAMMING THE EPIGENOME BY INDUCED PLURIPOTENCY

The use of cell extracts to reprogram cells to pluripotency was quite attractive because it did not rely on nuclear transfer, which often seemed to make a mess of reprogramming, or on cell fusion, which produced cells with odd chromosome counts that were liable to changes every time the hybrid cell divided. However, the extract method suffered from extensive cell death (cells do not like having holes poked in their membrane by SLO) and from the fact that not all laboratories were able to reproduce the results. With sufficient research effort, these problems might have been overcome, but these procedures were pushed into the background to some extent by the development of induced pluripotency. We hinted at this technique in the section on somatic cell nuclear transfer but did not indicate the effect that this breakthrough has had on ESC biology in the past few years.

The field of ESC research was shaken in 2006 by a paper from the laboratory of Shinya Yamanaka that described the generation of induced pluripotent stem cells (iPSCs). That publication detailed the reprogramming of a mouse somatic cell genome to one resembling a pluripotent cell genome. The reprogramming procedure has been heralded as one of the most important discoveries in cell biology and may in future provide a limitless source of patient-specific stem cells. Moreover, the protocol was exquisitely simple. The retroviral transduction of just four genes (*Oct4*, *Sox2*, *Klf4*, and c-*Myc*) reprogrammed mouse embryonic fibroblasts (MEFs) and adult tail tip fibroblasts (TTFs) into cells that showed characteristics reminiscent of mouse ESCs (mESCs). The resulting cells became known as **induced pluripotent stem cells**, or **iPSCs**, and were similar to mESCs in terms of morphology, growth rate, gene expression, and the epigenetic status of many of their genes. Microarray-based transcriptome analyses confirmed their similarity to mESCs but also highlighted some differences. However, the differences were largely glossed over once iPSC pluripotency was confirmed by the formation of teratomas. The initial iPSC lines generated in Yamanaka's laboratory were unable to contribute to chimeric animals when injected into mouse blastocysts (perhaps a more robust test of pluripotency) and did not fully silence the expression of the transgenes used to induce pluripotency. Consequently, it became clear that these cells were not going to be of much use in a clinical arena, but their development was nonetheless an important step forward.

Many of the problems with the first iPSC lines were possibly due to the method of their derivation. Initially, an antibiotic resistance gene was knocked into the *Fbx15* locus, to permit selection of reprogrammed cells on their attainment of pluripotency. However, when the antibiotic resistance gene was coupled to the pluripotency factor Nanog, whose expression is thought to be tightly linked to the attainment of pluripotency, more stable iPSC clones were obtained. Importantly, these iPSCs showed contribution to chimeric animals with germ-line transmission, another key test for the pluripotency of iPSCs. However, it was also reported that up to

20% of chimeric mice generated showed the formation of tumors attributed to the reactivation of the c-*Myc* transgene, so these cells were not perfect either.

The next major breakthrough came with the generation of human iPSCs. Yamanaka's group extended their earlier work and showed that adult dermal fibroblasts could be reprogrammed by retroviral transduction of *OCT4*, *SOX2*, *KLF4*, and *C-MYC* with good efficiency, but only after the cells were transduced with a mouse retrovirus receptor. Cells with hESC-like morphology were picked and expanded, and showed all the hallmarks of pluripotency, including teratoma formation, appropriate gene expression, and the ability to differentiate into multiple tissues. James Thomson's laboratory found that a slightly different set of genes—*OCT4*, *SOX2*, *NANOG*, and *LIN28*—were sufficient to allow iPSC generation from fetal and adult fibroblasts via lentiviral, rather than retroviral, transduction; this allowed the transduction of non-dividing cells, which was not possible with the use of retroviruses. The iPSCs generated showed all the signs of a pluripotent cell type, and the data further suggested that *Lin28* was not required, but aided by increasing the frequency of reprogramming events. These two studies used similar but distinct methods, which underscores the validity of iPSC technology and puts to rest questions of reproducibility while also showing redundancy in which genes can affect the reprogramming process. Closely following these results came the generation of iPSCs from another laboratory using a similar protocol. The addition of a ROCK inhibitor, which can aid hESC growth after separation into single cells (hESCs normally grow as colonies and do not like being separated), enhanced the formation of iPSCs from hESC-derived embryonic fibroblasts; importantly, iPSCs were also generated in the absence of *C-MYC* or *KLF4*, although at a lower efficiency.

From a therapeutic aspect and also to assess the potential of a terminally differentiated cell type to be reprogrammed to pluripotency, it was important to establish that various cell types can be used for iPSC generation. It is still not fully understood whether all cells can be targets for iPSC generation, although studies have shown that not just fibroblasts can be reprogrammed. Mouse pancreatic beta cells and adult mouse liver and stomach cells were shown to be relevant targets for iPSC generation. All cell types were deemed pluripotent, and iPSCs contributed to cells of all three germ layers in chimeric embryos, although only one animal survived to birth. Interestingly, iPSCs from liver and stomach were less tumorigenic than MEF-derived iPSCs, although perinatal death was higher, and when c-*Myc* was removed, only a 20–40% reduction in efficiency was observed, in comparison with a 90% decrease in MEF reprogramming. Terminally differentiated mature B cells from an iPSC-derived chimeric mouse could only be reprogrammed after cellular "sensitization" by expression of the myeloid transcription factor C/EBP-α. C/EBP-α can disrupt the function of Pax5, a transcription factor understood to have a major role in mature B-cell development and function. This suggests that not all terminally differentiated cells could be reprogrammed using the same factors. An elevated efficiency and speed of reprogramming of human keratinocytes in comparison with fibroblasts was also observed. The advantage of using keratinocytes was further demonstrated by the generation of iPSCs from cells isolated from single plucked human hairs, thus bypassing the requirement for skin biopsies to harvest fibroblasts.

It has even been suggested that the technique used to generate iPSCs can be used to alter cell phenotype without having to revert to a pluripotent state. Specifically, pancreatic exocrine cells were converted to cells closely resembling pancreatic beta cells by the retroviral transduction of three genes, *NGN3*, *PDX1*, and *MAFA*, which are all involved in beta-cell development. Because beta cells are the insulin-secreting cells destroyed in diabetes, this study is of obvious therapeutic interest.

Epigenetic reprogramming occurs during iPSC derivation

Derivation of iPSCs is a gradual process that extends over weeks in tissue culture. It begins with the initial transduction of fibroblasts with viruses carrying three or four factors, and culminates in the isolation of iPSCs manifesting most of the features that ESCs have, albeit with very low efficiency. The low number of iPSC colonies relative to the number of transduced cells put into the culture suggests that the reprogramming process is far from universal. However, it does require DNA demethylation of the promoters of the somatic cell's own copies of *Oct4*, *Nanog*, and so on. Together with the products of the newly activated genes, the virally expressed factors might reconstitute the core elements of the ESC-specific protein interaction network and recruit other transcription factors and chromatin modifiers to induce more stable and global changes. This requirement for demethylation (and probably other epigenetic changes) means that expression of the indigenous *Oct4* is a relatively late event in iPSC derivation, and because of the slow kinetics of reprogramming in this system it is likely that such demethylation is passive in nature, in contrast to the active demethylation that seems to predominate in nuclear transfer and normal fertilization.

This presents us with a problem if we are to understand the precise mechanism used to demethylate the genome during iPSC derivation, because in all probability there is a parallel requirement to methylate the promoters of genes specific to the somatic transcriptome. This means that simple exclusion of DNMT3A or DNMT3B from the nucleus, the method used by the pre-implantation embryo to demethylate the maternal genome passively (see Chapter 11), is unlikely to contribute to passive demethylation in iPSC derivation. To repress the fibroblast phenotype and keep iPSCs from differentiation, down-regulation of genes encoding lineage specification factors is required, with *de novo* methylation at lineage-specific loci possibly representing an additional mechanism. The role of methylation in reprogramming can be evaluated in fibroblasts that are deficient in DNMT3A or DNMT3B, the enzymes catalyzing *de novo* methylation.

The slow reestablishment of the pluripotent state also implies that the process might be largely stochastic in nature: it may well rely on random events occurring within the genome of the transduced cell that happen to expose a binding site for *Oct4* or other transgenes. It seems likely that the presence of the transgenes is able to "kick-start" certain pluripotency-associated molecular mechanisms, but only in a minority of the cells. So what other factors are needed to make this particular minority special? There are several possibilities, the most likely being the following: c-*Myc* induces the expression of genes involved in proliferation and self-renewal, and because it can also mediate global histone acetylation, it may perhaps mediate chromatin decondensation to allow the functions of DNA-binding transcription factors. *Klf4* may function to down-regulate p53 transcription—which has been shown to regulate *Nanog* expression—and also to inhibit ESC differentiation. *Lin28* is involved in RNA processing, and it modulates the expression of *Let7* microRNAs in stem cells to maintain pluripotency in mESCs.

Researchers intent on understanding the molecular mechanisms behind iPSC generation have begun to make use of "secondary systems" in which iPSCs are generated by the lentiviral transduction of doxycycline-inducible transgenes and are then injected into blastocysts to form chimeric mice from which fibroblasts, or other cell types, carrying the doxycycline-inducible transgenes can be isolated. Induction of these cells with doxycycline permits the reprogramming of a homogeneous cell type, which allows the induction process to be controlled so that it can be understood more fully. Importantly, induction with doxycycline also permits the self-selection of reprogrammed cells after the removal of doxycycline, because only cells with activated endogenous pluripotency genes will grow effectively. Stage-specific embryonic antigen 1 (SSEA1) and alkaline phosphatase (AP) activities have been shown to be early markers of the reprogramming process, while endogenous *Oct4* and *Nanog* expression and telomerase activity were only found in fully reprogrammed cells. Longer transgene expression was linked to enhanced reprogramming, but continued transgene expression ablated the differentiation capacity of iPSCs.

In a similar study, iPSC generation via a secondary system (for example using lentiviral transduction and doxycycline induction) was reported to be 20–50-fold more efficient than iPSC generation via the "primary" system (that is, the original protocol that relied on retroviral transduction), and multiple cell types could be used. In more detail, the secondary reprogramming system operates by generating iPSCs carrying doxycycline-inducible versions of the reprogramming genes *OCT4*, *SOX2*, *KLF4*, and C-*MYC*. Mice are created from these iPSCs by a process called tetraploid complementation; fibroblasts can then be harvested from the new mice when they are born. Treatment of these fibroblasts with doxycycline reprograms them and produces iPSC colonies with greater efficiency than observed from the first lentiviral transduction that gave rise to the initial iPSC line. This finding suggests that iPSCs can be generated from cells with different developmental origins and epigenetic states by using a secondary system. Interestingly, a DNA methyltransferase inhibitor, 5-aza-deoxycytidine, was also shown to boost the reprogramming efficiency, whereas a histone deacetylase inhibitor, trichostatin A, did not. This suggests that DNA methylation is more important to the reprogramming process, although the effects of modulating histone methylation were not studied. The timescale of reprogramming of the secondary cells (9–13 days) was noted to be similar to that of directly reprogrammed cells (10–14 days), suggesting that direct viral integration is not an essential part of the reprogramming process and, furthermore, that reprogramming is driven by stochastic epigenetic events requiring a minimum time of transgene expression.

A key issue originating from studies of this type is that most cells that appear to initiate reprogramming fail to become iPSCs at some point during the process. These cells seem to remain in an intermediate reprogramming state, suggesting that stochastic events are required for cells to become completely reprogrammed. Such partly reprogrammed clones were studied to understand the barriers to the reprogramming process; they showed incomplete activation of pluripotency-related genes, incomplete deactivation of genes related to the original cell phenotype, and incomplete epigenetic reprogramming. Interestingly, it was also reported that inhibition of DNMT1 led to the full reprogramming of these stable, partly reprogrammed cells. It therefore seems that altering the mechanism of maintenance methylation may be necessary for the completion of iPSC reprogramming, but quite how this operates is not yet known.

A more detailed analysis of reprogramming at different stages was undertaken in mouse cells, with transcription factor binding, gene expression, and histone methylation being examined at different stages of reprogramming. c-*Myc* was observed to be very important to early reprogramming, especially for the silencing of fibroblast-specific genes, contrary to its usually perceived function as a transcriptional activator. Furthermore, the researchers proposed that the lack of Oct4, Sox2, and Klf4 co-binding in partly reprogrammed cells contributes to the lack of full reprogramming. This lack of binding was suggested to be due to the lack of expression of Nanog, which allows the targeting of reprogramming factors.

Making iPSCs safe for clinical application

Initial iPSC studies used retroviruses for gene transfer into target cells and for their ability to be silenced, but alongside the inability of retroviruses to infect non-dividing cells it was noted that silencing was not maintained in iPSCs. Constitutive lentiviral use, in which transgene silencing is poor, was superseded by inducible lentiviral methods, with the aim being full silencing of transgene expression once the pluripotent state had been attained. However, the common problem with these vector types is the possibility of integrative mutations, or reactivation of transgenes, which has been shown to lead to tumorigenesis. Studies on viral integration seem to demonstrate that there are possible "hot-spots" for viral integration, but most seem to be random.

Solutions to these problems have been the center of intense recent research, and multiple techniques have been found to enhance iPSC technology. The use of a polycistronic transcript, in which all transgenes are present in a single linear form, could theoretically reduce the number of integrations required for iPSC generation to just one, and this has been done to generate iPSCs from both MEFs and TTFs with good efficiency (0.5–1%). Most of the iPSCs showed only one integration site and full transgene silencing. A further study was able to show that human neonatal foreskin fibroblasts could be reprogrammed similarly. The iPSCs were generated with relatively low efficiency, although they did express common pluripotency markers, formed teratomas with cells representative of all three primordial germ layers, and showed some differentiation potential.

The use of adenoviral transgene transduction, which does not lead to transgene integration, was studied next. Hepatocytes, which are easily infected by adenovirus, were successfully reprogrammed with no integration, and the iPSCs generated showed endogenous expression of pluripotency-associated genes. The adeno-iPSCs also formed teratomas in nude mice, and some chimerism and germ-line transmission was observed. However, this was only observed in one cell type, the efficiency of reprogramming was extremely low (0.001–0.0001%), and 23% of the iPSC lines were tetraploid. A similar study generated iPSCs by using a plasmid-based approach, which did not require the use of a virus—although, again, this was only in one cell type and at very low efficiency.

Recent studies have solved problems associated with the techniques outlined above by the use of plasmid-based transduction of transgenes using the *piggyBac* (PB) transposition system and Cre–*lox* technology, which permits the seamless integration and subsequent excision of transgenes. The method seems to be simple (it uses plasmids and accessible transfection products), can be employed with any cell type (it is not restricted by a cell's tendency to be infected by virus), and permits accurate transgene removal subsequent to iPSC generation. In the first study, stable iPSCs were generated from MEFs and human embryonic fibroblasts (HEFs) by using doxycycline-inducible transgenes and the *piggyBac* system. Seamless removal of the transgenes from iPSCs was observed after transient expression of transposase. Further, pluripotency of the MEF-derived iPSCs was confirmed via a tetraploid embryo complementation assay. The second study used a single plasmid with a 2A-peptide-linked reprogramming cassette, flanked by *loxP* sites, to generate iPSCs in MEFs, with subsequent removal of the transgene by *Cre* expression.

The single plasmid cassette was then combined with *piggyBac* technology to induce the stable formation of iPSCs from HEFs, although removal of the plasmid was not reported. This second technique, although still initially requiring transgene integration, does remove the possibility of transgene re-expression and associated tumorigenic consequences. This enhanced technique was then used in a therapeutic context, with the generation of disease-specific transgene-free human iPSCs. Fibroblasts were reprogrammed into iPSCs, and the transgenes were subsequently excised through expression of Cre recombinase, moving the major therapeutic goals of iPSC technology significantly closer. After excision of the viral transgenes, the expression profile of these iPSCs was more similar to that of hESCs than of previously generated iPSC lines. This suggests that a small amount of "leaky" expression from transgenes is still enough to perturb the gene expression profile of such iPSCs.

Updated methods for iPSC generation use vectors that cannot integrate the reprogramming transgenes into the genome of the target cell. A typical example of this method is the commercially available CytoTune™ system from Life Technologies, which uses vectors based on replication-incompetent Sendai virus to deliver the reprogramming transgenes. Sendai virus completes its life cycle in the target cell's cytoplasm and does not enter the nucleus at any time. So, in theory, the transgene RNA should not be capable of chromosomal integration but should only be translated into functional protein in the cytoplasm. The protein transcription factors made in this way will be capable of entering the nucleus to initiate the reprogramming process, but without replication the viral vector will eventually be eliminated from the cell population by progressive dilution at each mitosis.

The aim of this chapter has been to show that despite the clear need to maintain epigenetic modification patterns to preserve cell identity and function, there are artificial means by which the epigenome can be reprogrammed. This chapter also concludes our description of the cellular processes influenced by the epigenetic control of gene expression. In the next chapter we examine how these cellular processes may influence functions of the whole organism, particularly how disease may develop when epigenetic control goes awry.

KEY CONCEPTS

- Epigenetic modification patterns established during the cellular differentiation processes needed for embryonic development are highly stable.

- There are some unusual (and unnatural) methods that may be used to reprogram the epigenetic modification pattern.

- In very special circumstances, it is possible to reprogram the epigenome back to a pluripotent state that represents the very earliest stage of development.

- These techniques used to be difficult to apply and produced cells of doubtful utility, but the development of protocols to make induced pluripotent stem cells has greatly improved our ability to make pluripotent cells.

FURTHER READING

Ang YS, Gaspar-Maia A, Lemischka IR & Bernstein E (2011) Stem cells and reprogramming: breaking the epigenetic barrier? *Trends Pharmacol Sci* **32**:394–401 (doi:10.1016/j.tips.2011.03.002).

Bian Y, Alberio R, Allegrucci C et al. (2009) Epigenetic marks in somatic chromatin are remodelled to resemble pluripotent nuclei by amphibian oocyte extracts. *Epigenetics* **4**:194–202 (doi:10.4161/epi.4.3.8787).

Brero A, Hao R, Schieker M et al. (2009) Reprogramming of active and repressive histone modifications following nuclear transfer with rabbit mesenchymal stem cells and adult fibroblasts. *Cloning Stem Cells* **11**:319–329 (doi:10.1089/clo.2008.0083).

Corry GN, Tanasijevic B, Barry ER et al. (2009) Epigenetic regulatory mechanisms during preimplantation development. *Birth Defects Res C Embryo Today* **87**:297–313 (doi:10.1002/bdrc.20165).

Dowey SN, Huang X, Chou BK et al. (2012) Generation of integration-free human induced pluripotent stem cells from postnatal blood mononuclear cells by plasmid vector expression. *Nat Protoc* **7**:2013–2021 (doi:10.1038/nprot.2012.121).

Hajkova P (2010) Epigenetic reprogramming — taking a lesson from the embryo. *Curr Opin Cell Biol* **22**:342–350 (doi:10.1016/j.ceb.2010.04.011).

Inui A (ed.) (2006) Epigenetic Risks of Cloning. Taylor & Francis.

Li M, Liu GH & Izpisua Belmonte JC (2012) Navigating the epigenetic landscape of pluripotent stem cells. *Nat Rev Mol Cell Biol* **13**:524–535 (doi:10.1038/nrm3393).

Liang G & Zhang Y (2013) Embryonic stem cell and induced pluripotent stem cell: an epigenetic perspective. *Cell Res* **23**:49–69 (doi:10.1038/cr.2012.175).

Maruotti J, Dai XP, Brochard V et al. (2010) Nuclear transfer-derived epiblast stem cells are transcriptionally and epigenetically distinguishable from their fertilized-derived counterparts. *Stem Cells* **28**:743–752 (doi:10.1002/stem.400).

Mason K, Liu Z, Aguirre-Lavin T & Beaujean N (2012) Chromatin and epigenetic modifications during early mammalian development. *Anim Reprod Sci* **134**:45–55 (doi:10.1016/j.anireprosci.2012.08.010).

Mattout A, Biran A & Meshorer E (2011) Global epigenetic changes during somatic cell reprogramming to iPS cells. *J Mol Cell Biol* **3**:341–350 (doi:10.1093/jmcb/mjr028).

Murphey P, Yamazaki Y, McMahan CA et al. (2009) Epigenetic regulation of genetic integrity is reprogrammed during cloning. *Proc Natl Acad Sci USA* **106**:4731–4735 (doi:10.1073/pnas.0900687106).

Niemann H, Carnwath JW, Herrmann D et al. (2010) DNA methylation patterns reflect epigenetic reprogramming in bovine embryos. *Cell Reprogram* **12**:33–42 (doi:10.1089/cell.2009.0063).

Okita K, Nakagawa M, Hyenjong H et al. (2008) Generation of mouse induced pluripotent stem cells without viral vectors. *Science* **322**:949–953 (doi:10.1126/science.1164270).

Ruiz S, Diep D, Gore A et al. (2012) Identification of a specific reprogramming-associated epigenetic signature in human induced pluripotent stem cells. *Proc Natl Acad Sci USA* **109**:16196–16201 (doi:10.1073/pnas.1202352109).

Sawai K, Takahashi M, Moriyasu S et al. (2010) Changes in the DNA methylation status of bovine embryos from the blastocyst to elongated stage derived from somatic cell nuclear transfer. *Cell Reprogram* **12**:15–22 (doi:10.1089/clo.2009.0039).

Sridharan R & Plath K (2008) Illuminating the black box of reprogramming. *Cell Stem Cell* **2**:295–297 (doi:10.1016/j.stem.2008.03.015).

Sutovsky P (ed.) (2007) Somatic Cell Nuclear Transfer. Springer.

Takahashi K, Okita K, Nakagawa M & Yamanaka S (2007) Induction of pluripotent stem cells from fibroblast cultures. *Nat Protoc* **2**:3081–3089 (doi:10.1038/nprot.2007.418).

Takahashi K, Tanabe K, Ohnuki M et al. (2007) Induction of pluripotent stem cells from adult human fibroblasts by defined factors. *Cell* **131**:861–872 (doi:10.1016/j.cell.2007.11.019).

Tobin SC & Kim K (2012) Generating pluripotent stem cells: differential epigenetic changes during cellular reprogramming. *FEBS Lett* **586**:2874–2881 (doi:10.1016/j.febslet.2012.07.024).

Wang K, Chen Y, Chang EA et al. (2009) Dynamic epigenetic regulation of the Oct4 and Nanog regulatory regions during neural differentiation in rhesus nuclear transfer embryonic stem cells. *Cloning Stem Cells* **11**:483–496 (doi:10.1089/clo.2009.0019).

Yamanaka S (2008) Induction of pluripotent stem cells from mouse fibroblasts by four transcription factors. *Cell Prolif* **41**(Suppl 1):51–56 (doi:10.1111/j.1365-2184.2008.00493.x).

THE EPIGENETIC BASIS OF DISEASE

SECTION 3

14

EPIGENETIC PREDISPOSITION TO DISEASE AND IMPRINTING-BASED DISORDERS

From the preceding chapters we can appreciate that epigenetic control of gene expression is a versatile mechanism by which the information content of the genome can be used in a selective manner to define cellular phenotypes and respond to the environmental influences that cells experience during their lives. Similarly, by now it should also be apparent that perturbations in this mechanism could lead to changes in cell function that, if they are of sufficient magnitude or persist over a long enough timeframe, could alter the function of whole organ systems in a manner that we would recognize as a diseased state. The basis of many complex diseases is a form of such internal dysregulation.

The word "disease" covers many problems that afflict the body, but in general we use it to describe conditions that cause disruption of the body's normal functions. Illness and sickness are often synonymous terms, although the latter can often be used to describe the temporary means employed by the body in an attempt to rid itself of more serious dysfunction-causing agents. Examples of this could be the vomiting induced by the ingestion of noxious substances, or inflammation in response to infection by certain pathogenic organisms. These are not considered to be diseases in their own right, although they do cause some departure of the body from its apparently normal functions. In any case, the timescale of these processes is probably too short to influence all but the most transitory of epigenetic changes, so we will not consider them further in this chapter.

Diseases may be considered under the general heading of infectious or non-infectious, meaning that they are caused by the presence of pathogenic organisms in the body or by other factors, respectively. This allows us to distinguish those problems that are induced by the effects of some microorganisms on the body's organ systems from those that are caused by internal problems such as changes in the expression levels of certain genes or the loss of specific types of cells. The distinctions are not perfect and there are often overlaps between infectious and systemic diseases, the most obvious examples of this being infection by specific viruses inducing some cells to become cancerous.

This leads us to a useful distinction that may help us to understand how epigenetics might contribute to the multiplicity of ways in which bodily dysfunctions can arise, which is attack from outside versus attack from within. Most of the diseases that our ancestors had to deal with (or, for that matter, frequently died from) were bacterial, viral, or the result of some other form of opportunistic pathogen colonizing the body. The modes of transmission of the pathogens, such as person to person contact, airborne aerosols, contamination of food or water, or the bites of other animals (such as dogs or fleas) were merely mechanical matters, and the basic problems arose because of one type of organism invading another. Our immune systems are actually remarkably good at fending off this onslaught on a daily basis, and the added benefits of the medical and pharmaceutical advances of the past 150 years mean that external attack is no longer the main scourge of humankind.

As a species we lead longer lives than we have ever done, and this is attributable mostly to the public health advances of the nineteenth and twentieth centuries. Cleaner water, vaccinations, and antibiotics have made the probability of death from bacterial infection rather remote and something that generally only happens to the very old or persons with compromised immune function. However, not to be outdone so easily, nature has presented us with other difficulties—or, more likely, such difficulties have come to the fore because the impact of the older, "external attack" diseases has been reduced. Our main problems now are diseases that seem to arise from within ourselves. Conditions such as cardiovascular illness, cancer, diabetes, immune dysfunctions, and general alteration or decrease in the functioning of our organs as we age are the prevalent killers of our times. They are not new diseases. In all probability the cellular and/or biochemical problems that cause them have always been with us and they will have always contributed to the death of some of the individuals in a population, but these are generally diseases of the old and they would have been less apparent in a population that tended to die at a younger age.

Diseases such as those of the cardiovascular system are referred to here as internal attack problems because although there are several well-defined lifestyle options that can predispose someone to suffer from a heart attack, having high blood pressure, or developing cancer, these are only factors that swing the balance of probability in favor of the disease; they do not guarantee that you will get it. For example, the link between smoking and lung cancer is probably about as strong and well-characterized as it is going to get with our current levels of understanding, but there are plenty of individuals who, despite a lifetime of tobacco use, resolutely refuse to develop the disease. Conversely, many nonsmokers develop lung cancer, although in much lower numbers than smokers. These predisposing factors only become apparent at the level of whole populations, and the real issue seems to be the varying ways in which individual humans cope with the environmental toxins and other substances to which they are exposed throughout their lives.

14.1 PREDISPOSITION TO DISEASE

Life-course epidemiology seeks to explain disease

Understanding the factors that can lead to diseases is of interest to epidemiologists, who attempt to determine precise relationships between biological agents or other environmental exposures, intentional or otherwise, and different types of diseases. Epidemiology enjoyed some considerable successes in the nineteenth and twentieth centuries, such as linking observations of a greater number of cholera cases to particular water supplies in London in the 1850s, but it has not been so successful in predicting who will develop the more complex "internal attack" diseases of the present day.

Ischemic heart disease is a good example of this failing. Despite the identification of several risk factors such as obesity, smoking, high concentrations of high-density lipoprotein (HDL) cholesterol in the blood, and hypertension, it has proven difficult to provide an accurate individual risk beyond noting that there is a 1 in 6 chance that a seemingly healthy male will develop the disease within 5 years—even if that person is doing all the wrong things such as eating a high-fat diet and not exercising. The prediction rates for the opposite scenario, in which the individual is aware of risk factors and attempts to do something about them, are not much better. On a population level, regular exercise probably does reduce the risk of heart disease, but this is still subject to an enormous level of personal variation; a poignant example that underlines this point is the untimely death of James F. Fixx, widely hailed as the instigator of "jogging," who died of a massive heart attack at an age of only 52, despite his long-term claim that active people live longer. Two of his coronary arteries were sufficiently blocked that they would have warranted a bypass operation!

The common response to cases of this type is that the cause of the disease must have been genetic. The apparent failure of epidemiology to predict the risk of developing diseases has led to some impressive recent attempts to correlate variations in the genomes of individuals with the disease they suffer, but these studies have produced mixed results when applied to risk prediction. Most of the so-called **genome-wide association studies (GWAS)** have attempted to link individual variations in the genome—such as single nucleotide polymorphisms (SNPs), deletions, insertions, and variations in copy number—to disease; this approach followed from earlier genetic linkage examinations. One of the great advantages of GWAS is that enormous amounts of data can be generated, particularly since the advent of biobanks, which contain tissue or DNA samples from very large numbers of people. However, apart from the clear identification of small numbers of SNPs with conditions such as age-related macular degeneration (a form of damage to the retina leading to vision loss in the elderly) and the linking of a few genes to rheumatoid arthritis and hypertension, the success of GWAS has been limited. The main problem with this approach seems to be that most of the SNP variations are linked to only a very small increase in disease probability and are therefore of minimal predictive value.

So what else could cause the variations in the risk of developing disease? Epidemiologists have frequently studied siblings to see whether the "environmental" factors they are exposed to affect their health, and have termed these exposures the "shared environment." These can be influences such as diet or the age of the mother, but studies in animals such as guinea-pigs suggest that these have a minor effect on the phenotypes of offspring. As early as 1920, breeding experiments in which tangible environmental influences—such as maternal health (insofar as this could be defined), weather, or type of feed—were identical, guinea-pig littermates still showed variations in coat patterning that could not be explained by genetic inheritance alone. This implies that the "shared environment" had little influence. Additional studies in which a single pre-implantation animal embryo was split into two separate embryos that were subsequently allowed to develop and raised as separate individuals showed considerable phenotypic variation between the genetically identical offspring. This parallels many studies on human monozygotic (identical) twins, some of whom go on to develop disease whereas their siblings do not, despite their isogenic nature and an exposure to an apparently similar set of environmental factors during growth. These and numerous other examples demonstrate a significant contribution to phenotypic variation from seemingly random or stochastic events.

Epigenetics may be the basis of stochastic variation in disease

The deviations from genetic traits observed in experimental animals do seem to be mostly independent from one another, suggesting they may be due to non-stable characteristics; that is, something that is present in one animal (or even a single cell) but not another. The best candidate for this type of random variation is epigenetic modification of the genome, because this is a principal method of controlling the flow of information in and out of the genome. Epigenetic control of gene expression can establish two seemingly independent transcriptomes (the set of RNAs in a cell) from two identical genomes.

So how does epigenetics fit into our current understanding of disease? The answer to this question is far from simple, and a major part of our current inability to provide clear reasons why epigenetics may or may not cause a specific disease is probably due to our lack of understanding of the contribution of epigenetic control. However, there is another, more subtle, problem, namely that epigenetic alterations do not affect all cells similarly or permanently. Nonetheless, one of the principal reasons why epigenetic control of gene expression might contribute to disease is that even though DNA methylation and histone modification patterns are reasonably stable, they are a lot more plastic than the underlying sequence information of DNA itself. It is not difficult to imagine that time or environmental influences could cause these patterns to change, even in quite subtle ways that might change cell function. The acquisition and accumulation of DNA mutations may not be as rapid.

One of the problems in trying to link epigenetics to disease states arises from the likelihood that individual cells probably experience the random alterations to epigenetic modification patterns in different ways—that is to say, one cell may readily acquire a damaging epimutation but its neighbors may escape this. Worse still, the epimutation may be reversed at some later stage while the neighboring cells may acquire it, or even some other type of epimutations may replace previous types. In short, it is extremely difficult to predict how changes in the epigenetic modification patterns of individual cells will contribute to the onset of disease. The same problem exists as we observed for GWAS, so that at best the outcomes of epigenetic change at a population level are probabilistic rather than predetermined, and some epigenetic events that predispose an individual to disease may simply be viewed as bad luck. The key advantage of applying our increasing knowledge of the epigenetic control of gene expression is that it allows us to understand the basic mechanisms of disease, and although the random nature of the events may not help us to predict which individuals are more likely to suffer from specific conditions, the more plastic (that is, reversible) nature of epigenetic modification may allow us to develop treatment strategies that will correct the epimutation and perhaps influence the outcome of the disease.

In brief, the involvement of epigenetics in disease is not clearly understood at present, despite increasing evidence for its contribution to a wide variety of conditions. In the next section we look at various health problems, beginning with those for which the epigenetic contribution has been investigated and established in some detail.

14.2 IMPRINTING-BASED DISORDERS

Among the diseases with a known epigenetic component are imprinting-based disorders. We introduced the concept of gene imprinting in Chapter 11, but we need to reiterate this idea before we can understand how imprinting aberrations can lead to diseases. Genomic imprinting is regulated by DNA methylation and causes genes to be expressed from only one of the parental alleles inherited by the offspring. This also leads to exclusivity of function between the genes inherited from the male and female parents, especially during embryonic development. This was demonstrated by a number of experiments performed in the early 1980s in which the male pronucleus of a newly fertilized mouse embryo was replaced with the female pronucleus of a different embryo. After implantation into pseudopregnant female mice, these bimaternal or parthenogenetic embryos were unable to develop beyond the mid-gestation stage. A similar effect was observed for bipaternal or androgenetic embryos (two male pronuclei together), suggesting that the genomes from male and female gametes are distinguished in some manner that represses certain genes that are required for successful embryonic development. These genes can only be expressed by a gamete genome of the opposite sex, creating an absolute requirement for the genomes of both gametes to generate viable offspring. The mechanism that distinguishes these genes is epigenetic, as we have seen in Chapter 11.

The reasons for the failure of parthenogenetic and androgenetic embryos can help us to understand the functions of maternally and paternally imprinted genes. The general problem with parthenogenetic embryos seems to be an inability to develop extra-embryonic tissues (that is, parts of the placenta) required to support growth of the developing organism. Androgenetic embryos suffer from the opposite problem: they can generate the extra-embryonic tissues quite well, but embryonic structures (that is, parts of the body, such as internal organs) are underdeveloped, leading to early lethality. It is possible to isolate cells from parthenogenetic or androgenetic embryos and transplant them into embryos from normal fertilizations; in these chimeras, the transplanted cells show a reduced capacity to contribute to the formation of mesodermal and ectodermal structures (mainly those of the central nervous system). These developmental defects appear to originate exclusively from over- or underexpression of the relevant imprinted genes subject to epigenetic control.

This latter point has been demonstrated in a recent, elegant experiment in which somatic cells taken from parthenogenetic embryos were subjected to epigenetic reprogramming by the ectopic expression of four transcription factors (Oct4, Sox2, Klf4, and c-Myc) to generate induced pluripotent stem cells (iPSCs). After injection of these iPSCs into mouse blastocysts and subsequent implantation of the latter into pseudopregnant females, live offspring were born in which the iPSCs had contributed to all the structures of the complete organism, including those to which parthenogenetic cells do not normally contribute. This observation confirms that the inappropriate imprinting observed in parthenogenetic and androgenetic embryos is the sole barrier to the development of such embryos and, being epigenetic in nature, is therefore reversible by an effective reprogramming strategy. Further experiments in which the expression level of imprinted genes was manipulated by other methods confirm these observations.

Imprinting disorders can persist beyond embryogenesis

Recall from Chapter 11 that imprinting control regions (ICRs) regulate which of a pair of alleles is expressed. To recap in brief, imprinted genes tend to be found in clusters that have differentially methylated regions that constitute the ICRs. The ICR of a cluster can influence the three-dimensional structure of the cluster in a manner that makes individual genes residing there either capable or incapable of expression. In imprinting disorders, something goes awry with this process, and expression of both alleles becomes one of a number of possible consequences.

Expression from both alleles of many imprinted genes provides an excess of gene product that is often lethal to the developing embryo. However, this is not always the case, and survival of individuals carrying mutations that affect the imprinting of genes involved in a variety of diseases has been recorded. Many of these conditions are heritable and arise through errors accumulated during meiosis, such as the loss or addition of certain chromosomes. For example, if an extra chromosome carrying paternally imprinted genes is acquired by a female gamete, an embryo derived from fertilization of this gamete may be unable to repress the imprinted gene correctly, which could give rise to developmental problems. Alternatively, if the level of gene product is less than double, the embryo may survive but be diseased throughout its life. The range of epigenetic imprinting-related disorders and their impact on bodily functions is quite broad, as is evident from the description of several of these diseases below.

Prader–Willi and Angelman syndromes result from disruptions on chromosome 15

Prader–Willi syndrome and the related condition known as Angelman syndrome are caused by deletions of DNA sequence or inappropriate imprinting of a region of chromosome 15, specifically 15q11–q13 (**Figure 14.1**). This genomic region contains approximately 8 Mb of DNA sequence covering 100 genes in total and three breakpoints at which chromosomal damage may occur. The region contains a large cluster of imprinted genes and also a non-imprinted domain that exhibits biallelic expression of the genes encoded there.

Prader–Willi syndrome can arise after chromosomal deletions of three types. The type 1 deletion is approximately 6.6 Mb in size and represents a loss of DNA sequence between breakpoint 1 near the centromere and breakpoint 3; the slightly smaller type 2 deletion of 5.3 Mb occurs between breakpoints 2 and 3. Both of these deletions effectively eliminate most of the 15q11–q13 region. A smaller deletion between breakpoints 1 and 2 may also occur, resulting in the loss of the genes *GCP5*, *CYFIP1*, *NIPA1*, and *NIPA2*, although this region is not imprinted. Such deletions

Figure 14.1 Ideogram of chromosome 15, showing the order of protein-coding and noncoding genes and transcripts in the 15q11–q13 region. The locations of breakpoints (BP1 to BP3) for the typical type 1 and type 2 deletions involved in Prader–Willi and Angelman syndromes are shown. Also shown are maternally expressed genes/transcripts (red triangles, Angelman syndrome), paternally expressed genes/transcripts (blue rectangles, Prader–Willi syndrome), genes expressed on both chromosomes (green diamonds), genes with paternally biased expression (purple diamonds), genes with unconfirmed expression status (orange oval), imprinting control region (ICR), and small nucleolar RNAs (snoRNAs). [From Bittel DC & Butler MG (2005) *Expert. Rev. Mol. Med.* 7, 1. With permission from Cambridge University Press.]

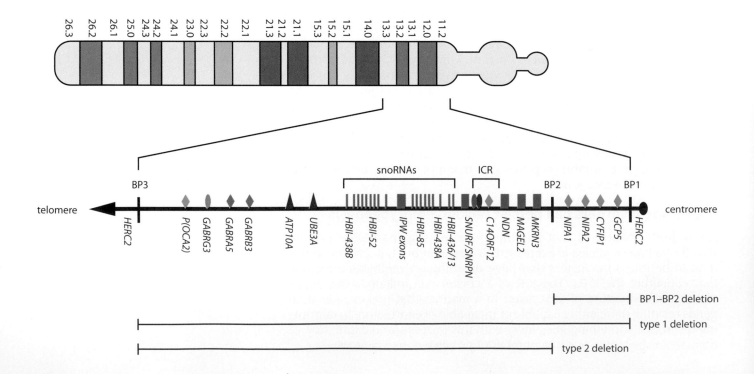

can result from an interstitial microdeletion of the paternally inherited 15q11–15q13 region (65–70% of cases of Prader–Willi syndrome).

Nondisjunction events that cause maternal uniparental disomy can also occur, resulting in oocytes with two maternal copies of chromosome 15. After fertilization, oocytes with this aberration would normally produce embryos with three copies of chromosome 15, which is usually lethal for the embryo. However, other events can occur in the early embryo that eliminate the sole chromosome 15 derived from the father, which leaves the two copies of maternal chromosome 15. This permits survival of the embryo but results in Prader–Willi syndrome. This problem accounts for approximately 20–30% of cases. The remaining cases generally result from imprinting problems, although the number of cases presented seems to be increasing in births resulting from artificial reproductive technologies such as *in vitro* fertilization.

Parental imprinting is present in 15q11–15q13 but is not universal, and we can identify four functional subsections of this region: (1) the section between breakpoints 1 and 2, which is not imprinted; (2) the Prader–Willi region, containing the protein-coding genes *MKRN3*, *MAGEL2*, *NDN*, and *SNURF/SNRPN*, which are expressed only from the paternal allele, and five small nucleolar organizer RNA genes (*HBII-436*, *HBII-13*, *HBII-438*, *HBII-85*, and *HBII-52*); (3) the Angelman syndrome region, containing the protein-coding genes *UBE3A* and *APT10A*, which are expressed only from the maternal allele; and (4) a non-imprinted region containing the genes *GABRB3*, *GABRA5*, *GABRG3*, *P(OCA2)*, and *HERC2*.

Several of the genes present in 15q11–15q13 affect brain development and function. For example, the function of *SNURF/SNRPN* is complex, but our current understanding indicates that exons 4–10 of this gene encode a protein that is involved in neuron-specific mRNA splicing. This gene is therefore likely to contribute to brain development and processes, although the precise nature of these is not clear. *NDN* is involved in the outgrowth of neuronal axons, particularly in the hypothalamus, and the gene *MAGEL2* is expressed in several regions of the brain, including the hypothalamus. This gene also seems to contribute to the measurement of circadian rhythms, the general structure of the brain, and fertility development. *MKRN3* is expressed in the central nervous system, but its function is not well understood. Despite this paucity of information, it seems likely that deletion or misexpression of these genes could result in poorly regulated neurological development and/or dysfunction of the adult brain in Prader–Willi syndrome.

Although genomic deletions are a major cause of Prader–Willi syndrome, the misexpression of the paternally imprinted genes in region 2 that results from a loss of paternal imprinting can generate similar, if not identical, symptoms to those of deletion events. Similarly, Angelman syndrome is caused by the loss of maternal imprinting in region 3. That both syndromes arise from the same 8 Mb stretch of DNA is a reflection of the unusual nature of the 15q11–15q13 region, namely that it has one ICR situated between the genes *NDN* and *SNURF/SNPRN*, as shown in Figure 14.1.

Most cases of Prader–Willi syndrome resulting from imprinting defects demonstrate a maternal-only DNA methylation pattern, despite the presence of both paternal and maternal alleles. These imprinting defects are thought to be random errors occurring during the process that installs the imprint pattern during spermatogenesis. A very small number of cases of Prader–Willi syndrome (15% of those caused by imprinting defects) arise from a microdeletion (from 7.5 to more than 100 kb) in the Prader–Willi ICR located at the 5′ end of the *SNRPN* gene and promoter. Approximately half of these are inherited from a non-diseased father (which he would be, because patients with Prader–Willi syndrome are sterile!) with a deletion in the imprinting control center of his maternally inherited chromosome 15. The other half arise from *de novo* deletions on the paternally inherited chromosome 15 that occurred during spermatogenesis or after fertilization.

The clinical manifestation of Prader–Willi syndrome is a complex neurodevelopmental disorder characterized by hypogonadism, feeding difficulties, short stature due to growth hormone deficiencies, and excessive appetite and food consumption that leads to early childhood obesity and an increased probability of developing type 2 diabetes. Prader–Willi syndrome is a relatively common condition by the standards of genetic diseases, with approximately 350,000–400,000 cases worldwide. Affected children have slightly lower than average birth weight, suggesting prenatal developmental problems, but Prader–Willi syndrome becomes evident after birth when the infant exhibits poor muscle tone (hypotonia) and an overall "floppy" appearance. The child frequently has a reduced or absent ability to suck and must be fed by artificial gavage; however, by 18 months of age, the feeding difficulties are replaced by insatiable appetite, which must be carefully controlled to avoid obesity. By adolescence, patients with uncontrolled Prader–Willi syndrome can weigh almost 200 kg. Such morbid obesity is coupled to a poor metabolic rate and generally reduced activity levels; patients with Prader–Willi syndrome tend to move around very little.

Angelman syndrome presents many clinical features that are similar to those of Prader–Willi syndrome. Typically, the affected individuals suffer from delayed development, speech impediments, reduced intellectual capability, epilepsy, abnormal brain activity (detected by an electroencephalogram), abnormal protrusion of the tongue, and hyperactivity. Once again, this disease is not uncommon in relation to other genetic diseases and affects 1 in 12,000 live births. As with Prader–Willi syndrome, most cases of Angelman syndrome result from deletions of parts of the 15q11–15q13 region of chromosome 15, such as the genes *UBE3A* and *APT10A*, or mutations of the coding sequences of these genes. Seventy-five percent of all cases result from deletion of approximately 6 Mb of the Angelman syndrome region, and 15% of cases arise after mutations in the coding sequence of the gene *UBE3A*.

UBE3A encodes an E3 ubiquitin ligase that has an important role in the function of neurons in the central nervous system. This gene is subject to paternal imprinting and is expressed only from the maternal allele in a neuron-specific manner. Selective mutation of the maternal allele accounts for most of the cases arising from *UBE3A* mutation, indicating that loss of *UBE3A* function is the principal cause of most Angelman syndrome phenotypes. Microdeletions that occur in the Angelman syndrome imprinting control center can disrupt the normal maternal expression of the *UBE3A* gene (2.4% of cases), whereas paternal uniparental disomy can result in the inheritance of two copies of the epigenetically silenced paternal allele (7.5% of cases).

Our understanding of the molecular and cellular defects leading to Angelman syndrome is considerably greater than for Prader–Willi syndrome. Mouse models of the disease have been created by deleting segments of the gene *Ube3a*, resulting in ataxia, epilepsy, and motor coordination problems. Studies of these mice suggest that many of the problems arise because the morphology of specific neurons is abnormal. The protein encoded by *Ube3a* localizes to the synapses, where it is thought to contribute to synaptogenesis and remodeling of the synapses to permit connections with other neurons (**Figure 14.2**).

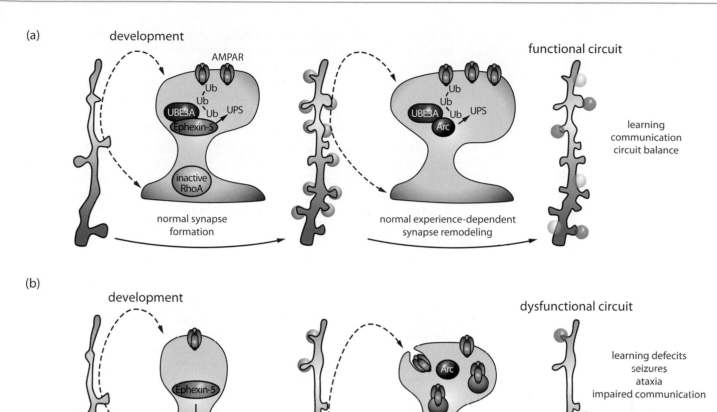

Figure 14.2 Schematic model illustrating the potential contribution of UBE3A to neuronal morphology and developing neural circuits. (a) Formation and experience-dependent remodeling of synapses. During synaptogenesis UBE3A ubiquitinates (Ub) and promotes the degradation of the RhoA guanine nucleotide exchange factor Ephexin-5 by ubiquitin proteasome system (UPS), leading to inactivation of RhoA and facilitating the formation of dendritic spines (highlighted in blue circles). UBE3A also ubiquitinates and promotes the degradation of Arc, encoded by an immediate-early gene, which facilitates experience-dependent remodeling of preexisting synapses by mediating the endocytosis of α-amino-3-hydroxy-5-methyl-4-isoxazole propionic acid receptor (AMPAR). This remodeling allows functional neural circuits to arise during development. Dendritic spines targeted for elimination (highlighted in red circles) and sites for the growth of new spines (highlighted in yellow circles) are shown. (b) Defects in UBE3A result in inhibition of formation and deficient experience-dependent remodeling of synapses. UBE3A deficiency leads to an accumulation of Ephexin-5 and Arc. Increased Ephexin-5 levels lead to enhanced active RhoA levels, which results in deficits in excitatory synapse formation. Inappropriately high accumulation of Arc leads to excessive endocytosis of AMPARs from glutamatergic synaptic sites and reduces excitatory synaptic transmission. This also increases the number of silent (AMPAR-lacking) synapses, which may subsequently be eliminated during experience-dependent synapse remodeling. The resulting synaptic and circuit dysfunction may underlie various Angelman syndrome phenotypes, including learning deficits, ataxia, seizures, and impaired social and communication skills. [From Mabb AM, Judson MC, Zylka MJ & Philpot BD (2011) *Trends Neurosci* 34, 293. With permission from Elsevier.]

The axons of most neurons in the central nervous system possess small membranous protrusions called dendritic spines, which increase the numbers of possible communication points between individual neurons. In mice with *Ube3a* defects, these spines seem to be misshapen and unable to alter their connections with other cells in response to external signals. Moreover, even synapses that appear morphologically normal may not be capable of transmitting signals. Examination of human post-mortem brain tissue indicates that similar defects may be the basis of the human Angelman syndrome.

UBE3A, the protein expressed by *UBE3A*, performs its function of marking other proteins for proteasomal degradation in most of the body's cells, so why should Angelman syndrome be a disease resulting largely, if not exclusively, from dysfunctions occurring in the brain? We have noted that expression occurs from the maternal allele only in the neurons of the central nervous system, but we still do not understand clearly why imprinted *UBE3A* should be restricted only to this organ system.

As with other genes, the imprinting of *UBE3A* depends on the establishment of DNA methylation that seems to be regulated by an ICR. The ICR for *UBE3A* lies upstream of the gene, in close proximity to the Prader–Willi syndrome control region (**Figure 14.3a**). However, there is more than one theory about how this region can control gene expression. Allele-specific DNA methylation of *UBE3A* is incomplete, and some of the 5′ sections are not methylated. This implies that parts of the gene may be capable of transcription even from the normally silent paternal allele (Figure 14.3b); however, these are prevented from actual translation into the gene product by a more abundant antisense transcript arising from the opposite strand of DNA to that used to make the normal mRNA. This so-called **antisense repression method** works because the RNA generated from the antisense DNA strand has the same sequence as the DNA in the sense strand, so it can outcompete the sense-strand DNA template for binding to RNA polymerase II and thereby prevent the synthesis of mRNA (Figure 14.3c). This method is referred to as the collision repression model, but it relies on transcription occurring in only one direction at any one time. This may not always happen, so an alternative repression model has been suggested, the **RNA–DNA interaction model**, which relies on interaction of the antisense RNA with the double-stranded DNA at the *UBE3A* locus to induce the deposition of histone modifications that alter the chromatin architecture and cause transcriptional elongation of the *UBE3A* mRNA to abort at these regions.

Beckwith–Wiedemann and Silver–Russell syndromes are consequences of disruptions of the *IGF-H19* locus

Beckwith–Wiedemann syndrome (BWS) causes children to overgrow and makes them more likely to develop tumors. It is quite rare, with only 1 in every 13,700 births being affected, but because accurate diagnosis can be difficult it is possible that this figure underestimates the true number of affected patients. The problems associated with this disease begin *in utero* when increased fetal growth rates become evident during the second half of pregnancy. This accelerated growth continues for the first few years of life, with the result that patients with BWS show a height and weight around the 97th centile for humans; however, the head sizes are generally disproportionate and are around the 50th centile for normal human heads. Quite often, patients have abnormally large and protruding tongues, making it difficult for them to eat. There are many other developmental defects associated with this condition, such as abdominal wall defects (for example hernia) and inappropriate organ sizes (particularly of the spleen, kidneys, and pancreas). The major problem is, however, cancer. There is a greatly increased probability of developing malignancies in the first 10 years of life, with the most common types being associated with the kidney and liver, such as Wilms tumor and hepatoblastoma, respectively. Other cancers typically associated with BWS are neuroblastoma and rhabdomyosarcoma (a tumor of striated muscle), although the list of tumor types is much more extensive than these examples.

As with Prader–Willi and Angelman syndromes, BWS can result from both genetic and epigenetic aberrations. The affected locus for BWS is found at chromosome 11p15.5, as shown in **Figure 14.4**. The 11p15.5 region

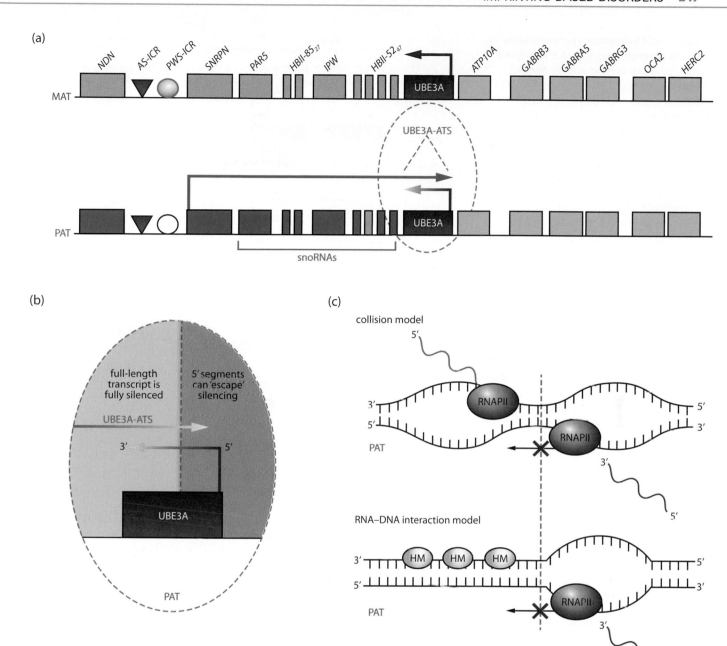

Figure 14.3 Possible mechanisms for UBE3A imprinting in the brain. (a) Map of the maternal (MAT) and paternal (PAT) human chromosome region 15q11–q13. Maternally expressed genes (lilac), paternally expressed genes (blue), non-imprinted genes (green), Prader–Willi syndrome (PWS) imprinting control region (PWS-ICR, circles), and Angelman syndrome (AS) imprinting control region (AS-ICR, triangles) are shown. Methylation (blue circle) at the maternal PWS-ICR globally represses expression of the surrounding genes (purple), including the UBE3A antisense (UBE3A-ATS) transcript; however, the maternal copy of UBE3A is expressed. Differential methylation (open circle) at the paternal PWS-ICR permits paternal gene expression, including the UBE3A-ATS transcript (blue arrow). UBE3A-ATS overlaps the paternal UBE3A locus, resulting in transcriptional silencing of UBE3A (red arrow fading to white). (b) The cutoff (dashed blue line) beyond which the silencing of UBE3A transcription by the UBE3A-ATS is incomplete: to the left of the line the antisense transcript competes with the sense transcript, resulting in silencing of full-length UBE3A sense transcripts; to the right, truncated paternal 5′ segments of the UBE3A sense transcript are produced. (c) Two hypothetical mechanisms of UBE3A-ATS/sense competition at the paternal allele: the collision model, in which transcription can only occur in one direction at a time and RNA polymerases (RNAPII) transcribing the sense strand (red) are competed off their template by oncoming complexes engaged in transcription of the antisense strand (blue), and the RNA–DNA interaction model, in which production of UBE3A-ATS induces histone modifications (HM) that modify chromatin structure along the UBE3A locus with transcriptional elongation of UBE3A prematurely aborted at these regions, resulting in truncated UBE3A sense transcripts. [From Mabb AM, Judson MC, Zylka MJ & Philpot BD (2011) *Trends Neurosci.* 34, 293. With permission from Elsevier.]

Figure 14.4 Schematic representation of the chromosome 11p15.5 imprinted region, which is functionally divided into two domains. Domain 1 contains two imprinted genes, *H19* and *IGF2*, and the *H19*-associated imprinting control region (ICR1). Domain 2 contains several imprinted genes including *KCNQ1*, *KCNQ1OT1*, and *CDKN1C* and the imprinting control region (ICR2) containing the *KCNQ1OT1* promoter. (a) ICR1 is usually methylated on the paternal chromosome and unmethylated on the maternal, with *H19* expressed from the maternal allele and *IGF2* expressed from the paternal allele. ICR2 is usually methylated on the maternal chromosome, and paternal expression of the *KCNQ1OT1* promoter regulates *in cis* the expression of the maternally imprinted genes in domain 2. KvDMR/ICR2 is the differentially methylated imprint control region. Two examples of imprinting alterations leading to Beckwith–Wiedemann syndrome (BWS) are (b) loss of methylation at the maternal ICR2 leading to reduced expression of *CDKN1C*, and (c) gain of methylation at the maternal ICR1, leading to biallelic expression of *IGF2*. Preferential expression of maternal (red) and paternal (blue) alleles, methylation (blue circles) at ICRs, and expressed (filled rectangles) and nonexpressed (open rectangles) genes are indicated. [From Weksberg R, Shuman C & Beckwith JB (2010) *Eur. J. Hum. Genet.* 18, 8. With permission from Macmillan Publishers Ltd.]

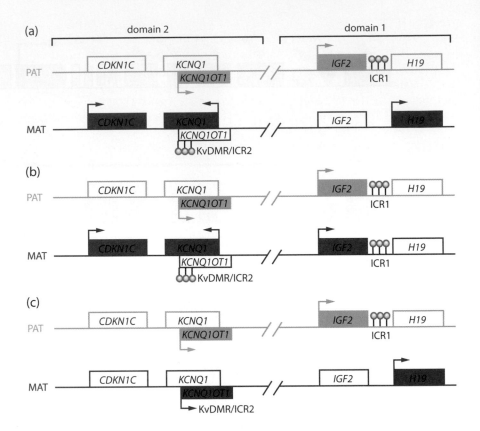

has two ICRs, one of which is located between the *IGF2* and *H19* coding sequences (ICR1) and a second positioned within the gene *KCNQ1* (ICR2). These differentially methylated regions are subject to sporadic changes; for example, 50% of patients with BWS show a loss of DNA methylation at ICR2, which reduces the expression levels of *CDKN1C*, whereas 5% of cases arise after a gain of methylation at ICR1, which causes biallelic expression of *IGF2*.

Silver–Russell syndrome has the opposite effect, namely intrauterine growth retardation, and is a much rarer condition than the other imprinting diseases described so far. Only 400 cases have been identified worldwide since the condition was first recognized in 1953. Patients typically present with intrauterine growth retardation, difficulty in feeding, failure to thrive, or postnatal growth retardation, suggesting some involvement of growth hormone deficiency.

The genetic causes of Silver–Russell syndrome are complex, and our understanding of these mechanisms is still far from complete. As one might expect, the disorder is associated with the abnormal regulation of growth control genes located on chromosomes 11 and 7, but there does seem to be significant involvement of loss of DNA methylation at the *IGF2–H19* region. Maternal uniparental disomy 7 (two copies of maternal chromosome 7) is also found in approximately 7% of cases.

Assisted reproductive technologies may increase the incidence of imprinting diseases

The number of diseases with a known imprinting problem is relatively small, and we have only attempted to describe the better-known examples. All of these conditions are still rare by the standards of other illnesses such as cardiovascular problems and cancers, but a disturbing trend in recent years has been the apparently higher incidence of BWS in children born after assisted reproduction (or *in vitro* fertilization; IVF). (The incidence of BWS in children resulting from IVF is 1 in 2500 as opposed to 1 in 13,700 in natural births.) It has been suggested that the culture media used to manipulate and support embryos outside the body may be sub-optimal and thus be responsible for the changes in DNA methylation in the imprinting centers associated with BWS. There have been further suggestions that other imprinting diseases such as Prader–Willi syndrome show increased incidence in children resulting from IVF, but the current data only show statistical significance for BWS.

14.3 EPIGENETICS OF MAJOR DISEASE GROUPS

Diseases arising from defective imprinting are quite unusual and affect only small numbers of patients; however, we have a considerable understanding of the impact of epigenetic mechanisms on their onset and progression. The pathology of more significant diseases such as cardiovascular problems, kidney dysfunction, and diabetes may also have epigenetic components, but these are not as well defined as those of imprinting diseases. Many of the ways in which these serious conditions develop have some common mechanistic threads; for example, hypertension can predispose to cardiovascular disease by favoring the formation of atherosclerotic plaques. Diseases of the kidney can lead to hypertension, whereas diabetes can affect kidney function to the extent of inducing nephropathy and leading to more atherosclerosis. In effect, all these conditions are interlinked, which complicates the identification of epigenetic mechanisms that may be unique to each of these diseases in isolation. We now review the evidence for epigenetic influences on the major diseases of the homeostatic systems that maintain blood flow, fluid/electrolyte balance, and glucose metabolism.

Cardiovascular disease is the major killer in high-income countries

Despite considerable advances in our ability to lessen the functional defects caused by cardiovascular problems, there is still a significant chance that citizens of Western nations will develop coronary heart disease or other cardiovascular conditions that are grouped under the general title of cardiovascular disease (CVD). Several variables have been identified that contribute to the development of this group of ailments, but this knowledge has not significantly reduced the tendency of Westerners to develop cardiovascular illnesses. To set this in context, a recent study showed that out of 2404 American and 8728 European males aged 40–59 years, 615 cases of coronary heart disease were diagnosed, 214 of which resulted in death or severe disability.

The emergence of CVD as a major public health problem during the 1940s encouraged research to understand the factors associated with this disease. As a result, several high-risk factors, such as hypertension, obesity, and diabetes, were identified and are commonly used today to assess the risk of CVD in individual patients. This knowledge has contributed to the substantial decline in CVD mortality observed in Western Europe and in the USA over the past 30 years; however, this trend is primarily due to improvements in quality of care and treatment, and much less so to the prevention of the disease itself. In short, we can treat the disease but we cannot prevent it from happening. Lifestyle changes in low-income and middle-income countries have resulted in an increased prevalence of obesity and diabetes, and consequently higher rates of CVD. Current recommendations for the treatment of CVD aim at reducing modifiable risk factors, such as high levels of low-density lipoprotein (LDL) cholesterol. However, as we noted at the beginning of this chapter, environmental and genetic effects can only explain a small part of the variability in CVD risk, even for well-established risk factors.

The basic problem in cardiovascular disease is atherosclerosis

CVD occurs because of "thickening" of the arterial blood vessel walls, which can reduce overall blood flow, cause inflammation of the vasculature, or block blood flow completely. These problems lead to a number of bodily dysfunctions ranging from peripheral vascular diseases (that is, the obstruction of arteries other than the aortic arch or coronary arteries) to strokes, myocardial infarction (heart attack), and cardiac ischemia (insufficient blood supply to the heart) due to occlusion of the coronary arteries. The common problem in all of these conditions is a greater or lesser degree of blood-supply restriction (ischemia) to one or several tissues or organs.

In the extensive research of heart disease, most attention has been paid to **atherosclerosis**, which is the accumulation of "plaques" based on lipids and cholesterol on the inner layer of large arteries. These restrict blood flow. However, another essential feature of arteries is their elasticity, which is needed to counteract the changes in pressure produced by the pumping of the heart. This elasticity declines during aging as a result of cross-linking of long-lived proteins—such as collagens, proteoglycans, and elastin—that form the tunica media (the middle layer) of blood vessels, and this is thought to be one reason for age-related increases in blood pressure. This latter condition is referred to as **arteriosclerosis** and can cause enormous damage by restricting blood flow over a longer term than atherosclerotic plaques.

So what actually causes the plaques to form? A complex and incompletely understood interaction exists between the critical cellular elements of the atherosclerotic lesion. These cellular elements are endothelial cells, smooth muscle cells, platelets, and leukocytes. The state of activation of the coagulation cascade, the fibrinolytic system, the migration and proliferation of smooth muscle cells, and cellular inflammation are complex and interrelated biological processes that contribute to atherogenesis and the clinical manifestations of atherosclerosis. The mechanism by which atherogenesis begins is uncertain, but there is considerable evidence to suggest that injury to the endothelial cells lining the arterial lumen causes a local inflammation that is the root cause of plaque formation. Many factors can injure the endothelium, but oxidized LDL, infective agents, xenobiotic toxins, and hyperglycemia have all been linked to CVD. The flow of blood through the arteries also seems to have an effect, because plaques characteristically occur in regions of branching and marked curvature at areas of geometric irregularity and where blood undergoes sudden changes in velocity and direction of flow. Decreased shear stress and turbulence may promote atherogenesis at these important sites within the coronary arteries, the major branches of the thoracic and abdominal aorta, and the large conduit vessels of the lower extremities.

Once the initial injury has occurred, circulating monocytes are attracted and infiltrate the intima (innermost layer) of the vessel wall, and these tissue macrophages act as scavenger cells, absorbing the LDL cholesterol and forming the characteristic **foam cells** (clumps of fat-filled macrophages) of early atherosclerosis. These activated macrophages produce numerous factors that are injurious to the endothelium, so a "vicious circle" of events begins from the quite normal response of the immune surveillance system to tissue damage. This is unfortunate, because rather than eliminating the LDL cholesterol from the injury site, the process actually serves to widen the lesion and allow the accumulation of additional lipids. This produces the earliest observable pathologic lesion, known as the fatty streak, which seems to provide an appropriate environment for the smooth muscle cells of the tunica media to proliferate and form much larger plaques, leading to vascular remodeling, progressive luminal narrowing, abnormalities in blood flow, and compromised oxygen supply to the target organ. The events that may occur after this stage are well documented and do not require further explanation here.

Epigenetic events may promote atherosclerosis by increasing known risk factors

We mentioned the Dutch hunger winter of 1944–45 in Chapter 1 as an example of how famine events can be used to study epigenetic inheritance. The outcomes of such studies strongly support the idea that exposure *in utero* to famine may contribute to the development of adverse metabolic phenotypes in the adult. For example, female children of starved mothers seem to have increased fat deposition and body mass index, whereas males suffer from detrimental changes in their levels of HDL, although these harmful effects may not be observed until the children are adults. Many of these children born during or shortly after the hunger winter were not examined until they had developed clinical symptoms, and this was up to 60 years after the events took place.

It is interesting that many hunger winter children have very different DNA methylation patterns than children who were not born to starving mothers. The affected genomic loci are those involved in growth and metabolism. This draws an interesting parallel with the imprinting disease mentioned in the last section, although there is little evidence that the starvation events actually led to an increased incidence of BWS (this may have been due to misdiagnosis, because BWS was not formally identified as an independent condition until 1964). Decreased DNA methylation on the gene *IGF2* was noted in the hunger winter children, coupled with increased methylation on genes such as *IL-10*, *LEP*, and *MEG3*. The persistence of such epimutations throughout life has been confirmed by animal studies. For example, it is conceivable that such events occurred only recently in the hunger winter children, as a result of aging or some other disease-related effect; however, restricting the food intake of pregnant mice produced offspring with similar DNA methylation effects to those of the hunger winter children, and these effects persisted throughout the lives of the mice. It is therefore likely that prenatal famine produces persistent changes in the DNA methylation patterns of humans.

Despite these interesting observations, because the analysis of the hunger winter children's DNA often took place after disease symptoms had become apparent, it is very difficult to prove that CVD arose in hunger winter children as a direct consequence of the epigenetic changes imposed by the mother's starvation. Animal models provide better information about the increasing risk of CVD after maternal starvation.

In mice subjected to intrauterine protein restriction, epigenetic changes were localized to specific genes that are known to increase the risk of CVD. As might be expected, well-established risk factors for CVD such as insulin resistance, hypertension, and obesity were indeed increased by this protein starvation. Additionally, at a molecular level, the offspring were found to have undergone DNA methylation changes in many genes whose normal function is to mediate lipid and fatty acid metabolism in the liver.

Unfortunately, there are few data from animal studies to indicate an increase in the actual occurrence of CVD as a result of these epigenetic and metabolic changes. There are known increases in the frequency of

CVD in hunger winter children, but we cannot be absolutely certain that the disease did not simply arise in some of these individuals because of a high-lipid diet, smoking, lack of exercise, or some other factor. The types of growth and lipid metabolism defects observed in these cases are likely to increase the probability of raised levels of HDL cholesterol that may predispose certain individuals to develop CVD, but there are probably many examples of hunger winter children who did not develop CVD. So, clearly, these epigenetic changes cannot be an overriding phenomenon that promotes atherosclerosis.

Additional evidence that atherosclerosis may have other causes comes from the seeming relationship between the condition and the serum levels of homocysteine, folic acid, and vitamins B_{12} and B_6. We came across these molecules as part of the one-carbon metabolism system, introduced in Chapter 5, by which DNA and protein can be methylated. We do not need to describe this system again, but it is worth reinforcing the fact that S-adenosylmethionine is the main methyl group donor and becomes S-adenosylhomocysteine after performing its methylation function. This molecule may be recycled by removing the adenosyl group to give homocysteine, which can then re-enter the methylation cycle (see Figure 5.1), or the homocysteine can be broken down further for complete removal from the body. The remethylation of homocysteine to methionine requires vitamin B_{12} and folic acid, so if these are not present at optimal levels, the homocysteine builds up in the bloodstream. A number of studies have implicated such raised homocysteine levels in the development of atherosclerosis.

How a buildup of homocysteine in the blood can cause injury to the endothelial cells in the arterial wall is not absolutely clear. It has been suggested that homocysteine may be toxic to endothelial cells in some way, but an alternative explanation is that an underlying dysfunction exists in the ability of these cells in the arterial intima to effect DNA methylation. This would imply that the higher homocysteine levels are simply an indicator that the methionine recycling/methylation system has been forced to slow down by the lack of folic acid and/or vitamin B_{12}. If DNA methylation is not as effective, it is possible that maintenance methylation of silenced genes may fail and they will begin to show inappropriate expression. This may lead to epimutations that change the endothelial cell phenotype in favor of the local inflammatory processes that begin the formation of an atherosclerotic plaque. After this point, the development of the plaque will depend on the types and concentrations of lipids in the blood of the affected individual.

Another key issue in plaque development seems to be the inappropriate proliferation of smooth muscle cells (SMCs). Unlike skeletal and cardiac muscle cells, which are terminally differentiated, SMCs within adult animals readily switch phenotypes in response to changes in local environmental cues (that is, they retain some developmental plasticity). For example, vascular SMCs express high levels of SMC-specific contractile proteins (to help them regulate the muscle tone of the artery) and do not generally proliferate, migrate, or secrete significant amounts of extracellular matrix. However, in response to extracellular cues released at sites of vascular injury or within atherosclerotic lesions, SMCs are known to show decreased expression of SMC-specific contractile proteins and increased migration, proliferation, and production of extracellular matrix components as well as matrix metalloproteases. It is conceivable that inappropriate DNA methylation could result in similar transcriptomic changes.

Evidence supporting this hypothesis is beginning to accumulate. Several studies have shown that demethylation of normally hypermethylated

CpG islands occurs in atherosclerotic arteries, and this seems to promote expression of pro-atherogenic genes. Microarray-based surveys have identified a range of affected genes, the most abundant classes being genes that express transcription factors, proteins involved in signal transduction, and proteins involved in organ homeostasis. Typically, many of these genes (roughly 40%) are involved in regulating the expression of other genes, implying that epigenetic changes are upstream events in the establishment of atherosclerosis. However, many other genes contribute to angiogenesis, modulation of the phenotype of SMCs, and inflammation. For example, vascular SMCs frequently change from a contractile, differentiated phenotype to an immature, secretory state after structural re-organization of their organelles. The differentiation marker gene *STMN*, which resides in the DNA hypomethylated regions, shows a corresponding increase in expression during this change in phenotype.

Naturally, it is very difficult to separate cause and effect in these studies. Moreover, the tissue samples taken from the atherosclerotic arteries will be a heterogeneous mixture of cell types, so we cannot be sure that the observed epigenetic changes do not simply reflect the enrichment of one particular type of cell over another. For example, if endothelial cells (for the sake of argument) were more numerous in the atherosclerotic arteries, we might be detecting an endothelial cell-specific pattern of epigenetic modifications. However, an interesting additional piece of information that supports the notion that the above-noted epigenetic changes are not simply cell-specific modifications but are evidence of epigenetic involvement in plaque formation is that if the stimulus provided by the plaque is removed, the SMCs revert to their normal phenotype. For example, studies in transgenic mice have demonstrated that SMC-specific contractile gene promoters are transcriptionally active within vascular SMCs under physiological conditions, whereas they are transcriptionally repressed in response to vascular injury and atherosclerotic plaque formation. After resolution of the injury, SMCs resume transcription of these genes and regain their fully differentiated phenotype.

Nearly all of these contractility-related genes of SMCs contain an evolutionarily conserved CC[AT]6GG sequence called a CArG box within their promoters; the CArG box is a transcription factor binding site and is required for SMC gene transcription *in vivo* because of its ability to bind serum response factor. This factor seems to be able to induce contractile gene expression only in the SMCs, because findings indicate that these genes are repressed in heterochromatin in other cell types. The response of SMCs to injury and to the plaque microenvironment seems to be heterochromatinization (of which DNA methylation is a key factor) of the promoter of the gene *ERα* (which encodes estrogen receptor α), leading to the proliferation of SMCs in atherosclerotic lesions.

The SMCs of atherosclerotic plaques appear to show global DNA hypomethylation compared with their counterparts outside the lesion, so it is unknown why *ERα* should be singled out for hypermethylation. There is a steady age-dependent increase in *ER* promoter methylation, which may reach 99% methylation level in the elderly. Still, age-related global hypomethylation of contractility-related genes in SMCs is the dominant process, and *ERα* hypermethylation is not the only exception. Age-related promoter hypermethylation of the genes encoding c-Myc, c-Fos, IGF2, MYOD1, N33, HIC1, versican, PAX6, DBCCR1, E-cadherin, and P15 has also been reported. Such gene-specific hypermethylation may have profound effects on the pathogenesis of atherosclerosis.

From the results of these studies, we can at least suggest that epigenetic changes induced by environmental factors may be responsible for atherosclerosis.

Epigenetics has a role in the regulation of arterial hypertension

Hypertension refers to the condition in which arterial blood pressure rises above 140/90, and in most cases (more than 95%) the causes are unknown. Several lifestyle factors are linked to a greater incidence of such high blood pressure, for example smoking, high salt intake, lack of exercise, stress, or excess alcohol consumption, but as with atherosclerosis, making such lifestyle choices does not guarantee hypertension. It is generally accepted, however, that hypertension does predispose the individual to develop CVD.

Increased occurrence of hypertension is observed in rats whose mothers were subjected to a low-protein diet during pregnancy. The offspring have a birth weight below normal and show elevated systolic and diastolic blood pressures (measured by tail cuff methods or by carotid artery catheters) as early as 4 weeks of age. Expression of the angiotensin receptor gene is up-regulated in the adrenal gland as early as the first week of life, as a consequence of DNA hypomethylation in the gene's proximal promoter region. There are two possible explanations: the starvation events could have favored the development of the rat adrenal gland to include more zona glomerulosa cells, which are the only cells expressing the angiotensin receptor, or the expression level in those same cells could be higher. Given the hypomethylation of the gene promoter, the second explanation is probably more likely. If starvation can induce these changes, then dysfunction of the *S*-adenosylmethionine DNA methylation system may produce similar effects.

Hypertension increases with age

The starvation model described above is interesting enough in its own right, but many humans develop hypertension even though their mothers were not starved during pregnancy. In fact, most of us undergo some increase in blood pressure as we age. Aging is one of the most significant factors involved in the onset of hypertension and is one of the few possible causes for which we can predict the effect.

We noted earlier that the elasticity of the blood vessels declines during aging and that this is thought to be one reason for age-related increases in blood pressure. The arterial wall consists of three layers: the tunica intima, tunica media, and tunica externa (**Figure 14.5a**), but it is the intima and media that suffer the greatest age-related changes. The intima is composed of endothelial cells that are separated from the tunica media by an elastic fibre lamina, whereas the tunica media consists of

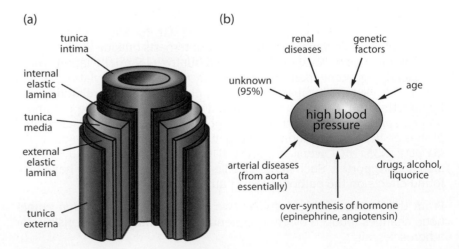

Figure 14.5 The arterial wall and causes of hypertension. (a) Structure of the wall of large arteries, showing the tunica intima, which is the principal site of damage that leads to atherosclerosis. (b) Mechanisms currently known to lead to hypertension.

SMCs connected by an extracellular matrix that contains many of the elastic proteins (such as collagen and elastin) that become progressively less elastic with age. Several other processes may also contribute to an increase in blood pressure (Figure 14.5b).

In large arteries, aging is characterized by a decrease in turnover of collagen and elastin and an increase in advanced glycation end products (AGEs) and cross-links. Elastic fibers undergo lysis and disorganization after their replacement by collagen and other matrix components. Moreover, lipids enter the elastin fibers, resulting in an influx of calcium ions that promotes the loss of elasticity yet further by increasing the activity of elastin-degrading elastase enzymes. These events cause the loss of elasticity and induce stiffening of the vascular wall, which probably contributes to an increase in blood pressure in the same way that squeezing water through a narrow solid tube gives higher pressures than using the same amount of force through a flexible hose, but there is little evidence of epigenetic regulation of this process.

Many other factors contribute to an alteration in blood pressure. Neurological control is a reasonably well-documented process for achieving blood pressure homeostasis, and melatonin-expressing neurons may provide input to the catecholaminergic neurons in the vasomotor centers of the rostral ventrolateral medulla (RVLM). This small area of the brainstem is one of the main regulators of sympathetic outflow to blood vessels, and RVLM dysfunction seems to be a mechanism for essential (that is, idiopathic) hypertension in humans. Treatment of experimental animals with nanomolar concentrations of melatonin causes global increases in the acetylation of histone H3, and because the vasomotor control sites in the area postrema of the brainstem are reported to contain high levels of melatonin receptors, it is possible that environmental factors may cause epigenetic modification of genes controlling the RVLM so that its trigger point for initiating vasoconstriction is set at a higher pressure level, with the overall effect of raising the blood pressure.

The cells of the artery may also contribute to the onset of hypertension. Endothelial cells have an important role in regulating vascular tone, permeability to a range of molecules, and, most importantly, the synthesis and release of the vasodilator nitric oxide. Nitric oxide promotes vasodilation by stimulating guanylate cyclase in the SMCs of the tunica media, so any changes in the expression level of nitric oxide synthetase and its isoforms may affect the ability of an artery to dilate. If the capacity to synthesize nitric oxide is reduced, an overall increase in blood pressure may result. However, it is interesting that an inducible isoform of nitric oxide synthetase (iNOS) seems to increase in expression level with age. This may be explained as a response to the decreasing effectiveness of nitric oxide synthesis through other isoforms of nitric oxide synthetase, but it has also been suggested that endothelial cells are protected from excessive concentrations of nitric oxide (which is toxic to them) by constant low-level synthesis by iNOS—in a sense, they develop tolerance to low levels of nitric oxide.

Another possible explanation for the change in vascular tone is the increasing degree of oxidative stress to which endothelial cells are subjected to as they age. We know that antioxidant enzyme systems such as superoxide dismutase function less efficiently in older tunica intima and that concentrations of superoxide ions are higher. Superoxide reacts with nitric oxide to produce the peroxynitrite ion, which, quite apart from creating another macromolecule-damaging reactive oxygen species, decreases the nitric oxide concentration. This could in turn reduce the vasodilation capability. The question then becomes whether the changes in gene expression mentioned above occur before and are a causative factor in the development of hypertension, or whether they are simply the cell's response to accumulated damage in the arterial wall.

Whether or not these changes are subject to epigenetic regulation is an even more important question. We know that the gp91^{phox} subunit of the NADPH oxidase enzyme complex, which generates superoxide in phagocytic cells, is controlled by the histone acetyltransferase GCN5 and that NADPH oxidases are important sources of reactive oxygen species in the vasculature. Several members of the NADPH oxidase family are detected in arterial cells: NOX1 is present in SMCs, NOX2 is found in endothelial cells, and NOX4 occurs in both cell types. This parallel expression of the NOX isoforms suggests that they may have different functions. For example, NOX1 and NOX2 localize to the plasma membrane, whereas NOX4 is more often found in the nucleus, the focal adhesion complexes, and the endoplasmic reticulum. NOX1 and NOX2 produce superoxide, whereas NOX4 produces predominantly hydrogen peroxide. Controlling the activity of the several NOX isoforms seems to be achieved partly through signal transduction mechanisms (such as TGF-β, which activates NOX4), but there also seems to be a strong epigenetic contribution because up-regulation of the microRNAs miR-146a and miR-25 seems to be able to induce NOX4 activity too. Moreover, overexpression or increased activity of NOX4 actually seems to reduce blood pressure by stimulating eNOS (endothelial nitric oxide synthase) as opposed to iNOS activity in the endothelial cells.

Other possible methods for brainstem control of blood pressure exist. For example, norepinephrine (noradrenaline) transporters are membrane proteins that conserve the catecholamine neurotransmitters norepinephrine and dopamine by transporting them back into the presynaptic neuron that released them. There may be co-release of epinephrine with norepinephrine at postganglionic sympathetic nerve endings as a result of increased activity of phenylethanolamine N-methyltransferase (PNMT), the enzyme that converts norepinephrine into epinephrine, primarily within the medulla of the adrenal gland. PNMT may be induced in brainstem catecholaminergic neurons, peripheral sympathetic neurons, and other neurons where epinephrine may be co-released, particularly during mental stress. PNMT is reported to act as a DNA methyltransferase and to be able to mimic the gene-silencing activity of MECP2 (see Section 5.2). This is thought to create a method for establishing and stabilizing a mechanism to increase plasma levels of catecholamines, which produces a state similar to stress and thereby increases essential hypertension.

Other genomic regions have been implicated in hypertension. For example, the region of chromosome 9 containing the mixed-lineage leukemia and acute lymphoblastic leukemia loci produces a sequence-specific DNA-binding protein that binds to the promoter sequence of a gene encoding a protein involved in the amiloride-sensitive renal epithelial sodium channel (ENaC-α). The H3K79 methyltransferase DOT1 functions as part of a repressor complex for the mixed-lineage leukemia and acute lymphoblastic leukemia loci, and the DNA-binding protein is the factor directing the assembly of the H3K79 methylating repressive complex at the (ENaC-α) promoter that effects epigenetic silencing. This sodium channel is involved in the renin–angiotensin system, which regulates blood pressure and fluid balance by secreting renin from kidney cells into the bloodstream when blood volume is perceived to be low. Renin converts angiotensinogen into angiotensin I, which is subsequently converted into angiotensin II, a potent vasoconstrictor that effects an immediate increase in blood pressure. Aldosterone also increases water reabsorption by the kidney, to ensure that blood volume increases. If the renin–angiotensin system becomes too active, hypertension will result. Aldosterone normally disrupts binding of the DNA-binding protein to the ENaC-α promoter; DOT1 is therefore probably able to influence the expression of genes regulating sodium transport, which may impose some predilection toward hypertension. In a similar manner, the hydroxysteroid dehydrogenase-11β2 (HSD11B2) degrades the stress marker cortisol to hydrocortisone. This is necessary because cortisol is as effective as aldosterone in increasing blood pressure, but its concentration in the blood tends to be much higher. High levels of DNA methylation at the HSD11B2 promoter are often found in hypertension, implying that epigenetic dysregulation of this gene could be one method by which an increase in blood pressure occurs.

These links to epigenetic regulation are quite speculative, and many more data are needed to confirm whether they make significant contributions to hypertension. However, a much stronger case exists for maternal water deprivation in animal models. The offspring of such mothers do not show immediate increases in blood pressure, but administration of angiotensin II has a much greater effect on them than on offspring from normally treated mothers. These data suggest that even short periods of water deprivation may affect the behavior of the renin–angiotensin system in a manner that disrupts vasorepressor responsiveness into adulthood. Specific epigenetic mechanisms for this process are not known, but likely targets are the genes of the renin–angiotensin system themselves, in addition to genes encoding proteins regulating osmolality and sodium transport.

Cardiac hypertrophy and heart failure also have an epigenetic component

A variety of conditions, including pressure or volume overload in the cardiovascular system and remodeling of the left ventricle of the heart after ischemic damage, result in heart failure, which is characterized by a reduction in contractile ability and a decrease in the number of viable myocytes in the heart. Treatment of heart failure remains problematic, and this condition is thus still one of the leading causes of human death.

Epigenetic status has been linked to cardiac hypertrophy and heart failure. Among a range of histone post-translational modifications monitored in human heart-failure patients, H3K4 trimethylation and H3K9 trimethylation seem to be affected the most, with several clustered regions of the genome showing extensive and abnormal accumulations of these modifications. These clusters seem to contain genes that are central to the function of cardiomyocytes or at least are significant loci for the development of chronic heart failure. This was deduced from the finding that the genes positioned close to the clusters were primarily those that encode components of signaling pathways related to cardiac function. The H3K4me3 modification was, for instance, associated with the genes RYR2, CACNA2D1, and CACNB2, whose products participate directly in the regulation of intracellular calcium concentration and in muscle contraction.

In a morphometric postmortem analysis of atherosclerosis in fetuses and children, it was demonstrated that high levels of cholesterol in the maternal blood were associated with a higher incidence of atherosclerotic lesions during the fetal period and a faster progression of these atherosclerotic lesions after birth, even if the offspring had normal cholesterol levels. The exact mechanism by which this can predispose the individual to develop atherosclerosis is unknown, but it may be related to epigenetic programming of the smooth muscle cells during their development, with the consequence that these individuals already have at birth inappropriate epigenetic modifications at genomic loci linked to cardiovascular function.

Histone deacetylation seems to be of particular importance for these processes, but this is most probably not a universal phenomenon because only a restricted number of histone-modifying enzymes are involved. HDAC5 and HDAC9, for example, seem to have anti-hypertrophic activity associated with repression of myocyte enhancer factor 2. Knockout of HDAC5 and HDAC9 in mice results in a much greater incidence of hypertrophy. Class I HDACs (such as HDAC2) are pro-hypertrophic by virtue of down-regulating expression of the gene encoding phosphatidylinositol-3,4,5-triphosphate phosphatase, which modulates activity of the cell-growth-promoting phosphoinositide 3-kinase–Akt–glycogen synthase kinase 3 (PI3K–Akt–GSK3) pathway.

Deacetylation of histones is usually stabilized by histone methylation, but few results are available to indicate the possible roles of this process in heart failure. Recent work has suggested that the arginine histone methyltransferases PRMT1 and PRMT3 can effect H3K9 trimethylation on genes that become repressed during cardiac hypertrophy. However, further genome-wide studies have shown that histone modifications such as trimethylated H3K4 and H3K9 do not always associate with active or inactive genes in cardiomyocytes that expand during hypertrophy, so it is difficult to determine a precise epigenetic mechanism for the inappropriate enlargement of a damaged heart.

Epigenetic drift may contribute to cardiovascular disease

The mechanisms and dysfunctions that may contribute to the onset of CVD that we have described so far can occur because of the environmental influences encountered by the mother during pregnancy. However, as we have noted, many people develop CVD whose mothers were apparently well nourished. We should therefore perhaps assume that the epigenetic mechanisms that may be involved are subject to spontaneous change—in short, *CVD just happens*. In addition, epigenetic changes are, of course, not restricted to the prenatal period, as studies of monozygotic twins suggest. Such twins experience drift in their epigenetic modification patterns as a function of lifestyle habits, other behavioral traits, and, above all, time. A question of potentially enormous significance is this: If CVD has a strong epigenetic input, can we reverse the harmful epigenetic modification patterns to reduce if not eliminate the damage that leads to atherosclerosis in the first place?

In at least a partial answer to this question, targeting enzymes such as histone acetyltransferases and histone methyltransferases may allow us to modulate the transcriptional regulation of genes involved in atherogenesis, inflammation, the proliferation of SMCs, and aberrant formation of extracellular matrix in the tunica media. It may even be possible to prevent the formation of at least some of the endothelial cell dysfunctions that start the process of plaque formation. To some extent, we may be doing this already. Statins, a well-known class of 3-hydroxy-3-methylglutaryl-coenzyme A inhibitors used to treat high levels of cholesterol, can function by increasing the nuclear concentration of HDAC7, which deacetylates histones at the promoter of the cholesterol 7-α monooxygenase gene. Naturally, HDAC7 activity is not specific to this gene, and doubtless many other targets will be identified that are affected by statin therapy, not all of them necessarily in a beneficial way. Future epigenetic therapeutic agents will therefore require a high degree of specificity. To create such agents we shall need a much more detailed knowledge of the genes controlling atherogenesis and precisely how epigenetic regulation contributes to their failure to maintain a pristine, healthy tunica intima.

14.4 EPIGENETICS OF KIDNEY DISEASE

Chronic kidney disease results in the progressive loss of renal function until little or no activity remains, a condition referred to as **uremia**. The early stages are often without distinct symptoms, and the only indicator is an increase in serum creatine concentrations or the presence of protein in the urine, which would generally be detected during health screens for other complaints. As kidney function decreases, several systemic problems occur: (1) blood pressure rises as a result of inefficient removal of excess fluid; (2) urea accumulates, with toxic effects such as lethargy, pericarditis, and encephalopathy; (3) blood electrolytes increase; and (4) excretion of phosphates, sulfates, and uric acid is reduced, leading to metabolic acidosis, which can interfere with the kinetics of certain enzymatic reactions and disrupt cardiac and neuronal membrane functions. Treatment of early-stage disease generally involves angiotensin-converting enzyme inhibitors to control blood pressure; these can slow disease progression but do not prevent the ultimate demise of the kidney. Dialysis is a well-established method of replacing kidney function *ex vivo*, but renal transplantation is the only long-term therapeutic option for progressive kidney disease at present. The problem is alarmingly common; a study performed by the US Centers for Disease Control and Prevention estimates that 16% of adults over 20 years of age have some form of the disease, and UK estimates suggest that 8.8% of the British population have symptoms. Naturally, these estimates do not include those individuals who have undergone functional changes in the kidney that are not detectable but are nevertheless the first steps in chronic kidney disease development.

Most of the problems leading to chronic kidney disease have a complex pathogenesis and involve many candidate genes. With 20 distinct cell types present in the adult kidney, the scope for disease is enormous. The kidney is composed of very large numbers of nephrons (**Figure 14.6**), whose job is to remove waste from the bloodstream by squeezing much of the liquid components across the walls of a network of fine capillaries (called the **glomerulus**) while retaining the cells in the blood. The liquid fraction is then sent off down a convoluted tube where water can be reabsorbed into a network of capillaries, and the resulting, more concentrated, fluid (urine) is passed into the collecting ducts for delivery to the bladder. The efficiency of this whole process, often measured as the glomerular filtration rate, can be influenced by many factors, but the most important of these are: (1) the number of nephrons present in the whole kidney; (2) the functionality of the blood vessels flowing in and out of the kidney (the renal artery and renal vein, the efferent and afferent arterioles, and so on)—by this we mean the presence or absence of obstructions such as may result from atherosclerosis; (3) the status of the ureter (normal versus blocked or damaged by kidney stones, for example); and (4) the functional capacity of the cells that make up the separate sections of the nephron.

Given its function as a selectively permeable barrier between the bloodstream and a waste excretion system, it is not surprising that most of the cell types that form the kidney are epithelial or endothelium-like cells. For example, the Bowman's capsule is composed largely of podocytes or visceral epithelial cells, whereas the proximal tubule is lined by epithelial

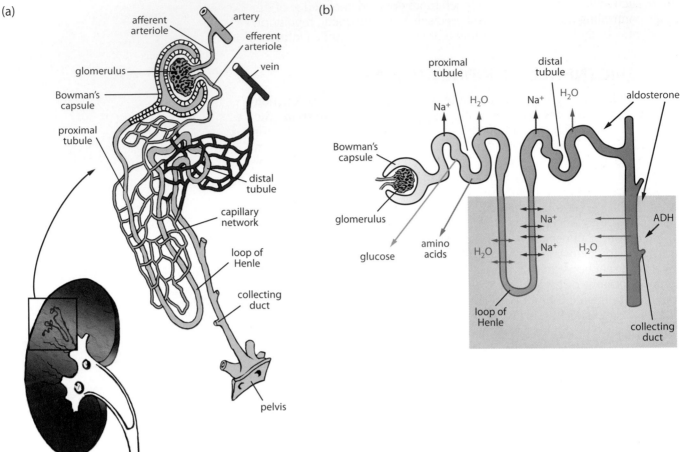

Figure 14.6 Kidney structure and function.
(a) Schematic structure of the nephron.
(b) Functions of the different regions of the proximal tubule, loop of Henle, distal tubule, and collecting duct. Nephric filtrate gathers in the Bowman's capsule and then flows into the proximal tubule, where glucose and amino acids are reabsorbed. Sodium ions can also be extracted, but this is under the control of angiotensin II. Phosphate reabsorption is controlled by parathyroid hormone, and the fine tuning of sodium reabsorption is controlled in the distal tubule by the action of aldosterone. Permeability of the collecting duct to water is altered by antidiuretic hormone. (Adapted from Kimball's Biology Pages. Courtesy of John W Kimball. http://users.rcn.com/jkimball. ma.ultranet/BiologyPages.)

cells with large numbers of microvilli to increase their surface area and facilitate their reabsorption function. The loop of Henle and distal tubules are composed of simple cuboidal and squamous epithelial cells.

The capillaries that surround the tubules of the nephron consist, as do most capillaries, of a single layer of flattened endothelial cells without the muscular or adventitial layers that we see in the arteries. This means that they do not suffer from atherosclerosis in the way that large or medium-sized arteries do, but damage to or dysfunction of the endothelial cells can still disrupt the flow of blood to tissues—this is referred to as **microangiopathy**. The kidney is one of the main organs to suffer from this type of damage, along with the retina and the brain. Such damage is frequently observed in diabetics, although a condition known as thrombotic microangiopathy can also arise.

The underlying pathogenesis of microangiopathy is considered to be injury to the endothelial cells, so it is perhaps not unreasonable to suggest that the misregulation of some of the components of the NADPH oxidase complex may damage the endothelial cells of the nephron in the same way as it was suggested to lead to endothelial damage in the larger arteries. Kidney endothelial cells produce nitric oxide in a similar manner to their large-vessel counterparts. This is partly controlled by levels of vascular endothelial growth factor (VEGF). It has been established that reductions in local VEGF concentrations in the podocytes of the Bowman's capsule can induce thrombotic microangiopathy, which can reduce the glomerular filtration rate if a sufficient number of nephrons are affected. If this problem persists, it can contribute to chronic kidney disease, and we can also envisage that epigenetic systems modulating the expression of NOX1, NOX2, and NOX4 subunits of NADPH oxidase may be able to influence the process.

In another parallel with the development of CVD, elevated levels of *S*-adenosylhomocysteine have been recorded in patients with chronic kidney disease and those with end-stage renal disease. Recall from the discussion of atherosclerosis that *S*-adenosylhomocysteine is a by-product of DNA methylation by *S*-adenosylmethionine and can result in a buildup of homocysteine in the blood, which can damage endothelial cells. Thus, it is reasonable to suggest that epigenetic modifications may be an important risk factor for progression of renal disease and CVD in patients with chronic kidney disease.

14.5 EPIGENETICS OF DIABETES

Diabetes mellitus is a significant health care problem that arises as a result of insufficient or absent synthesis of the hormone insulin by the beta cells of the islets of Langerhans in the pancreas. Type 1 diabetes results from complete loss of insulin synthesis, whereas type 2 arises when normally insulin-responsive systems such as the liver and skeletal muscles become less sensitive and the pancreas is unable to generate extra insulin to compensate for this decreased sensitivity. Insulin deficiency leads to chronic hyperglycemia, which predisposes the individual to CVD and renal failure.

Type 1 diabetes is thought to result from autoimmune destruction of the beta cells, although it is not clear how this destruction is mediated. This type of mechanism involves autoantibodies, which are immunoglobulins generated by the B cells of the immune system against antigens that are part of the self. Normally, the B cells that generate such antibodies are eliminated through clonal deletion; however, for reasons that are not fully understood, some of these escape and can attack specific cells or organs in the body. Once the beta cells have been eliminated, the patient is usually forced to take regular injections of insulin to regulate his or her glucose homeostasis. However, controlling hyperglycemia in this manner is imprecise, and the higher concentrations of glucose in the blood can affect the behavior of other cell types, such as the endothelial cells, whose damage can lead to atherosclerosis and microangiopathy.

Cardiovascular complications remain the major cause of morbidity and mortality in the diabetic population. Patients with type 1 or type 2 diabetes have a twofold to fourfold higher risk of CVD when compared with healthy individuals. There is strong evidence to suggest that exposure to high glucose levels is the major factor leading to these complications, and moreover, such exposure seems to involve a form of "metabolic memory" through which diabetic complications, particularly vascular events, continue to develop and progress, even in individuals who have returned to normal glycemic control after a period of transient hyperglycemia.

This was first observed during studies of retinopathy in diabetic dogs that were switched to good glycemic control after long periods of poor control (namely high blood glucose). The same numbers of dogs developed retinopathy regardless of whether they had been switched from poor to improved control or had been maintained with poor control throughout the study period, suggesting that some "memory" of the previously higher glucose levels had programmed the animals to develop retinopathy regardless of subsequent improvement in their glucose levels. It is possible, of course, that the retinopathy was a consequence of some undetected cellular damage that occurred during the high-glucose phase, but further studies showed that transient exposure to hyperglycemia induced epigenetic changes in the promoter of the nuclear factor κB (NFκB) subunit p65 in aortic endothelial cells both *in vitro* and in nondiabetic mice, leading to increased p65 gene expression. This is significant because activating the NFκB signaling pathway is known to trigger vascular inflammation, which is a major event in the progression of diabetic complications. The signaling leads to recruitment of monocytes and macrophages to the vessel and thus contributes to macrovasculature atherosclerosis. More recently it has been shown that high glucose levels promote the formation of foam cells, not only in macrophages but also in human cultured smooth muscle cells.

The above studies suggest an epigenetic involvement in the complications arising from type 1 diabetes that to some extent may also occur with the marginally less serious type 2 form of the condition. However, there is also evidence to suggest that the onset and progression of type 2 diabetes may themselves be influenced by epigenetic controls. A degree of heritability has long been recognized in type 2 diabetes: siblings of affected individuals are three times more likely to develop the condition than siblings of non-diseased individuals. This observation prompted many studies into the genetic inheritance of diabetes, but despite the identification of highly penetrant mutations in genes such as glucokinase, these were associated with only a small percentage of cases of type 2 diabetes and so were unlikely to be a universal cause of the disease. More detailed genome-wide association studies have identified further diabetic risk loci, but these still only explain approximately 10% of the familial risk. Added to this is the rapidly increasing incidence of type 2 diabetes in developed countries, a pace of development that is too rapid to be accounted for by the occurrence of new mutations alone. Taken together, these findings indicate that environmental factors such as diet and sedentary lifestyle, and gene–environment interactions are likely to have a significant role in the development of the disease.

Once again, the Dutch hunger winter has proven to be a useful source of information. Individuals born during or shortly after the famine have demonstrated that maternal and therefore intrauterine malnutrition and low birth weight, followed by an increase in body mass index during childhood, lead to an increased likelihood of developing type 2 diabetes later in life. This point has been investigated further in animal models, which show that intrauterine nutritional insufficiency results in adult animals with reduced beta-cell mass and decreased expression of the essential beta-cell transcription factor pancreatic and duodenal homeobox 1 (Pdx1), leading to a type 2 diabetes mellitus phenotype. Not surprisingly, acetylation of both histone H3 and H4 at the *Pdx1* locus is decreased in the fetal, juvenile (2 weeks), and adult (6 months) islets of Langerhans, and this change is accompanied by progressively lower H3K4me3 levels at the *Pdx1* locus. In addition, the transcription factor USF-1, which regulates *Pdx1* transcription, is no longer able to bind to its target site on the *Pdx1* promoter region in the starved offspring animals, suggesting that changes in the chromatin architecture of the promoter are denying access to proteins such as transcription factors. Treatment of isolated 2-week-old islets from such animals with the histone deacetylase inhibitor trichostatin A results in increased H3 acetylation and H3K4me3 enrichment, and permits the partial restoration of Pdx1 expression. Together, these results suggest that epigenetic changes at the *Pdx1* locus could be indicative of development of type 2 diabetes mellitus later in life and that reversal of epigenetic silencing at the *Pdx1* locus—for example, with a specific histone deacetylase inhibitor—might prevent late onset of the disease.

Modifications at the *Pdx1* locus may not be the only epigenetic changes that predispose to diabetes. The transcription factor hepatocyte nuclear factor 4a (Hnf4a) is also important for beta-cell replication in rodents in response to metabolic stress and glucose homeostasis, and mutations in HNF4a cause adult-onset diabetes in humans. It has been observed that a maternal low-protein diet and aging lead to epigenetic changes in the enhancer and the islet-specific P2 promoter region of the gene encoding Hnf4a, and in decreased Hnf4a expression levels in the offspring.

In summary, we have described a very limited range of diseases in which dysregulation of epigenetic control of gene expression may be a contributing causative factor. Our objective has not been to provide an exhaustive list of illnesses and precise details of the epigenetic aberrations that may give rise to them, but to establish the possibility that epigenetics may help us to understand the apparently random onset of some conditions. Furthermore, changes in epigenetic modification patterns may be the result of environmental influences that may occur long before the onset of disease without apparent effect at the time of exposure. Epigenetic records of events in the earlier lives of individual patients may make significant contributions to disease development in later life.

KEY CONCEPTS

- The immediate clinical effects of infection by microorganisms probably do not have an epigenetic component (although epigenetic modifications may result from the infection.)

- Diseases that result (at least in part) from dysregulation of the mechanisms that control epigenetic modification patterns are referred to as "internal attack" problems because they arise from the body's inability to maintain its cells, organs, and the information processed from its genome. The distinctions between internal and external attacks are not precise—for example, some types of cancer can arise from viral infections although most cancers occur because of improper regulation of cell growth, in which epigenetic control of gene expression is a major factor.

- Internal attack diseases are still influenced by external, environmental factors such as diet, exposure to chemical agents, stress, or ionizing radiation.

- The maintenance of epigenetic modification patterns is less effective as organisms get older, but exposure to environmental factors early in life may predispose an individual to develop disease at a later stage.

- The epigenetic modification patterns controlling gene imprinting can be changed, causing the loss of monoallelic gene expression. This leads to several well-characterized diseases.

- Epigenetic modifications have been implicated in other diseases that are thought to have complex multifactorial causes.

FURTHER READING

Berdasco M & Esteller M (2013) Genetic syndromes caused by mutations in epigenetic genes. *Hum Genet* **132**:359–383 (doi:10.1007/s00439-013-1271-x).

Braun MC & Doris PA (2012) Mendelian and trans-generational inheritance in hypertensive renal disease. *Ann Med* **44** (Suppl 1):S65–S73 (doi:10.3109/07853890.2012.665473).

Court F, Martin-Trujillo A, Romanelli V et al. (2013) Genome-wide allelic methylation analysis reveals disease-specific susceptibility to multiple methylation defects in imprinting syndromes. *Hum Mutat* **34**:595–602 (doi:10.1002/humu.22276).

Demars J, Le Bouc Y, El-Osta A & Gicquel C (2011) Epigenetic and genetic mechanisms of abnormal 11p15 genomic imprinting in Silver-Russell and Beckwith-Wiedemann syndromes. *Curr Med Chem* **18**:1740–1750.

Demars J, Shmela ME, Rossignol S et al. (2010) Analysis of the IGF2/H19 imprinting control region uncovers new genetic defects, including mutations of OCT-binding sequences, in patients with 11p15 fetal growth disorders. *Hum Mol Genet* **19**:803–814 (doi:10.1093/hmg/ddp549).

Drong AW, Lindgren CM & McCarthy MI (2012) The genetic and epigenetic basis of type 2 diabetes and obesity. *Clin Pharmacol Ther* **92**:707–715 (doi:10.1038/clpt.2012.149).

Ekström TJ & Stenvinkel P (2009) The epigenetic conductor: a genomic orchestrator in chronic kidney disease complications? *J Nephrol* **22**:442–449.

Girardot M, Cavaillé J & Feil R (2012) Small regulatory RNAs controlled by genomic imprinting and their contribution to human disease. *Epigenetics* **7**:1341–1348 (doi:10.4161/epi.22884).

Haffner MC, Pellakuru LG, Ghosh S et al. (2013) Tight correlation of 5-hydroxymethylcytosine and Polycomb marks in health and disease. *Cell Cycle* **12**:1835–1841.

Hanover JA, Krause MW & Love DC (2012) Bittersweet memories: linking metabolism to epigenetics through O-GlcNAcylation. *Nat Rev Mol Cell Biol* **13**:312–321 (doi:10.1038/nrm3334).

Issa JP (2013) The myelodysplastic syndrome as a prototypical epigenetic disease. *Blood* **121**:3811–3817 (doi:10.1182/blood-2013-02-451757).

Keating ST & El-Osta A (2013) Glycemic memories and the epigenetic component of diabetic nephropathy. *Curr Diab Rep* **13**:574–581 (doi:10.1007/s11892-013-0383-y).

Kim GH, Ryan JJ & Archer SL (2013) The role of redox signaling in epigenetics and cardiovascular disease. *Antioxid Redox Signal* **18**:1920–1936 (doi:10.1089/ars.2012.4926).

Koerner MV & Barlow DP (2010) Genomic imprinting-an epigenetic gene-regulatory model. *Curr Opin Genet Dev* **20**:164–170 (doi:10.1016/j.gde.2010.01.009).

Lehnen H, Zechner U & Haaf T (2013) Epigenetics of gestational diabetes mellitus and offspring health: the time for action is in early stages of life. *Mol Hum Reprod* **19**:415–422 (doi:10.1093/molehr/gat020).

Nistala R, Hayden MR, Demarco VG et al. (2011) Prenatal programming and epigenetics in the genesis of the cardiorenal syndrome. *Cardiorenal Med* **1**:243–254 (doi:10.1159/000332756).

Nolan CJ, Damm P & Prentki M (2011) Type 2 diabetes across generations: from pathophysiology to prevention and management. *Lancet* **378**:169–181 (doi:10.1016/S0140-6736(11)60614-4).

Ntziachristos P, Mullenders J, Trimarchi T & Aifantis I (2013) Mechanisms of epigenetic regulation of leukemia onset and progression. *Adv Immunol* **117**:1–38 (doi:10.1016/B978-0-12-410524-9.00001-3).

Sandovici I, Hammerle CM, Ozanne SE & Constância M (2013) Developmental and environmental epigenetic programming of the endocrine pancreas: consequences for type 2 diabetes. *Cell Mol Life Sci* **70**:1575–1595 (doi:10.1007/s00018-013-1297-1).

Stapleton G, Schröder-Bäck P & Townend D (2013) Equity in public health: an epigenetic perspective. *Public Health Genomics* **16**:135–144.

Turunen MP, Aavik E & Ylä-Herttuala S (2009) Epigenetics and atherosclerosis. *Biochim Biophys Acta* **1790**:886–891 (doi:10.1016/j.bbagen.2009.02.008).

Villeneuve LM & Natarajan R (2010) The role of epigenetics in the pathology of diabetic complications. *Am J Physiol Renal Physiol* **299**:F14–F25 (doi:10.1152/ajprenal.00200.2010).

Webster AL, Yan MS & Marsden PA (2013) Epigenetics and cardiovascular disease. *Can J Cardiol* **29**:46–57 (doi:10.1016/j.cjca.2012.10.023).

Wierda RJ, Geutskens SB, Jukema JW et al. (2010) Epigenetics in atherosclerosis and inflammation. *J Cell Mol Med* **14**:1225–1240 (doi:10.1111/j.1582-4934.2010.01022.x).

Xu SS, Alam S & Margariti A (2012) Epigenetics in vascular disease—therapeutic potential of new agents. *Curr Vasc Pharmacol*.

15

EPIGENETICS OF MEMORY, NEURODEGENERATION, AND MENTAL HEALTH

Whether or not epigenetics has a role in the individual's state of mind is probably the most difficult decision about cause and effect we will encounter, largely because our understanding of the cellular mechanisms that mediate consciousness and memory are at best rudimentary. In essence, the retention of memories is the basis of our personal identities, because even if we have the ability to fire neuronal impulses in specific patterns to respond to external stimuli or to maintain heart rate, respiration, or intestinal peristalsis, for example, if we cannot retain memory of our surroundings for more than a few seconds we will not be able to form a model of the world around us. Building a picture of the world is a critical determinant of how we process information at a higher level—in short, how we think. Consequently, understanding the molecular basis of memory is crucial to interpreting possible epigenetic influences on our mental state.

15.1 MEMORY

Memory formation relies on specific regions of the brain

Some definitions of the types of memory that may be stored in the human brain are helpful at this point. **Short-term memory**, also known as **primary** or **active memory**, represents the processing of current information. This can involve stimuli from the outside world such as visual signals, sounds, smells, and other sensory information, and also the input of thoughts from what we refer to as our conscious mind. Short-term memory is thought to be processed (at least in part) by supragranular pyramidal neurons in the dorsolateral prefrontal cortex (see **Box 15.1** for a fuller description of the regions of the brain), which is a highly developed primate brain structure that is essential to our complex behavior patterns.

Signaling between the cells relies on the ionic transport characteristics of their cytoplasmic membranes, with the flow of ions allowing the membranes to depolarize and generate a potential difference that is the basis of the nerve cells' electrophysiological action. Short-term memories last at most for 20–30 seconds, but their activity is essential to the establishment of longer-term memories that can be recalled. It is important to recognize that what we refer to as a "memory"—by which we mean, for example, an image of a particular location or the sounds that make up a piece of music—will involve the coordinated depolarizations and repolarization of a large number of neurons acting together in a very specific pattern. A different memory is likely to use a completely different pattern of cells, and our understanding of how such patterns form and relate to

Box 15.1 The regions of the human brain

The principal regions of the human brain are shown in **Figure 1a**. The largest structures are the frontal lobes and the prefrontal, premotor, and primary motor cortexes, which form part of the cerebral cortex and are thought to be the sites for so-called higher brain functions such as thought, voluntary decision making, and language. The parietal lobe processes information from the environment (such as sound), and the functions of this region help us to be spatially aware of our environment. The occipital lobe, consisting of the visual association area and visual cortex, is the primary region processing visual information. Functions of the temporal lobe region are concerned with object recognition, learning and memory, and emotional reactions. The cerebellum is located at the base of the brain and is associated with the coordination of voluntary movements, motor learning, balance, and posture,

among others. The cerebellum contains more neurons than all the other regions of the brain despite being only 10% of total brain mass. The brainstem is located just in from the cerebellum and has important roles in homeostasis such as the maintenance of blood pressure, breathing, and heart rate. The limbic system (Figure 1b) is involved memory, learning, emotion, neuroendocrine function, and autonomic activities.

Figure 1 Regions of the human brain. (a) The principal regions of the human brain. Regions within the frontal lobe (blue and green), parietal lobe (purple), occipital lobe (orange and yellow), and temporal lobe (brown) are indicated. (b) The internal structures of the limbic system. (Courtesy of Peter Junaidy, 123RF.com)

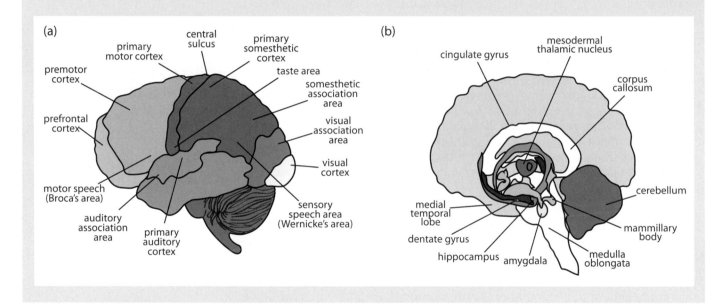

specific memory events is not clear at present. However, the molecular changes brought about in individual neurons in response to stimuli, both *in vivo* and *in vitro*, have been the subject of intensive investigation.

The electrophysiological activity of the neurons in short-term memory seems to be essential to the formation of long-term memories that psychologists would refer to as the **subconscious**—a set of information that is outside our current self-aware or conscious state but may be called into the short-term or working memory whenever it is needed. The formation of long-term memory seems to rely on an area of the brain referred to as the hippocampus, because selective damage to this structure impairs the ability to form long-term memories. Patients with damage to the hippocampus find difficulty in forming new, consciously accessible memories, although some memories that were acquired long before the injury seem to be preserved. This suggests that some of the actual memory-retention function may take place in other regions of the brain and that the hippocampus may simply be the structure controlling the initial establishment of long-term memory. The process of forming long-term memory may not be rapid, because newer memories (that is, memories formed imme-diately before hippocampal injury) seem to be lost along with the ability to generate new ones once the hippocampus is destroyed.

Structural changes and plasticity of synapses could be the basis of long-term memory

It is believed that the formation of long-term memories in a range of spe-cies involves structural changes in the connections between individual neurons. These connections, the synapses (Box 15.2), function by trans-ferring neurotransmitters across the synaptic cleft (the gap between the two neurons), with the neurotransmitter molecules originating from the presynaptic neuron and activating receptors in the postsynaptic neuron. Studies have shown that alterations in the types of tasks that animals are required to perform to obtain food produce changes in the number and size of vesicles in the presynaptic terminal (see Box 15.2) and cytoskeletal changes in the postsynaptic neurons. The latter changes occur mostly in the dendritic spines (protrusions from the dendrites that connect with the axons of other neurons) (see Box 15.2) and involve cytoskeletal changes that lengthen the spine. The changes that occur are referred to by the global term "synaptic plasticity," which includes alterations in the abil-ity of presynaptic neurons to release neurotransmitters, the numbers of receptors for such neurotransmitters on the postsynaptic neurons, and the structural changes described above. One of the major theories put forward to explain memory retention is that specific synaptic connec-tions are reinforced by increasing the responsiveness of the postsynaptic neurons to stimulation by the presynaptic neurons, a process that has been referred to as **long-term potentiation**. This process could involve changes in the number of receptor complexes (see Box 15.2) on the postsynaptic neuron to lower the threshold amount of neurotransmitter needed to activate the neuron. Activation occurs when neurotransmit-ters cross the synaptic cleft and bind to ligand-gated ion channels and/or G-protein-coupled receptors on the postsynaptic neuron that control the flow of ions into and out of the neuron. The neurotransmitter recep-tor proteins are subject to continual replacement; this implies a potential problem with the idea that memory might be maintained by specific num-bers of receptor protein complexes. Without a mechanism to replicate the manner in which specific synaptic connections are reinforced with a high degree of precision, long-term memories would not be possible. As explained below, there is mounting evidence that epigenetic control of gene expression may be one such mechanism through which long-term memories are created and maintained.

Epigenetic control of synaptic plasticity may contribute to memory maintenance

The formation of long-term memories seems to rely on gene expression changes that are brought about in response to the changes in receptor number and/or sensitivity that alter the synaptic sensitivity. N-methyl-d-aspartate (NMDA) receptors are glutamate-gated ion channels that function in a wide variety of neurons. High-frequency patterns of synaptic activity lead to the activation of NMDA receptors and an influx of Ca^{2+} into the cytoplasm of postsynaptic neurons. These higher calcium concentrations trigger several signal transduction pathways whose molecular cascades converge on the activation of extracellular signal-regulated kinase (ERK). Activated ERK regulates the expression of a range of genes, including several genes that encode transcription factors. The transcription factors, in turn, can establish a highly coordinated pattern of gene expression that results in the formation and stabilization of long-term memory. Maintaining this expression pattern seems to be under a

Box 15.2 How neurons work

In general, there are three kinds of neurons: **motor neurons** (for conveying motor information), **sensory neurons** (for conveying sensory information), and **interneurons** (which convey information between different types of neurons). Neurons all typically consist of four parts (**Figure 1**). First, the **main cell body** houses the nucleus, other organelles such as the endoplasmic reticulum, and a substantial proportion of the mitochondria. This serves much the same function as in other cell types and is responsible for the general synthesis and recycling of proteins. It is the second and third parts that are central to neuron functions: these are the processes that extend away from the cell body and function to conduct signals to or from the cell body. Incoming signals from other neurons are (typically) received through the **dendrites**, whereas the outgoing signal to other neurons flows along the **axon**. A neuron may have many thousands of dendrites, but it will have only one axon. The fourth distinct part of a neuron—the **axon terminal**—lies at the end of the axon. This is the structure that generates neurotransmitters that are able to cross the gap, known as the **synapse**, between the axon terminal and dendrites of two neurons (**Figure 2**).

To achieve long-distance, rapid communication, neurons have evolved special abilities for sending electrical signals (action potentials) along axons. This mechanism, called **conduction**, is how the cell body of a neuron communicates with its own terminals. The action potential is generated within the main body of a neuron by the transport of ions across the cell's plasma membrane. Normally, the membrane potential of a neuron rests at −70 mV, but the influx and/or outflux of ions (through ion channels during neurotransmission) makes the inside of the target neuron more positive. When this depolarization of the membrane reaches a threshold level, an electrical signal is generated that propagates along the axon (but not back through the dendrites) until it reaches the

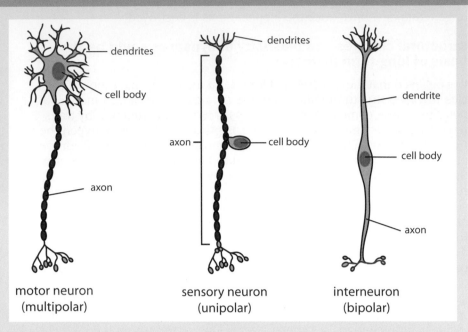

Figure 1 Basic types of neurons. Three principal types of neuron are shown along with relative positions of axons, nuclei, and dendrtites.

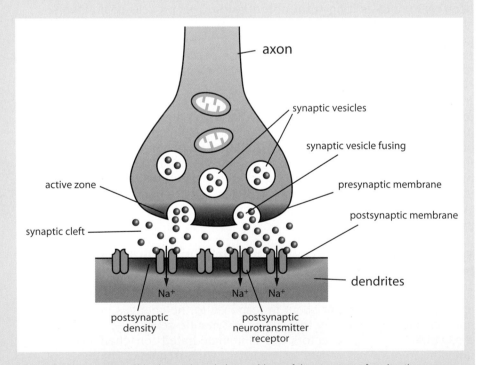

Figure 2 The synapse. This shows the relative positions of the structures forming the synapse.

axon terminal. An action potential travels along the axon quickly, moving at rates up to 150 m/s. Conduction ends at the axon terminal but triggers the synthesis of neurotransmitter chemicals, such as acetylcholine, that are released into the synapse and serve to pass the signal to any communicating neurons.

degree of epigenetic control, because several studies have suggested that the regulation of chromatin structure is important for facilitating long-term changes in neuronal function.

For example, exposure of experimental animals to pulses of light results in a change in chromatin structure in the neurons of the hippocampus (see Box 15.1). This change seems to involve increased acetylation of histone H4, while an alternative stimulus, that of the fear response, increases the acetylation of histone H3 but not that of H4. Interestingly, artificial modulation of histone acetylation through administration of HDAC inhibitors seems to elevate the level of synaptic sensitivity of hippocampal neurons, at least *in vitro*, and treatment of live animals with such HDAC inhibitors apparently leads to the enhanced formation of long-term memory. In many learning and memory experiments, acetylation has been shown to be a short-term modification whose levels peak 1–2 hours after the initial stimulus. It is highly likely that HDACs contribute to the ensuing decline in acetylation; therefore the success of HDAC inhibitors in enhancing long-term memory formation probably relies on slowing down this natural removal of the histone acetyl groups. In view of this, several studies have suggested that pharmacological HDAC inhibitors may be useful for treating the memory dysfunctions that accompany neurological diseases such as Alzheimer's disease.

Histone methylation is also implicated in the fear response. As little as 1 hour after introduction of a fear stimulus, H3K4me3 and H3K9me2 levels are higher in rat hippocampus. Although our knowledge of the specific genes affected by these changes is limited, the involvement of activating (H3K4me3) and repressive (H3K9me2) forms of histone methylation suggests that substantial changes in gene expression patterns are needed for memory formation. Another possibility is that a balance between the activating and repressive forms of histone methylation may be needed to set a threshold level of gene activity that functions to prevent the formation of memories of unimportant activities. For example, the H3K4me3 and H3K9me2 modifications may be present within the same gene promoter region (or other expression control elements), but it may be the precise amounts of each modification that control the expression level of that gene. If the concentration of a gene product is below a certain level, establishment of that particular neuron's contribution to the memory may not be possible. It is also possible that a system of this type could be "fine tuned" so as to increase or decrease the threshold for memory formation. Indeed, it may even be necessary to turn off the control while storage of useful memories is taking place, but the factors that might dictate the operation of controls of this type are unknown.

The interplay of multiple histone methylation marks in memory formation is underlined by changes that occur in the hours after fear conditioning in experimental animals. We have noted that H3K4 and H3K9 methylation is present in the hippocampus approximately 1 hour after stimulus, but regions of the brain thought to be involved in long-term memory formation under the control of the hippocampus do not show similar methylation changes until much later. For example, in the absence of further fear stimuli, increases in H3K4 methylation are absent from the dentate gyrus (see Box 15.1) until 48 hours after the initial stimulation.

It is important to distinguish between diseases that have a known, characterized cellular defect that can lead to neurodegeneration (such as Parkinson's disease and Alzheimer's disease) and those that do not seem to result from cell death at a gross level but nonetheless may arise from molecular dysfunctions in the central nervous system neurons. The latter class of conditions may be the cause of psychological illnesses if they interfere with the mechanisms through which memories are maintained, because such interference will disrupt the individual's ability to process and respond to information from the environment. Before examining the more complex dysfunctions (or at least more poorly defined ones) that lead to psychological illness, we will undertake a brief discussion of neurodegenerative diseases.

15.2 EPIGENETIC INVOLVEMENT IN NEURODEGENERATION

Most individuals undergo some degree of cognitive decline with age, ranging from occasional forgetfulness to large-scale loss of function associated with the destruction of significant numbers of central nervous system (CNS) neurons. The molecular basis of such neurodegenerative diseases is believed to be multifactorial, involving epigenetic, genetic, and environmental components. A principal risk factor for neurodegeneration is age. There is a high probability that a person who lives beyond 85 years will suffer from some degree of Alzheimer's disease, which is thought to result from a declining ability of CNS cell populations to respond with behavioral flexibility to environmental cues—that is, a decline in the phenotypic plasticity of these cells. Aging promotes a general decrease in DNA methylation, with hypermethylation at specific promoter regions of the genome in CNS neurons. These changes are similar to those associated with cancer. Age-related cognitive decline is known to be associated with changes in the plasticity of chromatin (for example the ability to change histone modifications to respond to environmental changes such as new activities or stress) in the cells of the hippocampus. Aging mice show deregulation of histone H4K12 acetylation (defined by an inability to maintain the same H4K12 acetylation patterns as the hippocampal cells of younger mice), and the accompanying cognitive failure is thought to be related to changes in the gene expression profile of the hippocampal neurons. Reestablishment of the acetylation levels restores both the gene expression pattern and cognitive ability to youthful levels.

Epigenetic alterations may contribute to the development of Alzheimer's disease

Alzheimer's disease (AD) is characterized by a range of progressive changes in the anatomy and function of the brain, with the most common feature being the accumulation of aggregates of the protein beta-amyloid (**Figure 15.1**). These aggregates, and the commonly associated neurofibrillary tangles, may have an impact on neuronal and synaptic function in certain regions of the brain and thereby lead to cognitive impairment. Alzheimer's disease is the most common form of dementia, affecting over 6% of the population aged more than 65 years, but the precise molecular events leading to beta-amyloid aggregate formation are not clearly understood. Mutations within the amyloid precursor protein (APP; a membrane protein present in neurons and possible regulator of synapse formation) and the proteins presenelin 1 and 2 (PSI and PSII—part of a gamma secretase enzyme complex that can cleave APP to generate the active form of the protein) have been linked to early-onset familial cases, but we have less information about how sporadic cases arise. It has been suggested that genes such as *Apolipoprotein E4* (the protein Apolipoprotein E4 is involved in lipid transport but its function in AD is unclear) may contribute to sporadic onset, but the data to support this are less convincing than the genes involved in familial AD. In either case, there is little we can do once AD has destroyed enough of the brain to cause significant cognitive impairment. None of the current treatment options affect the central molecular events that constitute the pathology of the disease.

Several studies of epigenetic alterations in the brains of patients with AD, obtained post mortem, indicate hypomethylation of DNA in the promoter region of the *APP* gene when compared with brains from healthy controls and from patients who had forms of dementia other than AD. Moreover, other studies showed that the *APP* promoter DNA undergoes

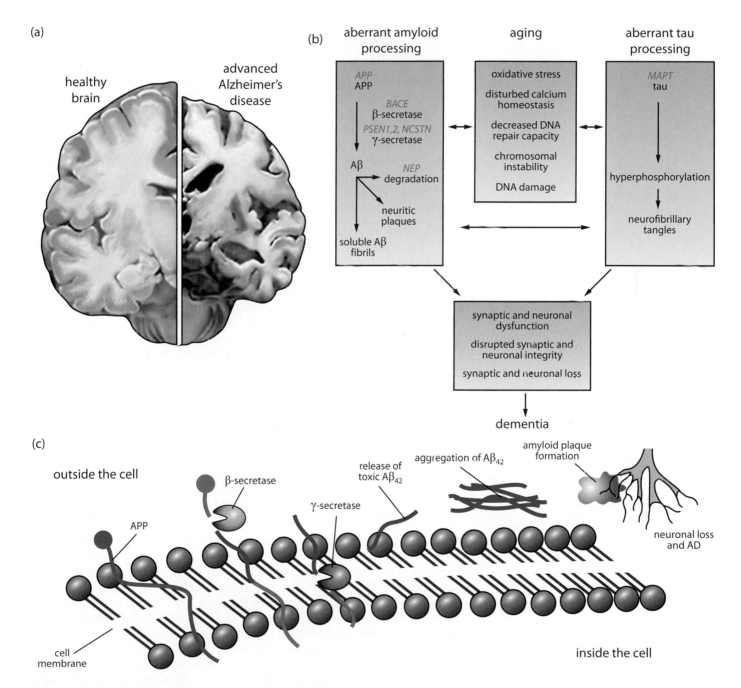

Figure 15.1 Mechanisms of Alzheimer's disease. (a) Extensive degradation of brain structure resulting from Alzheimer's disease. (b) Brain structure degradation is thought to result from the accumulation of amyloid plaques and fibrillary tangles arising from phosphorylated Tau protein. Aβ, beta-amyloid. (c) According to the amyloid hypothesis the amyloid precursor protein (APP) is cleaved by β-secretase and γ-secretase to form the peptide fragments of which the Aβ42 (the form of β-amyloid most commonly found in Alzheimer's disease) fragment is toxic to neurons. Moreover, these fragments can aggregate into amyloid plaques, which are a hallmark feature of Alzheimer's disease. Aggregation occurs when there is an imbalance between the accumulation and clearance of Aβ42 from the brain. [Adapted from Christensen DD (2007) *CNS Spectr.* 12, 113. With permission from Cambridge University Press.]

a progressive demethylation after the age of 70 years that may be related to the progressive deposition of amyloid in the brains of this age group. Another study showed that DNA methylation was decreased in the promoter region of the *Tau* protein gene in the parietal cortex, with a corresponding increase in the gene's expression level. The aggregation of Tau protein in neurofibrillary tangles is another contributing factor to AD, so dysregulation of the expression level of this protein could be a key event. We already noted in Chapter 14 that gradual hypomethylation of DNA accompanies aging, so it is conceivable that the progressive deficiency of the major methyl donor molecule, *S*-adenosylmethionine (AdoMet), could contribute to the hypomethylation of the *APP* and *Tau* promoters.

Some data exist suggesting that overexpression of the AD-linked gene encoding presenelin may be associated with decreased AdoMet levels, a point underlined by the parallel observation of higher levels of *S*-adenosylhomocysteine—the downstream metabolic product of AdoMet-mediated transmethylation reactions—in the brains of patients with AD. *S*-adenosylhomocysteine also inhibits the methyltransferases that utilize AdoMet, and methionine-*S*-adenosyltransferase (the enzyme that synthesizes AdoMet) is known to be expressed at lower levels in brains subject to neurodegeneration.

The AdoMet methylation system operates on many substrates within the cell, and we would therefore expect adverse changes within this system to affect the body in many different ways. In support of this idea, studies have linked folate deficiency (which adversely affects AdoMet concentrations) to many neurological and psychological disorders, including dementia, impaired cognition, depression, psychosis, AD, and Parkinson's disease. These observations suggest an epigenetic contribution to AD only insofar as they may be able to disrupt the maintenance of the DNA methylation patterns (and thus possibly the histone modification patterns) that control the expression of certain genes whose dysfunctions are thought to be causative factors in AD.

One of the underlying themes of this chapter is to attempt to explain the occurrence of disease in the light of events in the earlier life of the individual that may predispose that person to develop certain conditions. So what evidence is there for an epigenetic propensity to develop AD? The possibility that an epigenetic mechanism may apply at all comes from an examination of the influence of maternal behaviors that are known to induce long-term changes in behavioral and hypothalamic–pituitary–adrenal (HPA) axis responses to stress.

The **hypothalamic–pituitary–adrenal axis (HPA axis)** is a complex set of interactions between the hypothalamus, the pituitary gland, and the adrenal or suprarenal glands, which regulate things such as temperature, digestion, the immune system, mood, sexuality, and energy use. It is also a major part of the system that controls reactions to stress, trauma, and injury. Interestingly, rat pups that are more regularly licked and groomed by their mothers show lower levels of DNA methylation in exon 17 of the glucocorticoid receptor gene, a significant component of the HPA axis. Conversely, low levels of maternal care during early life led to an increase in DNA methylation and a decrease in histone acetylation in the same promoter region, which in turn increased the HPA responses to stress later in life. In humans, the post-mortem brains of suicide victims showed increased DNA methylation of the neuron-specific exon 17 of the glucocorticoid receptor gene, and this incidence also showed a strong association with abuse of those individuals as children. Abnormalities in the HPA axis have been described in a substantial proportion of patients with AD. Prolonged activation of the HPA axis generates higher plasma levels of cortisol, which is known to decrease the number of neurons in the hippocampus; because the hippocampus is a major site of pathological changes in AD, it is possible that AD is associated with HPA axis hyperactivity. Whether such hyperactivity is induced by epigenetic changes brought about by stressful experiences in early life is not clear. However, it is a tempting speculation, particularly because the incidence of cognitive decline is higher among those affected by the Dutch hunger winter (see Chapter 14 for more information about the latter).

There is some evidence that epigenetic mechanisms may contribute to Parkinson's disease

Parkinson's disease (PD) is the second most common neurodegenerative disease, but as yet no clear, direct evidence of epigenetic involvement in the processes leading to this disease has been established. PD is a progressive disorder with a range of neuropsychiatric symptoms for which neither preventative nor long-term remediation options are available. The condition arises when substantial numbers of neurons in the pars compacta region of the substantia nigra (see **Figure 15.2**) are destroyed (or at least functionally impaired) to the point that production of the neurotransmitter dopamine is greatly decreased. Loss of these neurons causes degeneration of the striatum, thus leading to pathological alterations in the basal ganglia motor circuit, which in turn produces the characteristic motor symptoms of PD, including tremor and postural instability. The molecular factors leading to the onset of PD are still poorly understood. We know that the incidence of the disease increases with age, which may imply some degree of epigenetic involvement, but there is also evidence for environmental and genetic contributions.

The best examples of environmental agents that affect the onset of PD are the synthetic opioid by-product and neurotoxin precursor 1-methyl-4-phenyl-1,2,3,6-tetrahydropyridine (MPTP), the herbicide paraquat, and the insecticide rotenone, all known to induce PD-like symptoms although a clear cause for these effects remains to be determined. More direct evidence of epigenetic involvement comes from the impact of therapeutic PD agents that are known to affect the epigenetic regulation of gene expression. One of the first-line treatment options for PD is administration of

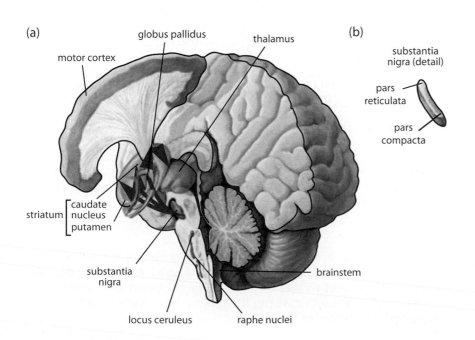

(a)
motor cortex
globus pallidus
thalamus
striatum { caudate nucleus / putamen
substantia nigra
locus ceruleus
raphe nuclei
brainstem

(b)
substantia nigra (detail)
pars reticulata
pars compacta

Figure 15.2 Pathophysiology of Parkinson's disease. (a) The brain regions affected by Parkinson's disease. (b) Detail of the substantia nigra, where the disappearance of pigmented dopaminergic neurons is an anatomopathologic aspect of Parkinson's disease.

the dopamine analog levodopa. In animal models in which PD-like symptoms are induced by MPTP, there is substantial decrease of H3K3me3 in the striatal neurons, whereas chronic levodopa therapy leads to the deacetylation of lysines 5, 8, 12, and 16 on histone H4. Another example is the use of monoamine oxidase inhibitors such as tranylcypromine (*trans*-2-phenylcyclopropylamine) to prevent the death of rat nigral dopaminergic neurons after the administration of toxins that would normally kill these cells. Tranylcypromine is an inhibitor of LSD1 (a flavin-dependent monoamine oxidase that acts as a lysine-specific histone demethylase) but has also been shown to affect the regulation of p53.

Genetic influences on PD development may also lead to alteration in epigenetic control of several downstream genes involved in PD. For example, in dopaminergic neurons (that is, neurons whose primary neurotransmitter is dopamine) of PD patients, misfolded α-synuclein is known to accumulate in cytoplasmic inclusions known as Lewy bodies, which may be one of the causes of PD-related neurodegeneration. The normal function of correctly folded α-synuclein is not clear. Some studies have shown the gene encoding α-synuclein (*SNCA*, short for synuclein, alpha (non A4 component of amyloid precursor)) to be under epigenetic control and to be hypomethylated in the brains of patients with PD when compared with undiseased controls. In addition, α-synuclein is known to decrease histone H3 acetylation by direct interaction with the nucleosomes and also by inhibition of histone acetyl transferases. Consistent with these findings, histone deacetylase inhibitors such as sodium butyrate are able to reduce the impact of α-synuclein toxicity in neurotoxin-induced PD models.

Additionally, the expression levels of NURR1 (nuclear receptor related 1 protein), a transcription factor central to dopaminergic neuron maintenance, are lower in PD patients. This gene product functions as part of the CoREST repressor complex, which also comprises the activities of HDACs, the G9a histone methyltransferase, and the histone demethylase LSD1. The range of potential targets for this epigenetic modification complex is large, and indeed there are many other genes that are targeted and potentially controlled by PD risk factor genes such as *PARK2* (which encodes PARKIN, a protein involved in the ubiquitin–proteasome system of protein degradation and which, if mutated, results in a familial form of PD). However, this merely suggests that some genetic disruption is affecting the ability of the systems that normally control the expression of downstream genes by epigenetic modification to perform their functions and does not imply any basic epigenetic cause of PD.

15.3 THE IMPACT OF EPIGENETIC CONTROL OF GENE EXPRESSION ON MENTAL HEALTH

Neurodegenerative diseases affect the CNS by altering the gross functionality of its neurons in a way that is ultimately lethal to them, which leads to loss of control over many bodily functions, including the wholesale destruction of memory and cognition. More insidious are those diseases that affect neuronal functions and interneuronal communications by disrupting the way in which the mind can interact with the environment. These so-called mental illnesses profoundly affect the individual's interaction with society.

We can gain some insight into possible impacts of epigenetic dysregulation on the workings of the mind from the influence of maternal deprivation on incidences of mental disorder in the offspring. Starvation is the primary mechanism for depriving the mother of nutrients during critical phases of development of her offspring. A study performed on children of

Chinese mothers subjected to food deprivation has provided interesting data on mental disorders. Between 1959 and 1961, food production in the Chinese province of Anhui (and many others, but Anhui was the worst in terms of the number of deaths) collapsed under the combined problems of floods, droughts, and poorly organized collective farming. During the famine years, the birth rates (per 1000) in Anhui decreased by approximately 80%, from 28.28 in 1958 and 20.97 in 1959 to 8.61 in 1960 and 11.06 in 1961. Among births that occurred during the famine years, the adjusted risk of developing schizophrenia in later life increased significantly, from 0.84% in 1959 to 2.15% in 1960 and 1.81% in 1961. These are not enormous increases in the probability of developing schizophrenia, and had it not been for the fact that this study confirmed similar findings from our other, by now familiar, example of starvation deprivation the Dutch hunger winter (see Chapter 14), the data might have been ignored.

Mental illness can be difficult to diagnose and treat effectively. To begin our investigation of this topic, we first need a clear idea of what we actually mean by mental illness. Mental health is a state of well-being in which an individual realizes his or her own abilities, can cope with the normal stresses of life, can work productively, and is able to make a contribution to his or her community. In this positive sense, mental health is the foundation for individual well-being and the effective functioning of a community.

Multiple social, psychological, and biological factors determine the level of mental health of a person at any point in time. For example, persistent socio-economic pressures are recognized risks to mental health for individuals and communities, with the clearest evidence being associated with indicators of poverty, including low levels of education. Poor mental health is also associated with rapid social change, stressful work conditions, gender discrimination, social exclusion, unhealthy lifestyle, risks of violence, physical ill-health, and violations of human rights.

There are also specific psychological and personality factors that make people vulnerable to mental disorders. Additionally, there are some biological causes of mental disorders, including genetic factors and imbalances in chemicals in the brain. It is conceivable that epigenetic control of gene expression may contribute to all of the factors listed above, but again, the evidence supporting such involvement is only now emerging. We will now review some of these data obtained in studies of specific psychological illnesses.

Disruption of epigenetic regulation may explain some features of bipolar disorder

Bipolar disorder is a condition characterized by periods of intense depression interspersed with periods of apparent elation and good mood or occasional irritability. The change from elation to depression can be very rapid. The condition is subdivided into three classes of disease. Type 1 bipolar disorder applies to individuals who have experienced at least one period of manic excitability and periods of major depression. This type used to be called manic depression. Bipolar disorder type 2 individuals display periods of high energy and impulsive behavior alternating with low-energy, depressive phases. The difference from type 1 is that the energetic periods are not as active (or manic). The third type is a much milder condition sometimes referred to as cyclothymia, which involves much more moderate mood swings. The manic phases of all three groups vary in duration, potentially lasting for months, and include symptoms of sleeplessness (or limited need for sleep), poor judgment, and lack of control over eating, drinking, drug use, temper, or physical activity. The depressive phases are characterized by intense sadness, eating problems (both excessive and reduced appetite), sleep disturbance, withdrawal from social contact, lack of energy, and suicidal tendencies.

Depression is thought to arise from the way in which the individual's brain processes information during times of stress. Evidence from epidemiological studies suggests that some degree of genetic predisposition may interact with environmental cues to potentiate the onset of depression. This belief arises from the occurrence of depression in some individuals exposed to stressful experiences while others seem to be more resilient. This variation may be related to the construction and function of neurotransmission systems, such as those mentioned at the beginning of this chapter. Some of the aberrations identified in these studies are slight alterations in the structure of the brain-derived neurotrophic factor (BDNF) as a result of single nucleotide polymorphisms in its gene's coding sequence and mutations in the serotonin transporter gene. The latter example involves variation in the length of a repeat sequence found in the 5' control region of the serotonin transporter gene, and individuals with the short form tend to show more depressive symptoms in response to stress.

These observations are interesting in their own right because disrupted epigenetic control of these (and other) depression-related genes may be responsible for producing symptoms even in persons not affected by the previously described mutations. Interesting evidence from a variety of species indicates that higher than normal levels of care from the parents (or conversely, increased stress) early in life influence the risk of depression in adulthood. Considerable amounts of data from human studies show that early childhood abuse or neglect increases the risk of a range of psychopathologies, but it is even more interesting that patterns of abuse and neglect may be transmitted from parent to child. The same seems to apply in nonhuman primates. Infant rhesus macaques who were removed from the care of a non-abusive mother and placed in the care of an abusive mother demonstrated similar abusive behavior toward their own young. Rats, too, demonstrate a range of care levels toward their offspring; in these cases, the expression of the glucocorticoid receptor in the hippocampus of the offspring is regulated by the amount of maternal licking and grooming. This regulation seems to be stable into adult life, and rats subjected to a higher level of care in the first week of life show greater resistance to and recovery from stress. Involvement of the glucocorticoid receptor implies the possible involvement of the HPA axis, which has been implicated in the development of AD. Several studies suggest that the expression of the glucocorticoid receptor is controlled by DNA methylation of the gene's 5'-control elements, because the offspring of mothers who provide greater levels of licking and grooming have lower DNA methylation than do the offspring of less caring mothers. These DNA methylation patterns are also stable into adulthood, but pharmacological intervention with HDAC inhibitors is able to reverse the epigenetic modification at the adult stage.

Several strategies have been developed to treat the symptoms of depression, including the administration of monoamine oxidase inhibitors, tricyclic antidepressant drugs, and the rather more extreme electroshock therapy, but there is still a paucity of information about how these treatments actually work to reduce depressive symptoms. Despite its reputation as a drastic treatment, electroshock therapy has been shown to increase acetylation of histone H4 in the promoter regions of the genes encoding both c-Fos and CREB (c-AMP regulatory element binding protein). The treatment also leads to H3 phosphorylation and acetylation in c-Fos, but these modifications generally require the repeated administration of electroshock therapy. These findings demonstrate that several long-lasting histone modifications occur in response to chronic electroshock therapy, leading to long-term effects on gene regulation.

So how does this relate to the specific condition of bipolar disorder? The condition seems to show a preference for maternal inheritance, but in cases in which it is transmitted from the father, the symptoms are much more intense. This mode of inheritance is strongly reminiscent of imprinting diseases (see Chapter 14), and paternally inherited cases of bipolar disorder are linked to chromosome 18q22. Searches for the relevant gene in this region identified *GNAL* (G-olf-α) as a potential candidate, but evidence of allele-specific methylation has yet to be found. In reality, the evidence for imprinting defects in bipolar syndrome is still sketchy. Conversely, some research findings—for example the successful treatment of bipolar symptoms with known epigenetic-modifying drugs such as valproic acid, and the antidepressant effect of the methyl group donor AdoMet—support the idea that bipolar disorder could nevertheless be epigenetic in its origins. The latter finding actually turns out to be more complex than perhaps would be implied by the involvement of AdoMet in DNA and protein methylation reactions, because administering AdoMet to bipolar patients is more likely to tip the balance between "mania" and "depression" in favor of mania. Giving methionine itself to bipolar patients also seems to work to some extent (because it is the precursor of AdoMet), but this also reverses the high DNA methylation levels of glucocorticoid receptors induced by low levels of early-life licking and grooming in rats, which is rather contradictory to the methylation-increasing effects of AdoMet.

In conclusion, the idea that epigenetic control of gene expression may have an impact on bipolar disorder is tantalizing, but the precise mechanisms by which this might occur are still far from clear.

Epigenetic regulation is a factor in major depressive disorder

A clear diagnosis of depression can be hard to reach, because most individuals suffer from low mood or sadness at some point in their lives, often as a result of unfortunate or stressful events. A depressive state may be said to occur when the low mood persists even after the stimulus for the normal emotion of sadness has been removed or when it appears for no obvious reason. However, it can be difficult to distinguish this latter form of depression from depression in which a psychological precipitating event is present. Certain symptoms need to be present before a patient's condition can be characterized as being a major depressive disorder (MDD). There should be characteristic sadness or irritability accompanied by sleep disturbances, loss of appetite, constipation, loss of the ability to experience pleasure, slowing of speech and actions, loss of sexual desire, and suicidal tendencies. Moreover, these changes should not be transitory and should last a minimum of 2 weeks, and they should not involve the periods of manic hyperactivity and euphoria that are characteristic of bipolar disorder. Despite these criteria, MDD is a heterogeneous condition in which the individual patient's response to treatment can be variable.

Major depressive disorder has some degree of heritability. Siblings born to parents with this condition have a threefold elevated risk of developing the disease, while the level of heritability in twins is 37%, which is lower than that of other mental disorders such as bipolar syndrome and schizophrenia. Nevertheless, this level of heritability is sufficient to have led to the search for genetic causes of MDD. Despite this, in every study of families affected by the disorder, the findings indicate MDD to be a complex genetic disease in which no single chromosomal locus has been implicated. The strongest evidence from genetic linkage studies suggests the involvement of chromosome 15q25–15q26 in early-onset depression; however, the attributable risk over the whole population for this locus is estimated to be fairly small, so it cannot be the sole cause of MDD. A few other polymorphisms in genes linked to the serotonin neurotransmitter system have been implicated. The serotonin transporter gene *5-HTTLPR* has a polymorphic variant in which the gene promoter is less active, the effect being that less serotonin is taken into presynaptic neurons and there is a consequent abnormal increase of serotonin in the synaptic cleft. There is evidence to suggest that this predisposes an individual to MDD, but other studies have indicated that it may only contribute to an anxious or pessimistic personality, which is not the same as actual depression.

Whether or not *5-HTTLPR* can be formally implicated in MDD, the serotonin link highlights an important mechanism believed to be a major root cause of depression, namely the increase or decrease in the supply of neurotransmitters in the brain. The availability of two of these neurotransmitters is controlled by the noradrenergic and serotonergic systems, which seem to be involved in the manifestation of MDD. Decreases in the availability of these neurotransmitters may also cause depression; many antidepressant drugs therefore counteract a lack of norepinephrine and serotonin by blocking their re-uptake by the presynaptic neurons. The effect of this is to increase the availability of these neurotransmitters to the postsynaptic neurons, thereby increasing the level of signaling through these cells. If MDD resulted from an excess of serotonin due to the *5-HTTLPR* polymorphic variant, antidepressant drugs of this type would not be effective. Monoamine oxidase is also part of this process, because this enzyme catabolizes norepineprine and serotonin in the presynaptic neuron; inhibitors of the enzyme can therefore increase neurotransmitter availability and thus are another important class of antidepressant drugs.

In fact, this latter observation has given rise to the **monoamine-deficiency hypothesis** of depression (**Box 15.3**), which is further supported by findings that brain monoamine oxidases are increased by 30% in patients suffering from MDD. Despite this, direct evidence of

Box 15.3 The monoamine deficiency and HPA axis hypotheses of depression

The monoamine hypothesis of depression postulates a deficiency in serotonin or norepinephrine neurotransmission in the brain. Monoaminergic neurotransmission is mediated by serotonin (5-hydroxytryptamine 1A [5-HT1A] and 5-hydroxytryptamine 1B [5-HT1B]) or norepinephrine (noradrenaline) released from presynaptic neurons.

Serotonin is synthesized in the presynaptic neurons and precise concentrations of this neurotransmitter are maintained by the destruction of re-internalized serotonin by monoamine oxidase (MAO; **Figure 1**). Increased levels and/or activity of monoamine oxidases in individuals with MDD may account for decreased concentrations

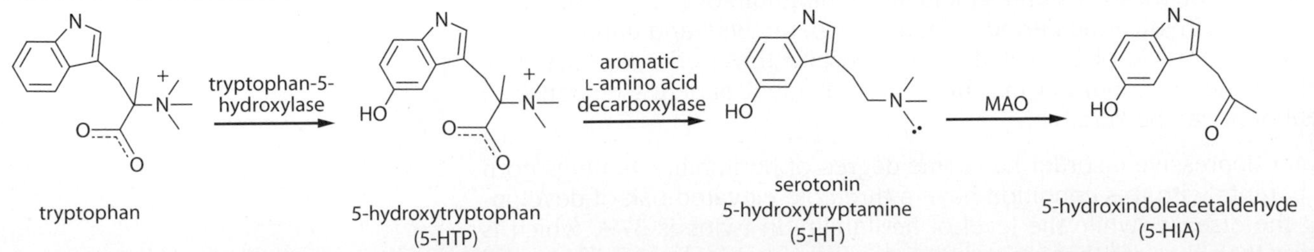

Figure 1 Formation of serotonin from tryptophan and its degradation by monoamine oxidase (MAO). The reaction needed to generate serotonin is shown with the enzymes required for each step.

Continued

Box 15.3 (*Continued*)

of serotonin that could be a causative factor of the disease. Findings in patients with depression that support the monoamine deficiency hypothesis include a relapse of depression with inhibition of tyrosine hydroxylase or depletion of dietary tryptophan, and an increased frequency of a mutation affecting the brain-specific form of tryptophan hydroxylase.

The HPA axis (**Figure 2**) hypothesis of depression postulates that abnormalities in the cortisol response to stress may underlie depression. Stressful stimuli perceived by the brain cause the hypothalamus to release ACTH (also known as corticotropin), a process mediated by corticotropin-releasing hormone; CRH); this induces the adrenal glands to produce glucocorticoids. Findings in patients with depression that support the HPA axis hypothesis include persistently raised levels of cortisol, an increased size of the pituitary gland and adrenal cortex, and reduced neurogenesis in the hippocampus.

Hyperactivity of the HPA axis is a well-documented phenomenon in MDD. This dysregulation is manifested by, among other things, cortisol hypersecretion, failure to suppress cortisol secretion after dexamethasone administration, exaggerated adrenal responses to endocrine challenges, and blunted ACTH response to CRH administration. This last observation has been interpreted as evidence of pituitary down-regulation of CRH receptors secondary to an increase in secretion of CRH. There is indeed good evidence of increased central drive, based on increased activity at the nadir of the circadian rhythm as well as more direct findings of elevated CRH in the CSF of depressed patients, and increased CRH immunoreactivity and mRNA levels in the paraventricular nucleus of the hypothalamus (PVN). Interestingly, post-mortem studies of suicide victims have also found evidence of chronic activation of HPA, such as adrenal hyperplasia, down-regulation of CRH receptors, and increases in mRNA for pro-opiomelanocortin (the precursor for ACTH) in the pituitary. It is not known whether these changes are due to the fact that a significant subset of suicide victims are patients with depressive disorders, to the stress surrounding the suicide itself, or to a neurobiological abnormality common to all suicide victims irrespective of diagnosis.

Figure 2 The HPA axis. To help cope with challenge, the hormonal stress response increases heart rate and respiration, energy reserves, and hormones like glucocorticoids, lower-priority functions, such as gastrointestinal function and reproduction are decreased. The immune system is stimulated in the short term but can be suppressed by prolonged, severe stressors.

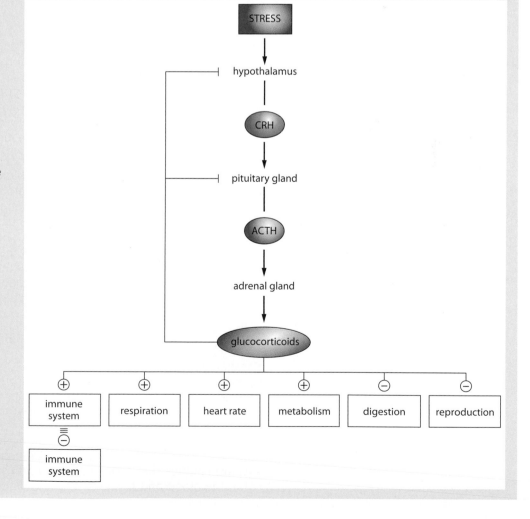

a defect in monoamine transmission as a cause of depression is still lacking. This has led to a recent trend of exploring the effects of this system on downstream components such as receptors for serotonin on the presynaptic and postsynaptic neurons. Some results suggest that the sensitivity of the serotonin 1A receptor is reduced in MDD patients and that the α-2-noradrenergic receptor seems to be more sensitive in MDD patients. Because the α-2-noradrenergic receptor modulates norepinephrine release by feedback inhibition in the postsynaptic neurons, the receptor's increased sensitivity is consistent with the lower than normal levels of norepinephrine found to be released in these cells, which may correlate with the symptoms of MDD.

It is also possible that disruption of the second messengers that are synthesized in response to primary neurotransmitter signals could generate symptoms similar to those that occur on the disruption of serotonin and norepinephrine regulation. In this regard, the levels of phosphatidylinositol and cyclic AMP in the post-mortem brains of MDD suicide victims have been found to be lower than those in non-MDD controls. This implies that downstream signaling from the monoamine oxidases may be able to induce symptoms of depression even without changes in the expression level of the underlying enzymes. However, the fact still remains that the strongest evidence for monoamine oxidase enzyme involvement in depression comes from animal models in which the genes encoding these enzymes or their downstream targets are knocked out. The serotonin re-uptake receptor knockout mouse is a much more anxious animal than control mice, and shows some similarities to humans who carry polymorphic variants of this gene.

Of note is the interesting effect that cocaine has on depression. Normally, as with amphetamines, this substance is able to induce the release of large quantities of monoamines by the presynaptic neurons (which presumably accounts for its stimulatory actions), but in patients with MDD the administration of cocaine frequently worsens the depressive state. It has been suggested that the ability of cocaine and amphetamines to deplete the presynaptic neurons of monoamines may be the factor that precipitates the "crash" into a much deeper depression in a brain that already has lower than normal levels of neurotransmitters.

So again we must ask the question of how does epigenetic regulation of gene expression relate to these possible causes of MDD. Naturally, it is possible that individual genes in the aforementioned neurotransmission pathways may be subject to epigenetic controls and that the ways in which these epigenetic states are regulated may change according to environmentally induced factors. The potential range of factors in a group of interacting cells as complex as the human brain is, of course, enormous because every nerve action potential, synapse formation, or modulation of the synapse sensitivity (the action of forming memories) is a potential stimulus for epigenetic modification of the neurotransmission system. Understanding the epigenetic modification patterns associated with every gene in every neuron in the brain is too complex a task to envisage. However, once more the HPA axis may be a significant factor.

For example, stressful situations normally stimulate the hypothalamus to synthesize corticotropin-releasing hormone (CRH; see Box 15.3). This molecule then stimulates the pituitary to release corticotropin into the bloodstream; in turn, corticotropin stimulates the adrenal cortex to release cortisol. In view of this, it is conceivable that epigenetic disruptions of the HPA axis, as discussed earlier, may be able to modulate the individual's response to stressful situations. In support of this hypothesis are findings that MDD patients tend to have higher levels of cortisol in the blood and higher levels of CRH in their cerebrospinal fluid. It is also interesting that adults with a history of childhood abuse show similar elevation of this molecule, but not all of these individuals go on to develop MDD. A possible difference between individuals with high levels of CRH but no MDD and those with full-blown MDD may be that the feedback mechanism for suppressing CRH synthesis by the hypothalamus in response to raised cortisol levels is mostly absent in severely depressed individuals. A drawback to this hypothesis is that many MDD patients do not seem to show imbalances of the HPA axis, but it may be that the imbalance occurred at some earlier stage in their lives and left a lasting epigenetic mark on other genes. Persistent early-life stress is known to cause hypermethylation of the promoter region of the gene encoding arginine vasopressin, which potentiates the actions of CRH under circumstances (such as stress) that demand sustained activation of the pituitary and adrenal glands.

15.4 SUMMARY

The principal message from this section on the impact of epigenetic control of gene expression on mental health is that there is a small but growing set of evidence favoring epigenetic causes for this group of conditions. Quite apart from epigenetic influences on the age-related neurodegeneration suffered by most individuals, there is evidence that some mental health problems can arise through changes in epigenetic regulation that are sometimes imposed on the genome relatively early in life. The range of such mental conditions is quite large, although much of this range probably reflects the criteria that we use to define the nature of these diseases.

In this discussion we have chosen to focus on just two of the major psychological illnesses, namely bipolar disorder and MDD. There are other conditions, such as schizophrenia, that are perhaps just as worthy of inclusion in this section; however, the evidence available from the limited number of studies of the epigenetic influences on mental illness suggests that many of the mechanisms by which epigenetic control could have an effect are similar. For example, we noted that changes in the HPA axis can affect Alzheimer's disease, bipolar disorder, and MDD. Changes in DNA methylation in the glucocorticoid receptor have also been implicated in depression, and because there is evidence suggesting this receptor's possible involvement in schizophrenia as well, there seems to be some value in the belief that a small set of parallel molecular processes can contribute to many mental conditions.

Perhaps the precise nature and symptoms of each condition depend on the type of psychological stimuli that cause some form of initial change in the synaptic connections between neurons and the rate at which they are able to signal to each other. The changes in gene expression needed to bring about this process of synaptic remodeling are very likely to be under epigenetic control. Therefore, if an individual has an underlying dysfunction in the systems used to control epigenetic modification in neurons, the initial synaptic connection change may be more damaging than it would be in individuals who do not suffer from an epigenetic control dysfunction. If so, the challenge is to identify the underlying epigenetic dysfunctions and their relationship to the symptoms displayed by those with mental illness.

KEY CONCEPTS

- Although not yet confirmed, memory formation may rely on the formation and stabilization of specific connections (synapses) between individual neurons.

- Epigenetic control of gene expression may be one factor controlling synaptic plasticity, namely the ability to form and maintain synaptic connections.

- Dysregulation of the epigenetic mechanisms controlling synaptic plasticity may contribute to mental health problems and cellular defects leading to neurodegenerative diseases.

FURTHER READING

Antonov I, Kandel ER & Hawkins RD (2010) Presynaptic and post-synaptic mechanisms of synaptic plasticity and metaplasticity during intermediate-term memory formation in *Aplysia*. *J Neurosci* 30:5781–5791 (doi:10.1523/JNEUROSCI.4947-09.2010).

Carasatorre M & Ramírez-Amaya V (2013) Network, cellular, and molecular mechanisms underlying longterm memory formation. *Curr Top Behav Neurosci* 15:73–115 (doi:10.1007/7854_2012_229).

Delgado PL (2000) Depression: the case for a monoamine deficiency. *J Clin Psychiatry* 61 (Suppl 6):7–11.

Dempster EL, Pidsley R, Schalkwyk LC et al. (2011) Disease-associated epigenetic changes in monozygotic twins discordant for schizophrenia and bipolar disorder. *Hum Mol Genet* 20:4786–4796 (doi:10.1093/hmg/ddr416).

Elgersma Y & Silva A (1999) Molecular mechanisms of synaptic plasticity and memory. *Curr Opin Neurobiol* 9:209–213 (doi:10.1016/S0959-4388(99)80029-4).

Gao Z, van Beugen BJ & De Zeeuw CI (2012) Distributed synergistic plasticity and cerebellar learning. *Nat Rev Neurosci* 13:619–635 (doi:10.1038/nrn3312).

Glanzman DL (2010) Common mechanisms of synaptic plasticity in vertebrates and invertebrates. *Curr Biol* 20:R31–R36 (doi:10.1016/j.cub.2009.10.023).

Hirschfeld RM (2000) History and evolution of the monoamine hypothesis of depression. *J Clin Psychiatry* 61 (Suppl 6):4–6.

Jarome TJ & Lubin FD (2013) Histone lysine methylation: critical regulator of memory and behavior. *Rev Neurosci* 27:1–13 (doi:10.1515/revneuro-2013-0008).

Josselyn SA & Frankland PW. mTORC2: actin on your memory. *Nat Neurosci* 16:379–380 (doi:10.1038/nn.3362).

Kerimoglu C, Agis-Balboa RC, Kranz A et al. (2013) Histone-methyltransferase MLL2 (KMT2B) is required for memory formation in mice. *J Neurosci* 33:3452–3464 (doi:10.1523/JNEUROSCI.3356-12.2013).

Kim MS, Akhtar MW, Adachi M et al. (2012) An essential role for histone deacetylase 4 in synaptic plasticity and memory formation. *J Neurosci* 32:10879–10886 (doi:10.1523/JNEUROSCI.2089-12.2012).

Kwok JB (2010) Role of epigenetics in Alzheimer's and Parkinson's disease. *Epigenomics* 2:671–682 (doi:10.2217/epi.10.43).

Labonté B & Turecki G (2011) The epigenetics of depression and suicide. In Brain, Behavior and Epigenetics (Petronis A & Mill J eds), pp 49–70. Springer.

Li Y, Meloni EG, Carlezon WA Jr et al. (2013) Learning and reconsolidation implicate different synaptic mechanisms. *Proc Natl Acad Sci USA* 110:4798–4803 (doi:10.1073/pnas.1217878110).

Liston C, Cichon JM, Jeanneteau F et al. (2013) Circadian glucocorticoid oscillations promote learning-dependent synapse formation and maintenance. *Nat Neurosci* 16:698–705 (doi:10.1038/nn.3387).

Marques SC, Oliveira CR, Pereira CM & Outeiro TF (2011) Epigenetics in neurodegeneration: a new layer of complexity. *Prog Neuropsychopharmacol Biol Psychiatry* 35:348–355 (doi:10.1016/j.pnpbp.2010.08.008).

Massart R, Mongeau R, Lanfumey L (2012) Beyond the monoaminergic hypothesis: neuroplasticity and epigenetic changes in a transgenic mouse model of depression. *Philos Trans R Soc Lond B Biol Sci* 367:2485–2494 (doi:10.1098/rstb.2012.0212).

Maze I, Noh KM & Allis CD (2013) Histone regulation in the CNS: basic principles of epigenetic plasticity. *Neuropsychopharmacology* 38:3–22 (doi:10.1038/npp.2012.124).

Mill J & Petronis A (2007) Molecular studies of major depressive disorder: the epigenetic perspective. *Mol Psychiatry* 12:799–814 (doi:10.1038/sj.mp.4001992).

Petronis A (2003) Epigenetics and bipolar disorder: new opportunities and challenges. *Am J Med Genet C Semin Med Genet* 123C:65–75 (10.1002/ajmg.c.20015).

Plaçais PY & Preat T (2013) To favor survival under food shortage, the brain disables costly memory. *Science* 339:440–442 (doi:10.1126/science.1226018).

Puckett RE & Lubin FD (2011) Epigenetic mechanisms in experience-driven memory formation and behavior. *Epigenomics* 3:649–664 (doi:10.2217/epi.11.86).

Rao JS, Keleshian VL, Klein S & Rapoport SI (2012) Epigenetic modifications in frontal cortex from Alzheimer's disease and bipolar disorder patients. *Transl Psychiatry* 2:e132 (doi: 10.1038/tp.2012.55).

Sando R 3rd, Gounko N, Pieraut S et al. (2012) HDAC4 governs a transcriptional program essential for synaptic plasticity and memory. *Cell* 151:821–834 (doi:10.1016/j.cell.2012.09.037).

16

EPIGENETICS OF CANCER

In the two previous chapters, we have seen that disruption of epigenetic modification patterns can contribute to the development of disease. The precise nature of the epigenetic contribution to conditions such as cardiovascular illnesses and mental disorders is a matter of some debate, and for these illnesses we can envisage alternative mechanisms to epigenetics to explain why a particular disease has developed in a specific patient. Naturally, this does not mean that the other explanations are true, but the point is that evidence for epigenetic involvement in some diseases is not overwhelming, and until more supporting data are collected, there is always the possibility that some diseases may not have an epigenetic component at all (although this seems increasingly unlikely). This is not the case with cancers, which almost without exception arise because of abnormal cell behavior that in turn arises from quantitative or qualitative changes in the genes expressed by those cells.

Cancer is a very serious disease. Although continued improvements in treatment have resulted in a decrease in the numbers of deaths from all forms of cancer, the USA still recorded more than 1.6 million new cases and nearly 600,000 cancer deaths in 2012. On the basis of rates of incidence of cancer in all anatomical sites between 2007 and 2009, it is estimated that 41% of individuals born within this period will be diagnosed with cancer of one form or another in their lifetimes.

So what do we mean by the term "cancer"? There are so many forms of the disease (more than 100, according to the US National Cancer Institute) that it is difficult to produce an overarching definition of what constitutes cancer, but all cancers involve the uncontrolled multiplication of abnormal cells to produce a tumor mass that can interfere with the function of the body's organs. The formation of abnormal tissue is referred to as neoplasia (the resulting tissue mass is a neoplasm), and the abnormal growth characteristics of the cells that give rise to neoplasms are known as dysplasia or metaplasia. It must be noted that dysplasia or metaplasia does not always give rise to neoplasms. The impact of such unrestricted growth need not be confined to the organ system in which the abnormal cells originated, because they can frequently migrate to other organs and begin their uncontrolled proliferation in new sites. This process, known as metastasis, is one of the features making cancers such a dangerous group of diseases, because tumors in relatively treatable locations, such as the skin, can give rise to tumors in organs that are much more difficult to treat, which greatly worsens the patient's chances of survival. Tumors that show this characteristic are classed as "malignant." Benign tumors can also occur, but these never enter into metastasis, can often be removed by surgery, and tend not to recur.

Cancers tend to be named according to the organ in which they originate (for example, skin cancer, or colon or liver cancer), but more useful descriptions come from the types of tissue in which certain types of tumor originate. Carcinomas, for example, originate from the skin or in the lining or covering tissues of the internal organs; the leukemias begin in the bone marrow and do not produce solid tumors; and lymphomas and myelomas—which also are not solid masses—originate in the immune system. Sarcomas develop in bone tissue or cartilage, fat, muscle, and connective or supportive tissues. Lastly, central nervous system cancers are those that arise in the tissues of the brain and spinal column.

16.1 UNCONTROLLED CELL REPLICATION

Loss of control of tissue homeostasis is a root cause of cancer

Homeostasis is the dynamic equilibrium of the internal environment of the human body. Tissue homeostasis is the process by which organs are able to replace dead or damaged cells while still maintaining the normal size and shape of the organ. More importantly, homeostasis also refers to the ability of the organ's tissues to maintain their environment within parameters that allow that organ to fulfill its function effectively.

The precise mechanisms that control organ size and shape are still unclear, but it is well established that cell turnover is the basic process by which the form and function of terminally differentiated tissues are maintained. In essence, the differentiated cells of a given organ are replaced at regular intervals, either by the division of other cells in that organ or by the differentiation and expansion of tissue-specific stem cells. The driving force behind the process is cell death: clearly, if too many cells die, they need to be replaced or the functional capacity of the organ will be diminished. However, our understanding of the signals that coordinate the growth of new cells is only poorly understood. One possible mechanism that may apply to the repair of organs such as the liver is contact inhibition (**Figure 16.1**). This process allows cells to divide until they come into contact with one another on all sides, after which they exit from the mitotic cell cycle.

It is quite likely that such a mechanism could contribute to controlling the size and shape of the liver, because damage is often repaired by the

Figure 16.1 Contact inhibition. Cells will grow as long as they have some part of their surface free from contact with other cells. As normal cells become denser, chemicals released by the cells cause them to stop dividing. Tumor cells have lost this characteristic and will not exit from the cell cycle even when densely packed. [Adapted from Lewis et al. (2004) Life, 5th Edition. With permission from McGraw Hill Higher Education.]

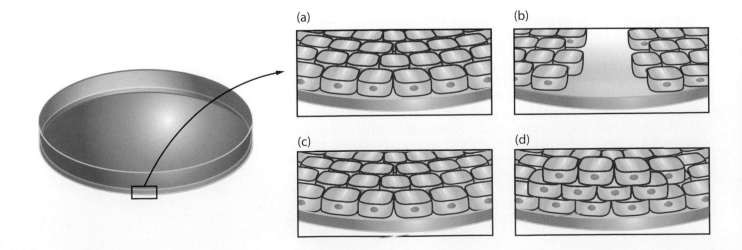

simple mitosis of existing hepatocytes. It is possible to remove up to 90% of the mass of the liver in experimental animals and, given time, the organ will regenerate itself by primarily using this cell division method. That is not to say that this is a simple process, because we do not understand the molecular cues that indicate the amount of organ damage or how the cells are able to recreate a shape with a high degree of similarity to the organ's structure before damage was introduced. The basic control system for contact inhibition probably causes dividing cells to exit from the cell cycle, but exactly how this can be used to remodel a whole organ is not known.

Repair and regeneration are a lot more complicated in organs that cannot or do not rely on the mitosis of existing terminally differentiated cells. Some organs contain large numbers of terminally differentiated cells that are incapable of mitosis (such as skeletal muscle, cardiac muscle, and central nervous system neurons), and these need to be replaced by the differentiation and proliferation of tissue-specific stem cells that normally reside within protected niches in these organs and only divide when called upon for repair. Once again, the signals that instruct such cells to effect damage repair are only partly known, and our knowledge of the repair process is also somewhat patchy. For example, we know a lot more about the regulation of the hematopoietic system than we do about the repair of a damaged cardiovascular system.

It is important to distinguish between normal levels of cell turnover and the regeneration that is needed to repair more significant damage. For example, regeneration is mostly triggered by a nonspecific insult (occurring through physical injury or disease) that causes acute damage and removes large amounts of tissue. The tissue usually responds by massive inflammation and the induction of signal transduction mechanisms that start the process of stem cell activation. In most cases, a reasonable repair can be made but with some limitations, such as scar formation and a decreasing regenerative capacity with age. This process is very different from normal cell turnover, which is characterized by the internally programmed removal of cells rather than a response to massive damage. In many ways, this turnover induced by disease resembles the developmental processes that produced an organ of specific size and shape during growth *in utero* but with greatly reduced regenerative capacity.

There is often a very poor correlation between the capacity of an organ to regenerate after injury and its ability to engage in "normal" cell turnover. For example, skeletal muscles regenerate reasonably well after physical damage despite having few mitotic cells, but the stem cells responsible for this repair (skeletal muscle satellite cells) are quite inactive in the uninjured muscle and evidence of "normal" turnover is scant. In contrast, there are examples of organ systems that have truly phenomenal turnover rates. The hematopoietic system supplies approximately 1 billion new blood cells every day to replace lost erythrocytes, lymphocytes, and macrophages that have been destroyed during the function of the immune system, and this replacement occurs by differentiation of the hematopoietic stem cells in the bone marrow. Similarly, it has been estimated that the cells of the intestinal epithelium are replaced every 5 days, and the continual loss of dead skin cells from the epidermis drives the replacement of dermal cells and their ongoing differentiation into epidermal tissue. In view of these data, it seems that cell turnover is a common feature of most if not all organ systems—it is just difficult to detect in some cases. The process is clearly of relevance to maintaining the health of the individual.

It would be possible at this stage to become embroiled in a detailed discussion of the contribution made by defective cell turnover to many different disease states, but because the emphasis of this chapter is cancer, it will probably suffice to say that cancer may be attributed to factors that tip the balance between cell loss and cell replacement toward a net increase in cell number. These factors include, but may not be restricted to, failure of the cell death mechanisms (apoptosis and necrosis) that normally remove cells exhibiting genetic damage, failure of the mechanisms that restrict the proliferation of differentiated cells (those that are still capable of mitosis), and dysregulation of stem cell activation and division.

Tissue homeostasis requires cell death

There are several ways for cells to die, but the best known are probably necrosis and apoptosis. Necrosis is considered to be the cell's version of "death by misadventure," whereas apoptosis is more often thought of as a highly regulated form of cellular suicide that has many applications in the body, ranging from the sculpting of organ morphology during embryonic development to the removal of cells that have acquired sufficient damage to their genomes to force their exit from the cell cycle. Necrosis on the whole is messy because it is largely unregulated. The cells swell, lyze, and release their components into the extracellular space, thus provoking inflammation, whereas apoptosis is a more discreet and controlled form of removal in which the dying cells are set aside for phagocytosis and recycling of useful materials.

It has been suggested that apoptosis is the primary mechanism for the elimination of cells during turnover, but this is difficult to prove for many tissues that have very slow rates of cell replacement. Added to this is the fact that many of the organ systems that display rapid tissue turnover have many different ways by which cells can be lost, such as mechanical removal of the intestinal epithelium, and so it becomes difficult to say whether apoptosis is the primary mechanism at work in these cases. Notwithstanding this, it is unlikely that necrosis would be the primary mechanism at work during cell turnover, for despite an observable increase in the occurrence of inflammation during aging, the body's organs are not subject to the large-scale disruptions that would occur if cells simply died and broke apart. On the balance of probability, apoptosis is the more likely mechanism by which cell removal can occur. Apart from any other considerations, strong links have been demonstrated between mutations and/or epimutations (such as changes in DNA methylation or histone modification patterns) in genes responsible for controlling apoptosis and many forms of cancer.

Loss of control of cell division is also known as cell transformation

Although the precise molecular mechanisms vary extensively with the type of cancer and, more importantly, the types of cells that are involved, the basic defect in cancer is that cell division can no longer be controlled and contact inhibition does not function properly. Contact inhibition can be observed for cells growing in culture, and the usual result is that normal cells form a confluent monolayer in their culture vessel. However, in many cancer cells this inhibition is absent, and if grown in culture they will simply pile up into mounds once the surface of the culture dish is covered. **Figure 16.2** shows the morphological differences between normal mouse fibroblasts and fibroblasts that have undergone a process called **transformation**. The transformed cells shown are not derived from a mouse tumor, but they will readily give rise to tumors if injected into other mice and so they are probably a good approximation to cancer cells.

(a)   (b)

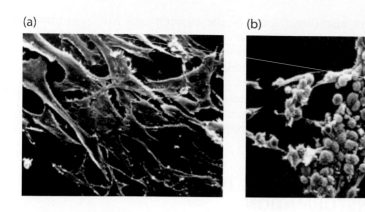

Figure 16.2 Comparison of normal and transformed mouse fibroblasts grown in culture. (a) The normal fibroblasts shown here form a confluent monolayer, a consequence of their contact inhibition. (b) These fibroblasts have undergone transformation and mound up on top of each other because they lack contact inhibition. (Courtesy of G. Steven Martin, University of California, Berkeley.)

Transformed cells will grow in culture under much simpler conditions than the nontransformed cell types they are related to. Essentially, they can make do with much poorer growth medium, nutrients, and growth factors. Another important feature is that they will go on growing indefinitely, provided that they get a steady supply of raw materials. This is sometimes referred to as immortality; indefinite growth is probably a more accurate way to describe them, but for all practical purposes, cancer cells can persist for a very long time. As an example, HeLa cells, the standard workhorse of many molecular biology investigations, were derived from a cervical tumor obtained from a woman who died in 1951 and have been in continuous culture ever since. Normal somatic cells do not survive indefinitely in culture because there is a limit to their proliferative capacity (the Hayflick limit), generally about 50 cell divisions before they enter into a quiescent state known as senescence. They do not die at this point; indeed, senescent cells can persist in the body for some time after they have stopped dividing, but they can no longer contribute to the growth or repair and regeneration of the tissues. The Hayflick limit has been viewed as a natural barrier to tumorigenesis, and although it does not seem to be very effective given the high incidence of cancer, the rate at which transformed cells are capable of growing suggests that without the imposition of some form of cell-division limit, tumors would have occurred at a frequency that could have prevented the evolution of large animals.

Dysfunctional genes are the basis of transformation

Somatic cells normally divide only when they must, to replace their dead or dysfunctional neighbors or to provide new cells that undergo rapid turnover, such as those in the blood or immune systems or the intestinal mucosa. This often requires tissue-specific stem cells, but a key step in the differentiation of these progenitors is their exit from the mitotic cell cycle and entrance into a state of quiescence usually equivalent to the G1 cell cycle phase. The same is true for cells such as fibroblasts, which are probably not renewed by stem cells but whose proliferation resumes in response to specific growth factors. In the case of fibroblasts, growth factors from the platelets damaged during wounding induce fibroblasts to re-enter the cell cycle, but most of these cells seem to also require epidermal growth factor (EGF), insulin-like growth factor (IGF-1), and transferrin. If the cells are grown in culture, removal of these factors causes quiescence. Replacing them allows the cells to resume mitosis.

As we saw in Chapter 10, the various signals that tell cells to begin or stop dividing impinge upon several cell cycle checkpoints that safeguard the integrity of the cell by controlling DNA replication and repair mechanisms and the elimination of defective proteins via the proteasome. The checkpoints can also direct a cell to die by apoptosis if it is too badly damaged to be repaired. Controlling this large array of functions is immensely complex, which presents many opportunities for dysfunction. Cancer can arise from the loss or disruption of one or more of these control mechanisms, because cancer is a disease of inappropriate cell proliferation. Fundamentally, all cancers permit the existence of too many cells. However, this cell number excess is linked in a vicious cycle with a reduction in sensitivity to signals that normally tell a cell to adhere, differentiate, or die.

16.2 CHANGES LEADING TO NEOPLASTIC TRANSFORMATION

Oncogenes and tumor suppressor genes are often altered during cancer progression

The changes leading to the loss of controls that govern cell number are usually mutations in specific genes or sets of genes, or major changes in the expression levels of these genes. The first genetic alterations shown to contribute to cancer development were gain-of-function mutations. These mutations define a set of **oncogenes**—genes that, when mutated or expressed at high levels, help turn a normal cell into a cancer cell. Oncogenes are mutant versions of normal cellular **proto-oncogenes**. The products of proto-oncogenes function in signal transduction pathways that promote cell proliferation. Transformation by individual oncogenes can be redundant (mutation of one of several genes will lead to transformation), or it can be specific to the cell type (mutations will transform some cell types but have no effect on others). This suggests that multiple, distinct pathways of genetic alteration lead to cancer, but that not all pathways have the same role in each cell type.

The change in an oncogene's function from normal to cancerous can be caused by a simple point mutation in the sequence of the gene. For example, a guanine-to-cytosine change in the *ras* oncogene, located on human chromosome 11, is frequently associated with bladder cancer. This simple change—which results in glycine at amino acid 12 being replaced by valine—drastically changes the function of the G protein encoded by the *ras* gene. Normally, the protein cycles from an inactive state to an active state by exchanging bound GDP for GTP and then back again to an inactive state by hydrolyzing the GTP to GDP. The mutation does not allow the hydrolysis of GTP, and the protein is continuously active. Because the signal delivered by the *ras* oncoprotein is thus delivered continuously, the cell continues to grow and divide.

Quite large numbers of oncogenes have been identified in the human genome, as shown in **Figure 16.3**, and **Table 16.1** shows that the protein products of many oncogenes are classified as growth factors, cell surface receptors involved in processing growth-inducing signals, or enzymes and transcription factors that contribute to growth control.

More recently, the significance of loss-of-function mutations in carcinogenesis has become increasingly apparent. Mutations in these so-called **tumor suppressor genes** were initially recognized to have a major role in inherited cancer susceptibility, and thus they have been included in the genes shown in Figure 16.3. Tumor suppressor genes (sometimes called *anti-oncogenes*) may therefore be defined as genes that protect the cell from at least one of the steps in the progression toward transformation and tumorigenesis. Unlike oncogenes, tumor suppressor genes generally follow the **two-hit hypothesis**, which implies that both alleles that code for a particular gene must be affected before alterations in cellular

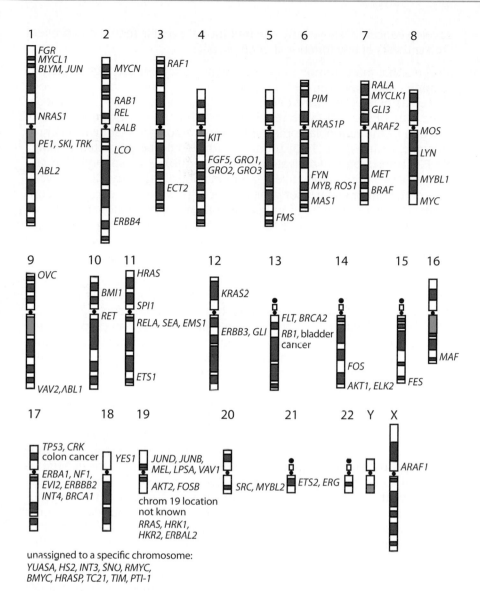

Figure 16.3 Chromosomal locations of several known oncogenes. Several genes known to function as oncogenes are shown. Gene symbols and approximate locations on human chromosomes are indicated. (Courtesy of Phillip McClean, http://www.ndsu.edu/pubweb/~mcclean/plsc431/cellcycle/cellcycl7.htm)

behavior are manifested. This means that heterozygous mutations may not produce an obvious phenotype, and therefore, unlike gain-of-function mutations, loss-of-function tumor suppressor mutations can be carried in the gene pool with no direct deleterious consequence. However, individuals heterozygous for tumor suppressor mutations are more likely to

TABLE 16.1 EXAMPLES OF SOME ONCOGENE PRODUCTS

Classification	Typical examples of class
Mitogens (growth factors)	C-sis (overexpressed in many types of tumour)
Receptor tyrosine kinases	Epidermal growth factor receptor Platelet-derived growth factor receptor Vascular endothelial growth factor receptor
Cytoplasmic tyrosine kinases	Src-family kinases
Cytoplasmic serine/threonine kinases	Raf kinase Cyclin-dependent kinases
Regulatory GTPases	Ras protein
Transcription factors	Myc protein

develop cancer, because only one mutational event is required to prevent the synthesis of any functional gene product.

It seems that many tumor suppressor gene mutations are highly likely to promote, and may even be required for, a large number of spontaneous as well as hereditary forms of cancer; indeed, many of these genes seem to be part of the cell cycle regulation machinery. For example, loss of function of the retinoblastoma tumor suppressor gene (shown on chromosome 13 in Figure 16.3) product, pRb, would be predicted to liberate E2F transcriptional activators without requiring phosphorylation and thus bypass a normal negative regulation controlling entry into the cycle. Loss of the tumor suppressor gene product p16 would have a similar consequence, liberating E2Fs by increasing pRb phosphorylation. In addition, cell cycle progression can be halted at several points by the tumor suppressor gene product p53, which is activated in response to checkpoints sensing DNA and possibly also chromosome damage; loss of p53 function would remove this brake to cycling.

Apart from the transmission of heterozygous mutations discussed above, most cancers arise from changes to the information contained in the DNA acquired throughout the person's life. Somatic mutations to the DNA sequence can be caused by ionizing radiation or mutagenic chemicals that find their way into the body. Additionally, quite often the process of DNA synthesis introduces errors into the nascent DNA strand that, because they are not immediately harmful to the cell, escape detection and become part of that cell's genome. Most mutations probably do not amount to much, but as we saw earlier, even a very small sequence change can substitute one amino acid for another in a protein and be sufficient to alter the protein's function drastically. These types of change are partly outside the scope of this book because they are genetic as opposed to epigenetic. Nevertheless, there are some significant mutational events, such as gene rearrangements, that can alter the epigenotype of the affected cell, and these will be included in our discussion of cancer epigenetics. Chromosomal rearrangements can potentially place an oncogene into a locus that is more permissive for expression of that gene, so these will also fall into our discussion. Other genetic faults such as gene duplication can affect gene expression levels, and although there is less evidence that this is based on epigenetic expression control, it will still be worth mentioning. However, point mutations may not be able to cause changes to the genetic architecture of the cell on a sufficiently large scale to be recognizable as an epigenetic cancer origin; thus we will ignore these for the most part.

Genomic instability is a common trait of cancer cells

When an oncogene is activated by a mutation, the structure of the encoded protein is changed in a way that enhances its transforming activity. Chromosomal translocations can also activate transcription factors in cancers by repositioning a gene from a locus where it is normally repressed to one where its activity is promoted, as we saw earlier. This repositioning can also incorporate the gene's coding sequences into other non-oncogenes to make fusion proteins whose properties differ greatly from those of the individual proteins. For example, in Ewing's sarcoma, the *EWS* gene is fused with one of several other genes, leading to altered transcriptional activity of the fused proteins. The EWS protein is an RNA-binding molecule that, when fused to a heterologous DNA-binding domain, can greatly stimulate gene transcription. Oncogene activation by genomic amplification, which usually occurs during tumor progression, is seen in the members of different oncogene families, including *MYC*, *CCND*, *EGFR*, and *FOS*. *MYC* is amplified in small-cell lung cancer, breast cancer, esophageal cancer, cervical cancer, ovarian cancer, and head and neck cancer.

Although the activation of specific oncogenes and disruptions of individual tumor suppressors alter the tumor phenotype in a specific manner, cumulative effects of such changes may be more apparent in tumors with high levels of genomic instability. Most cancers have an abnormal chromosomal content characterized by changes in chromosomal structure and number. Chromosomal aberrations are generally more numerous in malignant tumors than in benign ones, and the karyotypic complexity and cellular heterogeneity observed is often associated with poor prognosis. **Genomic instability** refers to a series of chromosomal changes occurring at an accelerated rate in cell populations derived from the same ancestral precursor and is a general term used to describe the processes that increase the rate of mutation, which in turn allows the cell to develop a more aggressive tumorigenic phenotype through adaptation to selection pressures present in the host tissue.

Genomic instability is generally classified into two major types: **microsatellite instability (MIN)** and **chromosomal instability (CIN)**. MIN involves simple DNA base changes that occur as a result of defects in the DNA repair processes, including base excision repair, mismatch repair, and nucleotide excision repair. CIN, in contrast, is characterized by grossly abnormal karyotypes, featuring both structural and numerical chromosome abnormalities. MIN and CIN mechanisms are generally found to be mutually exclusive and to produce different phenotypes, although recent findings suggest that there may be some overlap in these two pathways. One of the challenges facing cancer researchers today is to understand how cancer cells acquire genomes with such a high degree of genomic instability, and to determine in what way the genome and epigenome of cancer may be interacting to facilitate the occurrence of such instability.

Cancer cells frequently show major disruption in their DNA methylation profiles

Hypermethylation of CpG islands in gene promoters has been the most extensively studied area of research of DNA methylation in cancer, and this aberration probably contributes significantly to the loss of tumor suppressor gene function. Many genes with aberrant promoter hypermethylation have been identified in essentially all forms of cancer. Some of these susceptible genes include cell cycle regulators (*p16INK4a, p15INK4a, RB,* and *p14ARF*), DNA repair genes (*BRCA1, MGMT,* and *MLH1*), genes associated with apoptosis (*DAPK* and *TMS1*), hormonal regulation (*ER*), detoxification (*GSTP1*), metastasis (*CDH1*, which encodes E-cadherin, and *CD-44*), angiogenesis (*TSP-1* and *TIMP-3*), and many others. Although some genes such as *p16* are methylated in many cancers, other genes are methylated in specific types of cancer. An example is *GSTP1*, which is hypermethylated in more than 90% of prostate cancers but is largely unmethylated in acute leukemia. Transcriptional silencing via DNA hypermethylation can often be associated with a poor clinical outcome in several malignancies. For example, silencing of the cell cycle proteins CDKN2A and CDKN1A has been associated with poor clinical outcome in acute leukemias.

DNA hypomethylation also occurs frequently in many forms of cancer, ranging from solid tumors to hematological malignancies. Aberrant hypomethylation has been hypothesized to contribute to cancer progression by activating oncogenes such as *H-RAS, BORIS/CTCF, FGFR,* or *c-MYC*, or by retrotransposon activation or by increasing chromosome instability such as that observed in the rare, inherited immunodeficiency syndrome known as ICF syndrome.

One final way in which DNA methylation can contribute to cellular transformation is slightly less direct but nonetheless effective. Deamination of 5-methylcytosine (**Figure 16.4**) can be viewed as a mutagenic process, because it results in a change from cytosine to thymidine at that position. Normally, this thymidine would be detected as DNA damage and its removal during base excision repair would solve the problem, but there have been occasions when this type of point mutation has led to the functional inactivation of tumor suppressor genes such as *p53*.

The probability of this type of mutation naturally increases with the CpG island density. Analysis of DNA methylation in the coding regions of *BRCA1*, *RB1*, and *NF1* (all common oncogenes) showed prevalent CpG methylation, including those CpGs at mutational hotspots of these genes. In addition, hypomethylation of repetitive sequences may result in chromosomal and genetic instability, leading to further oncogenic events.

Impairment of DNA-repair mechanisms enhances cancer progression

A frequently observed feature of many cancer cells is a decrease in their capacity to repair damaged DNA, which leads in turn to an enhanced mutation rate. O^6-methylguanine-DNA methyltransferase (MGMT) is a DNA repair protein that protects cells against the toxic and carcinogenic effects of alkylating agents. The promoter of the *MGMT* gene is frequently methylated in cancers of the colon, lung, and lymphoid organs. The resulting silencing of *MGMT* that occurs in early tumorigenesis in humans (for example small adenomas of the colon) seems to predispose the cell to mutations such as G:C to A:T transitions in critical genes such as *K-ras* and *p53*. The DNA mismatch repair system, of which the MGMT protein is a critical component, recognizes and corrects base-pair errors in newly synthesized DNA, and disruption of this process increases the mutation rate as much as 100-fold. This type of mutation is particularly evident in regions of repeat DNA sequences such as microsatellites.

DNA repair genes other than *MGMT* have been implicated in cancer progression. *BRCA1* (breast cancer 1 early onset gene) is involved in the pathway that repairs double-strand breaks in DNA via homologous recombination—a repair normally performed in an almost error-free manner. Inactivation of *BRCA1*—again, via promoter methylation—leads to an increased rate of genomic rearrangements that can mutate tumor suppressor genes or oncogenes to the level at which control of cell proliferation and phenotype is lost.

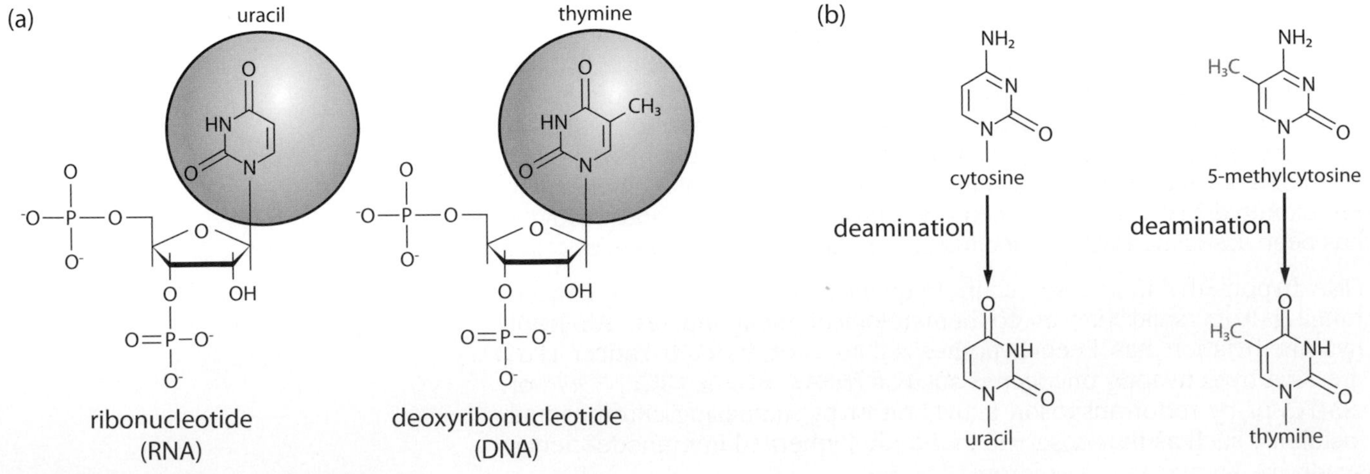

Figure 16.4 Cytosine deamination.
(a) Structural differences between RNA and DNA. (b) Spontaneous deamination of cytosine and 5-methylcytosine: the former deaminates to uracil, whereas the latter deaminates to give thymine. Both deamination events are mutagenic if unrepaired, leading to C:G to T:A changes in one daughter DNA strand after replication. [From Poole A, Penny D & Sjöberg B-M (2001) *Nat. Rev. Mol. Cell. Biol.* 2, 147. With permission from Macmillan Publishers Ltd.]

16.3 ABNORMAL PATTERNS OF DNA METHYLATION IN CANCER

DNA hypermethylation is typically mediated by DNMT1

Dysregulation of the epigenetic mechanisms that control cellular homeostasis is the basis of many forms of cancer, because genes involved in cell proliferation, differentiation, and survival are subject to tight but reversible epigenetic control under normal circumstances. The reasons leading to such dysregulation are of enormous interest because understanding these problems might allow us to develop better cancer treatments, but at present our knowledge of epigenetic dysregulation is no more complete than our comprehension of the normal state of affairs.

Abnormal patterns in DNA methylation were the first examples of epigenetic dysregulation to be characterized in human cancers. These abnormal methylations are thought to result either from the overexpression of DNMT or its aberrant recruitment, a consequence of mutations in the *DNMT* coding sequences that lead to structural alterations affecting the specificity of these proteins. Although abnormal DNA methylation affecting a variety of genes occurs in nearly every type of cancer that has been evaluated, some tumors show aberrant concurrent hypermethylation of numerous genes, a phenomenon known as the **CpG island methylator phenotype (CIMP)**.

CIMP was first described for a distinct subset of human colorectal carcinomas with high rates of concordant methylation of specific genes. Subsequently, CIMP has been reported to be in other human neoplasms, including tumors of the ovary, bladder, prostate, stomach, liver, pancreas, esophagus, and kidney, as well as neuroblastomas, leukemias, and lymphomas. CIMP is characterized by excessive DNMT3b expression and elevated total DNA methyltransferase activity, leading to the abnormal repression of many genes, with the unfortunate inclusion of many tumor suppressor genes. Studies performed on breast cancer cell lines suggest that genes without conventionally defined CpG islands may also be subject to enhanced methylation. Therefore, because the targets of aberrant methylation are not restricted to those genes with large CpG islands, it is probably more appropriate to refer to CIMP as a hypermethylator phenotype.

The fact that such hypermethylator phenotypes seem to include multiple independent loci reflects widespread deregulation of DNA methylation patterns and indicates that the cell may find it very difficult to control the activity of the excess DNA methyltransferases that are present. As we saw in Chapter 5, DNA methyltransferases are normally subject to a range of control mechanisms, such as the recruitment of DNMT to specific loci by transcription factors. However, the preferred enzyme involved in tumor suppressor gene hypermethylation seems to be the maintenance methylase DNMT1, which may be subject to a less exacting control system. This enzyme's preferred substrate is hemi-methylated DNA, as we have seen, but it may be that in cancer cells the mechanisms that direct DNMT1 to specific loci either do not work very well or the sheer excess of the enzyme allows it to overcome whatever control mechanisms normally perform this function.

Data from studies of the methylation status of 1184 CpG islands from 98 tumor samples (from different types of cancer) revealed that CpG island *de novo* methylation in tumor cells is widespread and differs between individual tumors and tumor types. An average of 608 (and potentially up to 4500) CpG islands were aberrantly hypermethylated in tumors in a nonrandom manner, indicating that certain CpG islands may be more susceptible to *de novo* methylation than others. Nevertheless, an important connecting factor for all of these samples was overexpression of DNA methyltransferase activity. Conversely, conditional DNMT1 knockouts generated in additional studies in certain cancer cell lines have been found to lead to mitotic arrest or cell death, demonstrating that DNMT1 is required for faithfully maintaining DNA methylation patterns in human cancer cells and is essential for their proliferation and survival. Furthermore, overexpression of Dnmt1 in mouse fibroblasts leads to transformation that is partly rescued by knockdown of expression by using small interfering RNAs directed against the Dnmt1 coding sequence. There is therefore little room for doubt that increased levels of this enzyme can contribute to the hypermethylation of CpG islands, although how this mechanism works is currently unclear. It is difficult to imagine that the mechanism simply relies on DNMT1 concentration in the cell, because the enzyme usually acts as part of a multiprotein complex; so, if the other members of these complexes do not show similar increases in expression, then explaining how the methylation function can be increased becomes problematic.

A known consequence of establishing blocks of 5-methylcytosine at specific genomic loci is that the region covered by methylation tends to spread over time. The vast majority of methylated DNA loci are established during embryonic development, with rapid deposition of 5-methylcytosine at the blastocyst stage followed by a more gradual methylation during the remainder of development. Once CpG sites within *cis*-acting loci have been established, they can act as foci for the spreading of methylation to distal CpG sites, because although DNMT1 is credited with the methylation of already hemi-methylated DNA, this is only a preference and not an absolute requirement. In fact, DNMT1 is approximately 15 times more likely to methylate double-stranded DNA with 5-methylcytosine on one strand, but that does not mean that it is incapable of *de novo* methylation. Such a capability, however small, allows for the possibility of methylation errors, and *in vivo* data suggest that methylation is maintained with a fidelity of approximately 95%, whereas *de novo* methylation occurs in around 5% of all attempts by DNMT1 to process DNA. If this is a universal occurrence, it implies that all cells should be subject to expansion of the methylated CpG islands with every cell division.

The possibility of such expansion with every round of DNA replication is supported by the observation that despite an overall decrease in the amount of DNA methylation in the genome with increasing age, most loci that tend to show increased methylation in cancers, such as colorectal cancer, also show similar methylation increases with age. It has therefore been suggested that the dominant cause of an increase in methylation in cancer is actually the methylation that arises in normal tissues as a function of age. This finding has been confirmed by several studies, and for malignancies of the epithelial tissues (the carcinomas), it is probably a universal phenomenon.

The mechanisms controlling DNA methylation are imperfect

Having established that methylation spreads naturally with age, what prevents it from covering the entire genome over time? CpG islands seem to have developed specific mechanisms to block such methylation spreading and this barrier is crucial in maintaining the methylation-free state. Therefore, one view of DNA methylation in predisposed genes is that of a struggle between methylation-promoting events and methylation-protecting events (**Figure 16.5**).

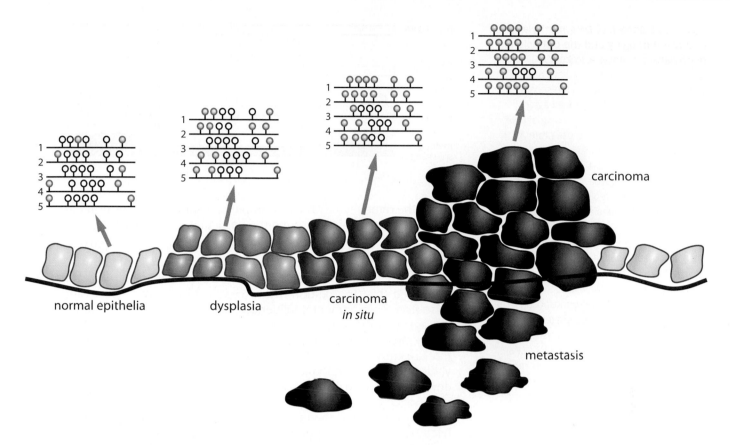

Bearing this in mind, a model of hypermethylation has been proposed that explains aspects of both age-related increases in methylation and neoplastic transformation: initially, *de novo* methylation and spreading is a replication-dependent phenomenon, and it is therefore crucially dependent on age. Aging tissues simply extend the patterns of DNA methylation that were deposited during embryogenesis, and methylation slowly extends toward promoters, piling up at the borders of the islands, where protection is presumed to be strong. For some genes, protection against the spreading of methylation is weak, either intrinsically or because protection is mediated by gene expression that can be reduced during aging, and this will show up as age-related promoter methylation in normal and neoplastic tissues. For other genes, protection is strong, and these genes will only become hypermethylated if there is strong selection for methylation, or if the barrier is somehow lost, possibly by a genetic event.

Given that there are probably a limited number of protective mechanisms, the inactivation of a single protecting gene might result in multiple genes being affected simultaneously, exactly as observed in CIMP. Further speculation has suggested that gene silencing may initiate with a spontaneous but perhaps random detachment of transcription factors required for the normal expression of the gene. This may be induced by environmental factors or by perturbations in the expression of other interacting genes. Regardless of how it happens, in the absence of the transcription factor that normally recruits epigenetic modifier proteins that maintain an open chromatin conformation, other repressive modifiers may bind the gene or its promoter and establish repressive modifications, such as DNA methylation. If the insulator or boundary to methylated DNA spreading is in part maintained by the transcriptionally active status of the gene, the methylation may be allowed to spread outside the normal limits. When transcription factor binding is reestablished, this extra methylation may be removed. However, if the residence time of the transcription factor is

Figure 16.5 An epigenetic model for cancer initiation and progression.
Five promoter CpG islands, as well as their neighboring CpG sites, critical to tumorigenesis are depicted here. In the initial neoplastic step, protection of some CpG island loci from aberrant DNA methylation is lost. *De novo* methylation occurs at the flanking CpG sites and progressively spreads into the core of a CpG island, resulting in silencing of the corresponding gene. This methylation spread may occur later in some other loci important for certain stages of neoplasm. In general, the density of methylated CpG sites within a locus as well as the number of methylated loci increase in more advanced stages of cancer. Methylated CpG dinucleotides (blue) and unmethylated CpG dinucleotides (white) are indicated. [Adapted from Nephew KP & Huang TH-M (2003) *Cancer Lett.* 190, 125. With permission from Elsevier.]

Figure 16.6 Spread of DNA methylation as a result of aging and the CpG island methylator phenotype. Naked CpG island DNA (top) is unmethylated (yellow) and coated by proteins (shown in the second panel as green ovals) that protect against the establishment and/or spreading of DNA methylation. The nature of these proteins is unknown, but they probably include transcription factors, co-activators, or similar molecules. During repeated rounds of the stem-cell mobilization and replication that accompany aging, DNA methyltransferases (red circles) are recruited to the borders of some CpG islands; the methyltransferases deposit methyl groups (blue) and create methylation pressure for these islands. The nature of this initial recruitment is unknown but is probably related to repetitive DNA sequences and/or retrotransposons. The balance of methylation pressure (circles) and methylation protection (ovals) is disrupted in the CpG island methylator phenotype (CIMP), resulting in the spread of methylation into the transcription start area and the triggering of the silencing cascade. Disruption of this balance is probably achieved through the loss of protective proteins (as indicated in the bottom panel), which could occur by mutations that inactivate these proteins or by mechanisms such as transcription factor loss or histone modifications that shut down expression of the encoding genes. Theoretically, this balance could also be disrupted by overactive *de novo* methylation pressure (circles), for example, by activating mutations in DNA methyltransferases. [Adapted from Issa JP (2004) *Nat. Rev. Cancer* 4, 988. With permission from Macmillan Publishers Ltd.]

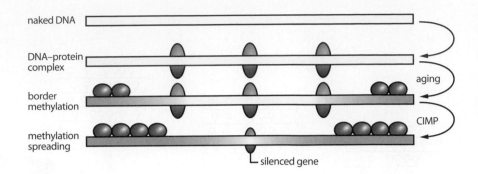

short, the removal may not be complete, resulting in long-term buildup of the methylated regions around the particular gene (**Figure 16.6**).

It is also possible that enhanced *de novo* methylation by DNMT1 might result from mutations in the structure of this enzyme that allow increases in its ability to bind and methylate naked DNA, but so far no studies have confirmed this. In addition, the excess load of 5-methylcytosine could arise from environmental influences, such as repeated exposure to "epimutagens"—for example, small molecules that methylate DNA independently of the DNA methyltransferases—or exposure to viral oncoproteins that are known to target the DNA methyltransferases and cause the methylation of certain genomic sequences. It is therefore conceivable that the development of a hypermethylator phenotype reflects the activities of carcinogenic molecules by enhancing an already increasing methylation load at certain susceptible loci. However, this is very difficult to confirm, because with the exception of cancer patients who have an obvious and well-recorded history of carcinogen exposure (such as industrial workers or smokers), the timing and epigenetic consequences of the exposure can be difficult to determine.

Other possible impacts of DNMT1 overexpression revolve around its other functions in the cell besides DNA methylation. DNMT1 levels are regulated with the cell cycle and are induced upon the cell's entry into S phase. DNMT1 regulates the expression of cell cycle proteins by its other regulatory functions and not through its DNA methylation activity. Once the mechanisms that coordinate DNMT1 and the cell cycle have been disrupted, DNMT1 exerts an oncogenic activity; however, in this context, overexpression of DNMT1 is probably not responsible for the aberrant methylation of at least some tumor suppressor genes.

DNMT1 binds to many other proteins, such as proliferating cell nuclear antigen (PCNA) and p53, which may also contribute to its cell-cycle-related functions (or arrest of the cell cycle if DNA damage is detected). Both alleles of the *p53* gene must be inactivated for there to be a failing response to DNA damage, and the inactivation usually occurs by mutation or methylation. Cells containing inactivated versions of the p53 protein or repressed *p53* gene can live on and develop into tumors. DNMT1 and p53 are known to cooperate in the methylation of promoters containing p53-binding sites, so it is possible that DNA hypomethylation of p53-responsive genes can be partly explained by p53 loss. This hypomethylation should not be affected by DNMT1 overexpression, because an alternative method of recruiting DNMT1 to the p53-binding site in the absence of active p53 protein is not known.

Elevation of signal transduction via the Ras pathway may also have an important role in DNMT1-mediated promoter methylation. The GTPase Ras mentioned earlier is probably one of the most commonly occurring oncogenes in human cancers. More than 30% of all cancers are associated with Ras activation, and although the initial activation is usually due to a DNA mutation event, oncogenic Ras also epigenetically represses an array of cancer-related genes involved in apoptosis, cell cycle regulation, DNA repair, and cellular differentiation. These include *Fas*, *P16*, and *TGFβ receptor type II*, and the epigenetic modification of the promoters of these genes is brought about by a group of Ras-associated epigenetic silencing effects (RESEs). RESEs comprise transcriptional regulators and DNA-binding proteins, proteins involved in histone and DNA modifications (such as DNMT1), and several polycomb group (PcG) proteins. Their activity is increased in cells transfected with the mutated form of Ras normally found in tumor-producing cells.

The key enzyme for most of the RESE complexes seems to be DNTM1, particularly in the down-regulation of tumor suppressor genes such as *Fas*, although several PcG proteins, including the histone methyltransferase EZH2, were among the identified RESEs. Other studies have shown that EZH2-containing polycomb complexes act as recruitment platforms for DNMTs and direct specific DNA methylation by marking the targeted promoters with methylated histone H3. EZH2 may therefore serve as a potential downstream effector through which Ras mediates target-specific epigenetic silencing. However, at the time of writing, it seems likely that the actual mechanism operating in cancer development is probably rather more complex, because simply transfecting many primary cell types with oncogenic Ras triggers their senescence rather than neoplastic transformation. This suggests that activation of the Ras-mediated epigenetic silencing program may require additional inputs or a specific cellular setting.

Abnormal DNA hypomethylation contributes to cancer formation and progression

Although more attention has been paid to hypermethylation as a means of inactivating tumor suppressor genes, there is also substantial evidence that abnormal DNA hypomethylation contributes to oncogenic transformation or tumor progression. Indeed, cancer-linked DNA hypomethylation is probably just as prevalent as cancer-associated DNA hypermethylation. The loss of 5-methylcytosine from specific promoter regions of oncogenes, thereby allowing their inappropriate expression, would be consistent with the gain-of-function hypothesis described earlier in this chapter, but the evidence suggests that DNA hypomethylation is more likely to promote tumorigenesis by affecting chromosomal stability. The hypomethylation of genomic sequences often exceeds hypermethylation in some tumor types, so that cells from these tumors typically display lower levels of 5-methylcytosine than do a variety of normal adult tissues. Detailed methylation analyses, using sodium bisulfite DNA sequencing, have revealed that hypomethylated CpGs in tumors are not distributed randomly but are instead clustered within defined regions. Moreover, cancer cells can display hypomethylation without any associated hypermethylated regions of the genome, although both forms of aberrant methylation are frequently found in the same cells. In view of this, it is likely that hypomethylation is a genuine contributor to carcinogenesis and not merely a by-product of the oncogenic transformation process. This should suggest caution in the use of small-molecule inhibitors of the DNA methyltransferases (such as 5-azacytidine) as cancer therapies, because they may actually increase the hypomethylation of some genomic regions and thereby lead to increased tumor progression.

The preference for clustered genomic loci means that cancer-associated DNA hypomethylation frequently involves repeated DNA elements, such as long interspersed nuclear elements (LINEs), whereas cancer-associated hypermethylation seems to occur predominantly in the nonrepeated regions of the genome. The satellite repeats are another common target for hypomethylation. This preference is potentially serious because these loci are normally held under tight repression as constitutive heterochromatin.

The centromeric satellite regions of chromosomes 1 and 16 seem to be commonly undermethylated in many types of cancer cells, but it is not immediately clear why this should occur. Such regions are often referred to as **microsatellites** (also known as **small tandem repeats** or **variable number tandem repeats**) and are generally composed of varying repeats of oligonucleotide sequences, differing in size up to pentanucleotides. They are scattered throughout the genome and are usually located outside, but occasionally within, coding sequences. Microsatellites are faithfully replicated in healthy cells, and such maintenance ensures that coding sequences remain within the appropriate reading frame. Insertion or deletion of one or more of the repeats can induce frameshift mutations in genes downstream of the microsatellite, which leads to aberrant gene function.

Clustered genes such as the HOX clusters are also known to lose methylation from key control sequences. An example of this, one that is commonly observed in many types of malignancy, is the up-regulation of *HOXA11* by a loss of methylation from its promoter region. Hypomethylation of some gene regions that have no apparent link to cancer has also been observed, and examples of this include the loss of methylation from the promoter of the β-globin gene in the epithelial cells (where this gene would not normally be expressed) of colon and breast adenocarcinomas. This effect is probably incidental to the general hypomethylation process, but because the β-globin gene is also a member of a gene cluster, future studies of this apparently non-cancer-related event may shed light on the basic mechanism behind hypomethylation in cancer cell types.

The consequences of DNA hypomethylation that can lead to cancer formation depend for the most part on the genomic loci that are affected by this aberration. Because the mammalian genome consists of relatively short unmethylated domains embedded in a matrix of long, stably methylated domains, these latter regions tend to be affected most often (a situation that seems to be made worse by the exceptional sensitivity of these domains to a loss of 5-methylcytosine). The principal problem for the affected cell is one of general instability of its genome. Demethylation of repetitive sequences located at centromeric, pericentromeric, and subtelomeric chromosomal regions may cause the induction of chromosomal abnormalities. For example, recent findings have demonstrated that DNA hypomethylation at the centromeric region causes permissive transcriptional activity at the centromere and the subsequent accumulation of small, minor satellite transcripts that impair centromeric architecture and function. Hypomethylation of LINEs, SINEs, transposons, and retroviral elements present in the genome activates them and, by increasing the possibility of their transposition to other loci, can enhance genomic instability. Similarly, hypomethylation of the subtelomeric regions can permit the transcription of previously silenced genes from this area.

Imprinted genes were introduced in Chapter 11, where we saw that allele-specific expression dependent on the parent of origin was crucial for several developmental processes. Loss of the DNA methylation that contributes to the silencing of one of these alleles of these genes is considered to be one of the most frequent alterations in cancer, although the mechanism by which this permits tumorigenesis is not completely understood.

The molecular mechanism of DNA hypomethylation is less clear than that leading to hypermethylation. One might imagine that if an excess of

DNA methyltransferase activity contributes to a hypermethylation phenotype, a deficit of these enzymes should logically lead to hypomethylation. Although it is true that mice carrying a hypomorphic *Dnmt1* allele, which reduces *Dnmt1* expression, show substantial genome-wide demethylation in all tissues, it is less likely that such a deficit contributes to cancer-related demethylation. The reason is that hypomethylation is frequently observed alongside hypermethylated promoters in cells that overexpress DNA methyltransferases, an indication that hypomethylation is not dependent on a deficit of DNMT enzymes. Although it is also true that these mice develop aggressive T-cell lymphomas, the above-noted observation of concurrent hypomethylation and hypermethylation of promoters casts some doubt on the validity of a model that relies on a deficit of DNMT enzymes as an explanation for sporadic hypomethylation in cancer cells.

Another possibility is a mechanism based on *S*-adenosylmethionine (AdoMet; see Chapter 5), which connects the one-carbon metabolic pathway to effective DNA methylation. It is well known that long-term exposure to an inadequate supply of methionine, choline, folic acid, or vitamin B_{12} results in a profound loss of cytosine methylation in the livers of male rats and mice. In addition, it is believed that the loss of DNA methylation induced by exposure to arsenic, diethanolamine, trichloroethylene, or alcohol is associated with perturbations in cellular homeostasis of AdoMet. However, disruptions in one-carbon metabolism would also be expected to induce global hypomethylation, which does not always occur in cancer, so this mechanism is unlikely to be the major contributor to the hypomethylation phenomenon.

The results of several studies have demonstrated that the presence of unrepaired lesions in DNA can lead to DNA hypomethylation. Specifically, the presence of 8-oxoguanine and 5-hydroxymethylcytosine in DNA—common DNA modifications resulting from oxidative damage—interferes with the binding of MECP2, a protein that recognizes methylated CpG islands. The consequence is a reduced effectiveness of maintenance methylation at that locus. The effect of oxidative DNA damage on passive loss of DNA methylation has also been used to explain the gradual decrease in global methylation that occurs with increasing age. This is consistent with the observation that oxidative lesions accumulate in the genome over time, but how would this effect be more prevalent in regions of DNA repeats?

Some studies have suggested that the microsatellite repeats have a higher mutation frequency after DNA damage than do nonrepetitive sequences in human cells, which may be one of the factors that enhances the loss of 5-methylcytosine. Research in which cells were treated with excess reactive oxygen species such as hydrogen peroxide suggests that oxygen-induced DNA lesions can affect the faithful maintenance of the microsatellite repeats more than would be expected for other, nonrepeat DNA. The implication of this is that the sequence of DNA bases in the microsatellites might react with free radicals of oxygen more readily than nonrepeated DNA. However, although there is some evidence that this may occur in the TTAGGG repeat sequence of telomeric DNA (guanine residues are particularly susceptible to oxidation), the base composition of microsatellites is not known to have any significant effect on their instability in cancer. This implies that microsatellite DNA may be no more susceptible to oxidation than other regions, so there must be other mechanisms that allow the buildup of oxidative lesions in these regions.

One possibility may be the way in which repeat DNA performs its DNA repair functions. Double-strand breaks, induced for example by ionizing radiation, present a significant threat to cell survival, whereas single-strand breaks are a feature of many types of oxidative insult, which induce the degradation of 2-deoxyribose, and are rapidly repaired in competent cells. However, repeat DNA seems to have some unusual structural properties. Observations in yeast suggest that the DNA of repeat sequences adopts a helix conformation that is different from that of normal B-DNA, meaning that the helix diameter, helix pitch, and number of bases per turn may be quite different. **Figure 16.7** shows the radical structural differences between B-DNA (the normal one) and two alternative conformations, A-DNA and Z-DNA. We can see that the accessibility of bases on the A and Z forms is substantially less than on the normal B form. As a consequence of this altered structure, lesions within repetitive sequences may be less efficient substrates for DNA repair enzymes, which could in turn allow the accumulation of oxidative lesions, such as 8-hydroxyguanine, that could affect DNA methylation.

In addition, it has been shown that the presence of 8-hydroxyguanine in a nascent (and unmethylated) DNA strand, one or two bases upstream of the target cytosine, significantly diminishes cytosine methylation. However, 8-hydroxyguanine in the methylated parent strand does not affect DNA methyltransferase activity toward target cytosines in the nascent strand. It would therefore seem that the effect of 8-hydroxyguanine on cytosine methylation is strand-specific and could affect DNA methylation patterns in rapidly dividing cell populations exposed to oxidative insult. Such alterations in DNA methylation could influence carcinogenesis by switching on the expression of growth-promoting or anti-apoptotic genes.

Oxidative stress has additional effects on epigenetic processes that impinge on cancer

It is worthwhile to examine the impact of oxidative stress on tumorigenesis in a little more detail before moving on to other epigenetic lesions involved in cancer. The hypermethylation of certain genes, discussed earlier, may also respond to increases in the concentrations of reactive oxygen species in cells, because such changes alter the overall cellular redox status. This can alter the structure of transcription factors and their

Figure 16.7 Structural conformations of (a) B-DNA , (b) A-DNA, and (c) Z-DNA. [From J Kendrew [ed] (1994) The Encyclopedia of Molecular Biology. Blackwell Science Ltd. With permission from John Wiley and Sons.]

(a)   (b)   (c)

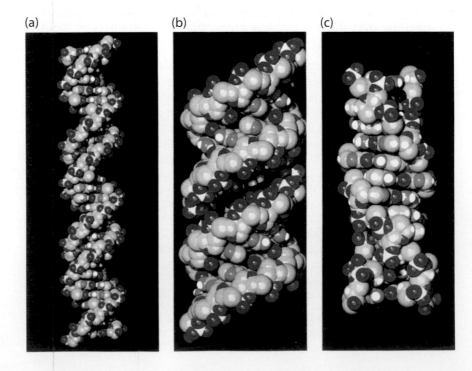

binding to cognate DNA sequences, often through modification of the thiol–disulfide status of cysteine. We saw earlier that decreasing the time during which transcription factors are present on their binding sites in gene promoters may allow repressive epigenetic modification complexes to initiate DNA methylation at these loci, so structural alterations that reduce the residence time still further may contribute to this process.

The cell has systems to cope with altered redox states. For example, it can enhance the synthesis of glutathione, which reacts with free radicals of oxygen to produce less harmful molecules; indeed, redox alterations can form part of the normal physiological responses of some types of cells. A more significant problem arises if oxidative lesions form on the DNA sequence of a promoter. Promoter regions in eukaryotic genes contain at least a TATA box (5′-TATAAA-3′) 30–90 bp from the transcription initiation start site, which is generally flanked by GC-rich sequences. Additional elements residing between 40 and 110 bp of the promoter region in some genes are a CAAT box and a GC box. A feature of several transcription factor-binding sites, including factors that are involved in gene responses to oxidative stress, is the presence of guanine-rich sequences, which are particularly susceptible to oxidation, as mentioned earlier. The presence of 8-hydroxyguanine in various transcription factor-binding sites modulates the accessibility of the site (mostly this reduces accessibility, but there are a few instances in which this is increased) to the transcription factor and therefore would be expected to alter not only gene expression but also the ability to recruit epigenetic modifier complexes that maintain an open chromatin conformation at that locus. The converse may also be true, so that in the absence of transcription factor binding, repressive complexes may "take over" and methylate the DNA. 8-Hydroxyguanine accumulation has been observed at the binding sites for Sp-1, AP-1, and NF-κB transcription factors. The consensus sequence recognized by Sp-1, for example, is highly guanine-rich (5′-GGGGCGGGG-3′) and should thus be a prime target for oxidative damage. Replacement of guanine residues in the core of this GC box (guanines 4 and 5) with 8-hydroxyguanine leads to a loss of more than 90% binding of Sp-1, with replacement of residues on either side of this core resulting in a loss of 70–80% binding. This positional specificity reflects the varying importance of interactions between amino acid residues in the zinc-finger domains of Sp-1 and DNA. Significant decreases in AP-1 and Sp-1 binding to their cognate target DNA sequences were also noted after the introduction of 8-hydroxyguanine. In contrast, binding of NF-κB was apparently unaffected by 8-hydroxyguanine in this study, which may reflect the fact that this transcription factor is intimately involved in the cellular response to oxidative stress.

Definite evidence that oxidative damage to promoter regions can lead to increases in DNA methylation is still needed, and the contribution of this form of damage to any of the DNA methylation changes discussed so far remains speculative. Nevertheless, these lesions are currently the best indications we have as to how cancer-related hypomethylation and hypermethylation might be formed in their respective parts of the genome.

The influence of microRNA on DNA methylation in cancer

Compared with our knowledge of oncogene versus tumor suppressor gene expression in cancer, our knowledge of the contribution made by microRNAs (miRNAs) to cancer is still rudimentary. However, this is an area of research that has undergone an enormous expansion in recent years, and several experiments suggest that microRNAs may function as a novel class of oncogenes and tumor suppressor genes. Those miRNAs whose expression is increased in tumors may be considered to be oncogenes. These oncogene miRNAs are known as **oncomirs** and promote tumor development by inhibiting tumor suppressor genes and/or genes that promote cell differentiation or apoptosis. As opposed to oncomirs, the expression of some miRNAs is decreased in cancerous cells. These types of miRNAs are likened to tumor suppressor genes. Tumor suppressor miRNAs usually prevent tumor development by inhibiting oncogenes and/or genes that impede cell differentiation or apoptosis.

miRNAs are very useful as diagnostic molecules for various tumor types, because the tissue from which a tumor originated can be identified and tumors classified solely on the basis of microRNA expression profiles. Moreover, the microRNA expression profile provides valuable information about the developmental lineage and differentiation state of a tumor. These profiles provide an enormous advantage over a diagnosis based on gene expression (that is, on mRNA) in that they permit the classification of poorly differentiated cancers. Furthermore, they may be used to define the probability of a patient's survival in a way that is very difficult with more traditional systems based on gene expression.

miRNA genes tend to be located within the introns of protein-encoding genes, but it is of interest that many of their locations (more than 50%) are found within the regions highlighted as cancer-linked loci (as determined via traditional linkage studies). This seems to suggest that their dysregulation may have an important role in cancer. They are also located in regions of heterozygosity loss, regions of amplification, and common breakpoint regions that are genetically altered in human cancers.

We saw in Chapter 8 how these small noncoding RNAs are able to control DNA methyltransferase activity and the methylation of specific target genes by recruiting chromatin remodeling complexes to regions of the genome that have recently been transcribed. **Figure 16.8** shows various ways in which miRNAs might be able to influence the chromatin architecture. From this figure it is clear that dysregulation of the system that allows miRNAs to be synthesized in a cell-specific manner could lead to inappropriate expression of oncogenes and thus to tumorigenesis. Moreover, as noted above, not only has recent research implicated miRNAs in cancer development but it has also led to the observation that some miRNA levels are reduced in various human cancers and to speculation that these miRNAs may normally function as tumor suppressors.

This concept is supported by the interesting observation that expression of several miRNAs (miR-29a, miR-29b, and miR-29c) is down-regulated in non-small-cell lung carcinomas and that enforced expression of miRNA-29s in lung cancer cell lines restores normal patterns of DNA methylation, which are associated with the re-expression of tumor suppressor genes. The restoration of normal methylation seems to inhibit the transformed phenotype of the lung cancer cell lines *in vitro*, and, at least in mice, administration of miRNA-29 (via injection of tumor cells pretransfected with the microRNA) seems to prevent the formation of lung carcinoma. Injection of non-small-cell carcinoma cell lines into mice would normally result in a high incidence of tumor formation in these animals. This is a fascinating result, because although this protocol has not yet been shown to eliminate an already existing tumor, it could pave the way for the development of miRNA-based cancer therapies.

miRNA-29 seems to function by targeting the expression of DNMT3A and DNMT3B via binding to complementary sites in 3'-untranslated regions of their genes (as we saw briefly in Chapter 8). It should be noted that miRNA-29 seems to have other targets as well, because its complementary binding site is present in the 3'-untranslated regions of several

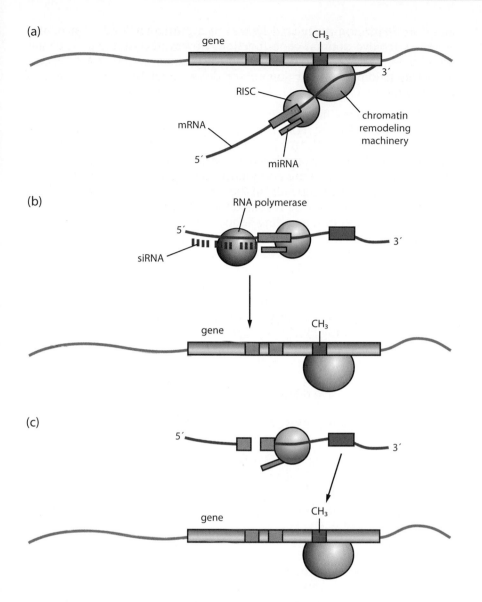

Figure 16.8 Mature miRNA is first produced from a precursor and exported from the nucleus. (a) The mature miRNA might return to the nucleus with an RNA-induced silencing complex (RISC) and base pair to a matching sequence (red) in the mRNA being produced. The RISC might then, via the mRNA, recruit chromatin-remodeling machinery to the DNA to achieve methylation. (b) Small interfering RNAs (siRNAs) might be produced by an RNA-dependent RNA polymerase that is directed by miRNAs or by miRNA-induced mRNA cleavage. The siRNAs might then directly or indirectly (via spreading) target a gene's region (blue here) for methylation. (c) The mRNA could be modified by an miRNA-induced mark (such as cleavage) and then be guided to the DNA to induce methylation. [From Ronemus M & Martienssen R (2005) *Nature* 433, 472. With permission from Macmillan Publishers Ltd.]

genes, such as those encoding B7H3 (an immunomodulatory glycoprotein present on the surface of many cancer cell types) and tristetraprolin (a regulator of epithelial cell polarity and metastasis).

16.4 HISTONE MODIFICATION PATTERNS AND CANCER

Disruption of the DNA methylation machinery is important in tumorigenesis, but as we have seen throughout the preceding chapters, no single epigenetic modification seems to act in isolation. Given the enormous importance of post-translational modifications of histones on the control of gene expression, it is essential to examine the impact that disruptions in the mechanisms that initiate and maintain these histone marks have on cancer development.

How does histone acetylation contribute to tumorigenesis?

To recap briefly, acetylation of specific lysines on the N-terminal polypeptide chain of histones H3 and H4 is frequently associated with actively transcribed genes or those that are at least in an accessible chromatin environment that is permissive for transcription. The balance between acetylated and non-acetylated lysines at a particular locus is synchronized to achieve optimal gene function. However, dysregulation of this balance can lead either to oncogene up-regulation (in the case of excessive acetylation) or tumor suppressor down-regulation (in the case of deacetylation).

Nucleosomes do not have to be completely acetylated to prevent higher-order chromatin folding. For instance, acetylation of 46% of the available lysines in positions 9, 14, 18, or 23 of histone H3 is sufficient for transcription. There is therefore considerable scope to increase acetylation levels during tumorigenesis. There is evidence to suggest that the level of gene transcription achieved depends on the amount of acetylation present. For example, some studies have revealed multistep gene activation in which histone acetylation of the coding sequence occurs first via the action of histone acetyltransferases of generally low specificity, followed by a more precise acetylation of regions in the gene promoter to initiate transcription. Loss of, or decreased control of, the histone acetylase systems that establish these patterns of histone modification could lead to changes in the precise level of gene transcription and affect the functionality of the cell. This may explain why a large number of acetylation-regulated genes that are also involved in cancer development have been identified. We saw in Chapter 10 that histone acetylation has a major role in the regulation of the cell cycle. It is therefore not surprising that many types of cancer show extensive disruption of cell cycle control that correlates with similarly extensive disruption of their histone acetylation patterns.

The HAT/HDAC balance requires dysregulation of other factors

In many types of cancer cells, genes encoding histone acetyltransferase enzymes are known to undergo translocations that could change their expression levels, which could in turn alter the ability of the cell to maintain an appropriate level of acetylation at tumor suppressor loci. Similarly, if HDAC levels are decreased, excessive acetylation may accumulate in oncogenes. However, there is little evidence that either HDACs or HATs have specificity for individual genes, so changes in the expression levels of both the HDACs and the HATs are less likely to be the major determinants behind cancer-related changes in acetylation. Significant under-expression of either enzyme class may lead to either hyperacetylation or hypoacetylation, but this is likely to be nonspecific and therefore may not be the most important mechanism leading to cancer-related epigenetic disruption.

HATs and HDACs usually operate as part of a multiprotein complex, and one of the principal cancer-associated genes, c-*myc*, is known to influence histone acetylation of its target genes by virtue of its protein product recruiting HATs preferentially to promoters that are already enriched for euchromatic marks such as H3K4me2 or H3K4me3 or acetylation of lysine 9 on histone H3. Once bound to its target promoters, Myc introduces further changes in chromatin. In particular, Myc interacts with a variety of histone modifiers—such as the HATs Tip60, GCN5/PCAF, and p300–CBP, or HAT-associated proteins such as TRRAP—resulting in the recruitment of transcription factors to chromatin and in local hyperacetylation of histones. Genes that are already repressed in constitutive heterochromatin probably cannot be up-regulated by c-*myc*. However, for facultative heterochromatin or euchromatic regions, inappropriate gene expression levels of c-*myc* targets may result from overexpression of this oncogene. This is particularly true for genes involved in the cell cycle, which are continually capable of expression in mitotically active cells but whose expression at an inappropriate level might remove control over cell division and establish continuous growth, a hallmark of neoplastic transformation.

Histone methylation contributes to tumorigenesis

Although there is evidence that histone acetylation serves to maintain an accessible chromatin conformation, the link between DNA methylation states and repressive histone methylations is more firmly established. In general, dimethylation and trimethylation of lysine 9 on histone H3 correlates with DNA methylation at that locus (and therefore gene repression), whereas acetylation of lysine 9 tends to be associated with gene activation and shows an inverse correlation with DNA methylation. We saw in Chapter 5 how a large array of different enzymes are capable of adding the methyl mark to either lysine or arginine residues at various positions of the histone N-terminal tails. Presumably, changes in the expression levels of one or combinations of these enzymes or dysregulation of their functions could lead to changes in the pattern of histone methylations that might predispose the cell to transformation.

Although evidence for aberrant DNA methylation and histone modification changes in cancers is accumulating, it remains unclear how aberrant DNA methylation and histone modifications interact at a certain locus, why some genes are particular targets of epigenetic machinery, and whether aberrant DNA methylation and histone modifications confer specific roles in cancer initiation and progression. Of increasing clarity is the apparent cross-talk between histone methylation and the establishment of DNA methylation at particular loci.

From a "traditional" viewpoint, histone modification was a downstream event from DNA methylation, because methylated-DNA-detecting proteins such as MECP2 were known to recruit complexes containing HDACs and methyltransferases. However, studies in fungi (*Neurospora crassa*) show that mutations of histone H3K9 methyltransferase decrease DNA methylation, indicating a simple linear model in which H3K9 methylation acts as an upstream epigenetic mark that signals to DNA methylation. In addition, the epigenetic repression of Oct4 during embryonic stem cell differentiation requires the histone methyltransferase G9a, which subsequently recruits DNMT3a and DNMT3b, which methylate appropriate cytosines. The interaction between G9a and the DNMTs depends more on its ankyrin subunit than on the catalytic SET domain, suggesting that DNA methylation on the promoter depends on the recruitment of G9a (especially the ankyrin motif) rather than on the histone methyltransferase activity itself. The SET domain can be mutated to eliminate its capacity to methylate lysine, but as long as the ankyrin domain binds, DNA methylation will still take place. These interactions between DNA methylation and histone H3K9 methylation currently fit a model in which these two changes form a reinforcing silencing loop or bidirectional cross-talk. Therefore, if one of these modifications is subject to aberrant changes during tumorigenesis, this may lead to corresponding dysregulation of the other factor.

An interesting observation arising from large-scale chromatin immunoprecipitation studies suggests that many of the genes that are more likely to be hypermethylated in cancers are targets of the polycomb repressive complexes, particularly PRC2, which potentially links H3K27 methylation and *de novo* DNA methylation at these loci. The histone methyltransferase associated with PRC2 is EZH2, and because EZH2 is also known to interact with DNMTs in some cases, this activity may be one mechanism responsible for enforcing the DNA methylation. It must be stressed that there are many instances where this link does not apply. For example, most genes enriched with the H3K27me3 mark in prostate cancers do not have CpG islands in their promoters, whereas polycomb complexes seem to prefer CpG-rich targets in normal cells. This could mean that cancer cells disrupt the polycomb silencing system in some way so that it targets tumor suppressors (among other genes) that it should not normally repress. If so, this could mean that either the polycomb complex in the cancer cell is structurally or functionally different, so that its target selection changes completely, or some other factor is involved in directing the polycomb complex to interact with a new set of genes. Alternatively, there may be some structural or sequence features of the polycomb-target genes that predispose their CpG islands to the spread of DNA methylation over time or in response to elevated levels of DNMT; these might even be some of the features needed to recruit PRC2 in the first place.

Bioinformatic analysis of CpG island sequences known to be prone to hypermethylation in cancer cell lines and primary tumors shows a strong association with embryonic targets of PRC2, and an even stronger association with the PRC2 subunit SUZ12. An estimate from genome-wide studies is that as many as 10% of the total number of genes in the human genome are PRC2 targets; only a fraction of these are further targeted for *de novo* DNA methylation in cancer cells. This would imply that both sequence information and SUZ12 occupancy (and perhaps other factors) are required to define a specific CpG island as being prone to aberrant hypermethylation. Any of these situations would lead to the epigenetic reprogramming of several genes and cause genes that are originally silenced by PcG to acquire DNA methylation as an added silencing mechanism. This epigenetic switch to DNA methylation-mediated repression reduces the epigenetic plasticity, locking the silencing of key regulators and contributing to tumorigenesis. Changes in the target specificity of PcG could also lead to the reactivation of oncogenes if these sequences were no longer subjected to H3K27 methylation and the associated DNA methylation.

16.5 EXAMPLES OF EPIGENETIC DYSREGULATION LEADING TO CANCER

Cancer is a diverse collection of illnesses that depend on the neoplastic transformation and expansion of an equally wide range of cell types. Although the contribution of changes in DNA methylation to the process of transformation is only poorly understood, our knowledge of this particular epigenetic aberration is far better than our knowledge of how histone modifications can contribute to carcinogenesis. One reason for this is the diversity of post-translational modifications that may be applied to the histone N-terminal tails and the combinatorial nature of the information that they provide to the cell. As a consequence of this complexity, most studies of histone modification in cancer so far have been largely descriptive. Not surprisingly, this makes it quite difficult to present a comprehensive discussion of the mechanisms by which histone modification may be involved in tumorigenesis without resorting to pure speculation. The best that can be achieved with the current state of knowledge is to provide a few examples that show the types of epigenetic changes associated with specific types of cancer.

Hematological malignances such as leukemia are good examples of epigenetic dysregulation

Essentially, this group of malignancies occurs because the systems controlling the types and number of blood cells that are formed in the bone marrow become disrupted. An overview of blood cell formation, known as hematopoiesis, is shown in **Figure 16.9**, but the complexity of the process is much greater than this simple diagram would imply.

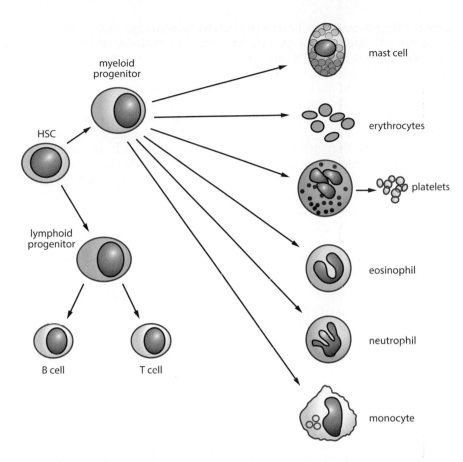

Figure 16.9 The hierarchy of blood cell development from hematopoietic stem cells (HSCs). HSCs can differentiate into common myeloid progenitors, which then differentiate into granulocyte/macrophage progenitors and megakaryocyte/erythrocyte progenitors. [From Speck NA & Gilliland DG (2002) *Nat. Rev. Cancer* 2, 502. With permission from Macmillan Publishers Ltd.]

Hematopoietic development is a delicately orchestrated process that results in the maturation of immature hematopoietic progenitor cells in the bone marrow into differentiated cellular components of peripheral blood, including neutrophils, red blood cells, and platelets. The requirement for tight control of blood cell production is emphasized by the short life of many of these cells in peripheral blood and by the large number of cells that are required for protection from infection, the proper delivery of oxygen, and hemostasis. For example, the half-life of the more than 1 billion neutrophils in peripheral blood is approximately 8–10 hours. Leukemia is, in essence, deranged and disordered hematopoiesis that results from mutations in hematopoietic progenitor cells; the mutations confer a proliferative and/or survival advantage on these cells and impair their hematopoietic differentiation. This leads to the division of hematological malignancies into either lymphoid or myeloid types, depending on the differentiation pathway and cell type affected. The lymphoid leukemias—acute lymphoblastic (ALL) and chronic lymphocytic (CLL)—affect the lymphocytes (mostly the B cells), whereas myeloid leukemias come in acute (AML) and chronic (CML) forms and affect a variety of blood cells that normally differentiate from myeloid stem cells.

The leukemias as a group are probably the twelfth most common form of neoplastic disease and the eleventh most common cause of cancer-related death. They can be treated with variable degrees of success by using radiotherapy or chemotherapeutic agents that kill the aberrant cells. Ablation of the bone marrow, followed by transplantation, is one option for treating some of these conditions. However, this requires a donor of matched or similar haplotype and even then requires lifelong immunosuppression to prevent rejection of the grafted bone marrow.

The pathogenesis of the various forms of leukemia differs, but one thing that leukemias have in common with all other cancers is an underlying genetic defect such as chromosomal translocations and other rearrangements, small deletions, amplifications, mutations, or loss or gain of entire chromosomes. Some of these abnormalities are apparent at the cytogenetic level, whereas others are detected only with molecular techniques. One of the better-characterized chromosomal aberrations is the Philadelphia chromosome (or Philadelphia translocation), which is associated with CML and results from a reciprocal translocation between chromosomes 9 and 22. The presence of this translocation is a highly sensitive test for CML, because 95% of people with CML have this abnormality. The result of the translocation is the *BCR–Abl* gene, which is located on the shorter chromosome 22. *BCR–Abl* is a chimeric oncogene that encodes the protein p210 or sometimes p185, and it functions as a constitutively active tyrosine kinase that activates a number of cell-cycle-related proteins.

DNA hypermethylation and hypomethylation contribute to the leukemic phenotype

DNA hypermethylation is observed in many forms of leukemia. Several research groups have shown that acute myeloid leukemia (AML) cells possess several methylation lesions, with 95% of all patients examined possessing at least one hypermethylated allele. Some results suggest that this correlates with increased DNA methyltransferase expression and/or activity, but other results contradict these findings. So, as in other forms of cancer, there is no clear definition of how such methylation changes are generated, but because they do appear in a nonrandom fashion (that is, hypermethylation occurs only on specific gene promoters), the causes are probably similar to those discussed earlier in this chapter.

Chromosome 11 (particularly the p arm) seems to be a methylation "hotspot" in AML, but the p15 and p16 cyclin-dependent kinase inhibitors also undergo frequent hypermethylation-based inactivation. Another characteristic feature of AML is that such methylation events tend to occur at later stages of disease progression, suggesting that some of the initial transformation events are required to establish the hypermethylator phenotype. One possibility is that oncogenic fusion proteins, which are a hallmark of AML, are able to recruit the DNA methyltransferases to gene promoters. This is supported by the observation that the PML–RAR fusion protein [the fusion gene is generated by a chromosomal translocation which places the retinoic acid receptor (RAR) in the promyelocytic leukemia (PML) locus] is able to methylate the *RARβ2* promoter, resulting in its repression. Hypermethylation has also been observed in the *HIC1*, *ER*, and *ABL1* genes during CML development, with similar involvement of *ER* and *HIC1* in acute lymphoblastic leukemia (ALL).

One genomic region that contributes a great deal to the development of many forms of leukemia comprises the *HOX* gene clusters, because they have crucial roles in the control of differentiation of adult hematopoietic cells. Several members of the *HOX* gene family, particularly from the *HOXA* cluster, are frequently inactivated by CpG island hypermethylation in leukemia, with inactivation of *HOXA4* and *HOXA5* being common to both myeloid and lymphoid malignancies. *HOXA5* is a known regulator of the myeloid differentiation pathway from hematopoietic progenitors (see Figure 16.9). If its expression is reduced below a certain threshold level, differentiation toward more terminal lineages such as granulocytes is blocked, although the myeloid progenitors still form. These accumulate in increasing numbers to produce the CML phenotype. Interestingly, ectopic re-expression of *HOXA5* in CML-derived cell lines causes them to re-start their differentiation program, although they seem unable to complete it. This inability points to the involvement of other genes that may have been inactivated as a consequence of CML. In addition to the frequent targeting of *HOXA4* and *HOXA5* in CML, hypermethylation of these genes was also often observed in AML, CLL, and childhood ALL and AML, suggesting a general role for the inactivation for these *HOX* genes in the development of human leukemia. Many other cancer cell types show *HOXA* cluster dysregulation, a finding consistent with their central role in the specification of cell identity and differentiation.

Inactivation of specific members of the *HOX* clusters in a cell-type-dependent manner relies mostly on repression by the polycomb group proteins. The observed DNA methylation occurring at the numerous CpG islands distributed throughout the clusters is therefore probably due to the recruitment of DNA methyltransferases by such complexes. We saw in Chapters 5 and 9 how the *HOX* clusters are subject to precise epigenetic controls over their spatio-temporal expression, but one of the factors contributing to this control mechanism that was not discussed in detail is the expression of small noncoding RNAs from the intergenic regions of these clusters. It has been suggested that selective expression of some of these noncoding RNAs in different cell types may be responsible for controlling the recruitment of histone-modifying enzyme systems such as polycomb group complexes, although how selective targeting of specific genes is achieved is not yet known. It may be that the patterns of histone modification enrichment have more to do with specific noncoding RNA activity than with direct control of specific *HOX* genes, although it is not yet clear whether the histone modification represses the noncoding RNA or whether the noncoding RNA is responsible for silencing itself by recruiting the polycomb complex to its own locus. Clearly, we do not yet know enough about this mode of regulation, but the cell-type context is crucial to the noncoding RNA expression pattern.

Figure 16.10 shows the differences in the *HOXA* clusters in fibroblasts derived from the lung and the skin of a human foot. Even this apparently identical cell type shows enormous differences in *HOXA* cluster

Figure 16.10 Epigenetic modification and expression from the *HOXA* cluster depends on anatomical location.
Occupancy of SUZ12, H3K27me3, and Pol II versus transcriptional activity over ~100 kb of the *HOXA* locus for primary lung (top) or foot (bottom) fibroblasts (Fb). For chromatin immunoprecipitation (ChIP) data, the $\log_2$ ratio of ChIP/input is plotted on the y axis. For RNA data, the hybridization intensity is shown on a linear scale. The dashed line highlights the boundary of opposite configurations of chromatin modifications and intergenic transcription. [From Rinn JL, Kertesz M, Wang JK et al. (2007) *Cell* 129, 1311. With permission from Elsevier.]

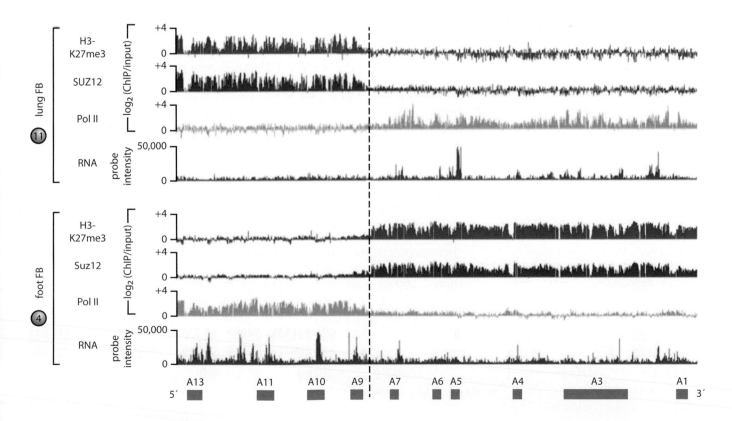

organization, which seems largely dependent on the cluster's anatomical location (remember that *HOX* genes establish the initial body plan). At the time of writing, similar data do not exist for the precise *HOXA* cluster structure in hematopoietic stem cells, but one can imagine that even small disruptions in the maintenance of this epigenetic control system could lead to re-expression or repression of *HOXA* genes that are not consistent with maintenance of the cell's identity or its ability to differentiate along a specific pathway. If its identity is lost, most of its functionality will be lost too.

On the basis of the discussion of hypermethylation-specific CpG islands, the *HOXA* cluster (and the other clusters, *HOXB* to *HOXD*) are "classic" PRC2 targets and, as highlighted in Figure 16.10, show continual occupancy by SUZ12 in many cell types throughout adult life. One could imagine that this would single them out for aberrant hypermethylation in response to DNMT1 overexpression, but curiously, none of the homeobox-type genes were identified by the original screen of CpG islands subjected to SUZ12 occupancy. However, it is possible that the clustered structure of the *HOX* genes confers some additional structural constraints that makes them sensitive to methylation changes.

Repression of *HOX* genes involved in the maintenance of hematopoietic cell identity and self-renewal is one contributing factor to leukemia, but the reaction of genes that are normally repressed in the regulation of this phenotype is also of importance. The histone methyltransferase MLL1 is part of a multiprotein complex that establishes the H3K4 trimethyl mark on genes undergoing active transcription, and it gets its name from Mixed Lineage Leukemia, because linkage studies highlighted its importance in the condition. Chromosomal translocations that fuse the MLL1 methyltransferase to one of more than 50 known partners are correlated with the occurrence of aggressive myeloid and lymphoid leukemias in infants and in patients with therapy-associated leukemia. Evidence suggests that native MLL1 regulates Hox genes in hematopoietic cells to establish cellular identity. Disruption of MLL1 function by chromosomal translocation results in Hox misregulation, coupled with the onset of leukemic phenotypes.

MLL1 normally binds near the transcriptional start sites of most genes occupied by RNA polymerase II. This behavior pattern may alter the chromatin architecture of a very wide range of genes after aberrant MLL1 expression, but the 3′ end of the *HOXA* cluster seems to do something rather different. Binding events within some genes occur in regions considerably upstream and downstream of the transcription start sites. The 3′ end of the *HOXA* cluster seems to be especially rich in this type of binding site, which may be a consequence of the organization that permits the co-linear expression of *HOX* genes. MLL1 occupies the genome far upstream (3775 bp) and downstream (2250 bp) of individual *HOX* genes within the *HoxA* cluster, but it also seems to target large regions encompassing *HoxA1* and the 5′-*HoxA* subcluster, including *HoxA7*, *HoxA9*, *HoxA10*, *HoxA11*, and *HoxA13* (**Figure 16.11**). In Cd133-expressing hematopoietic progenitor cells, the 3′-*HOXA* cluster is normally subject to H3K27me3-mediated gene repression (just as for the foot fibroblasts noted above) with limited H3K4 methylation. If the fusion proteins of

Figure 16.11 *HOX*-specific binding profile of MLL1. Schematic representation of the 130 kb *HOXA* cluster region on chromosome 7p15.2. Binding events for MLL1 (blue rectangles) and H3K4 trimethylation (red rectangles) are shown for each individual *HOX* gene probe. [From Guenther MG, Jenner RG, Chevalier B et al. (2005) *Proc. Natl. Acad. Sci. USA* 102, 8603. With permission from National Academy of Sciences.]

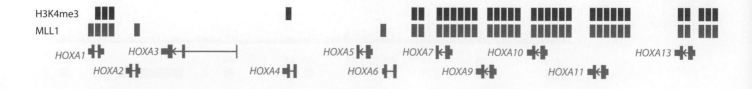

MLL1 are able to increase the enrichment of H3K4 methylation in this area, this may be able to de-repress the genes, leading to the phenotypic aberration that in turn can produce transformed leukemic cells.

How epigenetics contributes to lung cancers

Lung cancer is one of the most prevalent cancers and is the leading cause of cancer deaths in the world. So far, enormous progress has been made in understanding the molecular and cellular biology of lung cancers, with the strong link between lung cancers and smoking having been a major driving force behind research in this area. An important role is played in the development of this subset of cancers by oncogene up-regulation and tumor suppressor repression, and, once again, changes in DNA methylation and histone modification are central to this inappropriate gene-expression control.

So far, several known and putative tumor suppressor genes have been identified that are involved in the pathogenesis of lung cancer and are frequently inactivated by DNA methylation (**Table 16.2**). RARβ functions as a key retinoid receptor, mediating growth-control responses. Frequent loss of *RARβ* mRNA expression has been described in both primary non-small-cell lung cancers and bronchial biopsy specimens from heavy smokers.

The *adenomatous polyposis coli* (*APC*) gene is a tumor suppressor gene associated with both familial and sporadic colon cancer. However, it has also been reported that methylation of *APC* promoter 1A occurs frequently in lung cancers and correlates with a loss of *APC* expression (as revealed by reverse transcription PCR). In addition, genes involved in the cell cycle, such as *p16*, are frequently silenced, as are several cell surface glyco-protein genes such as *CHD1* and *CHD13* (which encode E-cadherin and H-cadherin, respectively), whose lack of expression may compromise homophilic cell adhesion and recognition. Other genes that are frequently silenced by promoter methylation are *MGMT*, the DNA-repair gene expressing O6-methylguanine-DNA methyltransferase, and *DAPK*, the apoptosis-associated gene expressing death-associated protein kinase.

In contrast to the situation in some other types of cancer, DNA methylation seems to be an early event in lung cancer development. For example, *p16* methylation was found in 75% of all precursor lesions that eventually gave rise to lung carcinoma, and the frequency of *p16* methylation

TABLE 16.2 TUMOR SUPPRESSOR GENES FREQUENTLY INACTIVATED BY DNA METHYLATION IN LUNG CANCERS

Gene	Non-small-cell lung carcinoma (%)	Small-cell lung carcinoma (%)
APC	49–96	15
CDH13	45–45	15
RARβ	40–43	45
FHIT	37	64
RASSF1A	30–40	79–85
TIMP-3	19–26	not applicable
P16	25–41	5
MGMT	16–27	16
DAPK	16–44	not applicable
CDH1	18–33	60
P14	6–8	not applicable
GSTP1	7–12	16

is known to increase during the progression from basal cell hyperplasia through squamous metaplasia to carcinoma *in situ*. Interestingly, the methylation states of the promoters of genes shown in Table 16.2 are useful markers of the probability of an individual developing either small-cell or non-small-cell lung carcinoma. Sputum samples obtained from smokers who had not developed cancer contain sufficient DNA to amplify the promoter regions using PCR and examine their methylation status. These studies show extensive methylation of key tumor suppressor genes in smokers and former smokers, whereas these levels are much lower in healthy nonsmoking controls. Moreover, in the case of the *p16* gene, the level of methylation was significantly associated with the number of years over which the individual had smoked and their smoking behavior (for example heavy or light) during that period.

It is worth discussing the link between tobacco smoke and lung cancer in a little more detail, given the wealth of data supporting the carcinogenic activity of the smoking habit and the very high incidence of smoking-related lung cancers. The active (or at least desirable) component of tobacco smoke, nicotine, is not a carcinogen, but each puff of each cigarette delivers a mixture of more than 60 established carcinogens, along with toxicants, tumor promoters, co-carcinogens, oxidants, free radicals, and inflammatory agents. The gas-phase constituents include 1,3-butadiene, ethylene oxide, benzene, and aldehydes. The particulate-phase constituents include polycyclic aromatic hydrocarbons (PAHs), the best known of which is benzo[*a*]pyrene (BaP), and tobacco-specific nitrosamines such as the potent lung carcinogen 4-(methylnitrosamino)-1-(3-pyridyl)-1-butanone (NNK). Consistent with the presence of these carcinogens, both the gas phase and the particulate phase of tobacco smoke can induce lung tumors in rodents on exposure by inhalation.

Tumor promoters, co-carcinogens, and toxicants all have various deleterious effects. Tumor promoters are not carcinogenic themselves, but they enhance the activity of carcinogens when given subsequently. The tumor-promoting activities of tobacco smoke and its condensate have been clearly demonstrated through administration both by inhalation and by application to mouse skin. These tumor promoters are only partly characterized, but extensive data indicate that they are found mainly in the weakly acidic fraction of tobacco smoke condensate. Co-carcinogens are also not carcinogenic themselves, but they enhance the activity of carcinogens when given concurrently. Catechol, methyl catechols, and certain PAHs are well-established co-carcinogens in tobacco smoke, as indicated by studies on mouse skin. One of the main toxicants in cigarette smoke, with a demonstrated relationship to lung carcinogenesis, is acrolein. Although it is not strongly carcinogenic itself, acrolein is highly toxic to cilia of the lung, thus impeding the clearance of tobacco smoke constituents. Acrolein also reacts directly with DNA and protein to produce adducts with potentially important consequences. Other toxicants in tobacco smoke include nitric oxide and poorly characterized free radicals, which may contribute to tumor promotion or co-carcinogenesis by causing oxidative damage.

The carcinogens and their metabolites bind to DNA, resulting in DNA adducts and subsequent somatic mutations. When these mutations occur in crucial genes, such as oncogenes and tumor suppressor genes, the result is a loss of normal cellular growth-control mechanisms, genomic instability, and cancer. The damage induced by smoking is truly impressive. DNA sequencing of 623 cancer-related genes revealed more than 1000 somatic mutations in 188 human lung adenocarcinomas, and 26 of these genes, including the tumor suppressor gene TP53 and the oncogene *KRAS*, were mutated at significantly higher frequencies. Alterations were commonly observed in genes of the MAPK-signaling, TP53-signaling, Wnt-signaling, cell cycle, and mTOR pathways. The multiple mutations caused by tobacco smoke carcinogens are also consistent with the concept of field cancerization. Although the lung is the first point of entry of these materials and therefore the most likely site for disease progression, tobacco smoke has adverse effects on nearly every aspect of human health (**Figure 16.12**).

The inclusion of the effects of smoking on the reproductive system is quite appropriate in this review of epigenetic contributions to disease, because there is evidence that exposure of pregnant female animals to tobacco smoke can alter the gene expression profile of the lungs of their offspring when they reach adulthood. It has been shown that tobacco smoke extract causes specific down-regulation of DNMT3B expression that leads to the demethylation of CpG islands in several potential oncogenes in lung epithelial cells and that a similar effect can be induced by small interfering RNA knockdown of the *DNMT3B* gene. In addition, it has been also demonstrated that tobacco exposure induces epigenetic changes through promoter methylation of tumor suppressor genes and proapoptotic genes. Increased expression of DNA methyltransferase 1

Figure 16.12 Tobacco smoke has adverse effects on nearly every aspect of human health. Nonmalignant cardiovascular and respiratory conditions and cancer account for most smoking-related deaths. Additional health risks associated with cigarette smoking are a weakened immune system, impaired wound healing, diabetes, decreased fertility in men and women, and an increased risk of sexually transmitted diseases. HPV, human papillomavirus. [From Stämpfli MR & Anderson GP (2009) *Nat. Rev. Immunol.* 9, 377. With permission from Macmillan Publishers Ltd.]

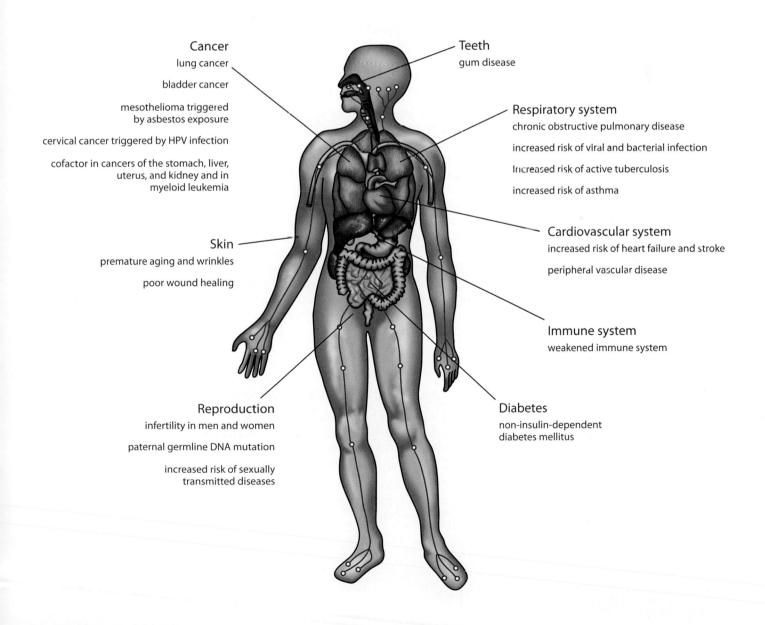

Cancer
lung cancer
bladder cancer
mesothelioma triggered by asbestos exposure
cervical cancer triggered by HPV infection
cofactor in cancers of the stomach, liver, uterus, and kidney and in myeloid leukemia

Skin
premature aging and wrinkles
poor wound healing

Reproduction
infertility in men and women
paternal germline DNA mutation
increased risk of sexually transmitted diseases

Teeth
gum disease

Respiratory system
chronic obstructive pulmonary disease
increased risk of viral and bacterial infection
increased risk of active tuberculosis
increased risk of asthma

Cardiovascular system
increased risk of heart failure and stroke
peripheral vascular disease

Immune system
weakened immune system

Diabetes
non-insulin-dependent diabetes mellitus

(DNMT1) has been linked to tobacco-caused hypermethylation of the tumor suppressor gene *RAS-association domain family 1A* (*RASSF1A*). We saw earlier that disruption of polycomb repressive complex function is a frequent cause of epigenetic changes leading to inappropriate gene expression in cancer. Tobacco smoke extract is known to engage polycomb repressive complexes to mediate the epigenetic silencing of *DKK1*, which normally serves to repress *WNT5a*, a gene involved in noncanonical WNT signaling, which is critical for distal airway development. Overexpression of *WNT5a* in non-small-cell lung carcinoma correlates strongly with enhanced proliferation of the transformed cells and a lower probability of patient survival. The components of tobacco smoke that cause the recruitment of polycomb proteins to the *DKK1* promoter are unknown, but such recruitment results in a dose-dependent increase in H3K27me3, which coincided with the recruitment of EZH2 and SUZ12 as well as Bmi1, a member of "maintenance" polycomb repressive complex 1.

Additional analysis confirmed the dose-dependent deacetylation of H4K16, which coincided with the recruitment of SirT1. A class III HDAC, SirT1 mediates H4K16 deacetylation, which is thought to be significantly disrupted in lung cancers. This observation, among others, has led to the suggestion that HDAC inhibitors may be useful in treating non-small-cell lung carcinoma (which shows the greatest levels of histone deacetylation at tumor suppressor loci). However, as with other types of cancer, this must be approached with caution because there is still the possibility of deacetylating histone H3 at oncogene loci or even deacetylating genes whose expression is not directly linked to the neoplastic transformation but may still produce a cytotoxic effect on otherwise healthy cells.

The two preceding examples give only a brief snapshot of the contributions made by the epigenetic regulation of gene expression to the development of cancer, and doubtless there are as many cancer-related variations to this regulatory mechanism as there are cellular aberrations that may potentially produce tumors in human organs. Most epigenetic problems are centered on changes in DNA methylation and histone post-translational modification that result either from altered expression of the enzyme systems that maintain these epigenetic states or through the expansion or contraction of epigenetically modified regions of DNA by mechanisms that are as yet unclear. Additional examples of cancers affecting humans and mammals are unnecessary: they would serve only to underline further the types of dysfunctions that have been already described. It is therefore hoped that the reader will have gained a useful, albeit incomplete, overview of epigenetics in cancer and, indeed, in disease in general.

KEY CONCEPTS

- Cancer arises when homeostatic control of cell growth fails.

- This loss of homeostatic control occurs when the expression of specific genes is dysregulated—these genes are known as tumor suppressors and oncogenes on the basis of their ability to restrict or promote tumor formation.

- Disruptions in the epigenetic mechanisms controlling the expression of tumor suppressors or oncogenes can contribute to the loss of homeostasis.

FURTHER READING

Cedar H & Bergman Y (2011) Epigenetics of haematopoietic cell development. *Nat Rev Immunol* **11**:478–488 (doi:10.1038/nri2991).

Cohen I, Poręba E, Kamieniarz K & Schneider R (2011) Histone modifiers in cancer: friends or foes? *Genes Cancer* **2**:631–647 (doi:10.1177/1947601911417176).

Ehrlich M & Lacey M (2013) DNA hypomethylation and hemimethylation in cancer. *Adv Exp Med Biol* **754**:31–56 (doi: 10.1007/978-1-4419-9967-2_2).

Goel A & Boland CR (2012) Epigenetics of colorectal cancer. *Gastroenterology* **143**:1442–1460.e1 (doi:10.1053/j.gastro.2012.09.032).

Henson DE & Albores-Saavedra (eds) (2001) Pathology of Incipient Neoplasia, 3rd ed. Oxford University Press.

Herceg Z & Ushijima T (2010) Introduction: epigenetics and cancer. *Adv Genet* **70**:1–23 (doi:10.1016/B978-0-12-380866-0.60001-0).

Hsiao SH, Huang TH & Leu YW (2009) Excavating relics of DNA methylation changes during the development of neoplasia. *Semin Cancer Biol* **19**:198–208 (doi:10.1016/j.semcancer.2009.02.015).

Jiang Y, Liu S, Chen X et al. (2013) Genome-wide distribution of DNA methylation and DNA demethylation and related chromatin regulators in cancer. *Biochim Biophys Acta* **1835**:155–163 (doi:10.1016/j.bbcan.2012.12.003).

Jovanovic J, Rønneberg JA, Tost J & Kristensen V (2010) The epigenetics of breast cancer. *Mol Oncol* **4**:242–254 (doi:10.1016/j.molonc.2010.04.002).

Katto J & Mahlknecht U (2011) Epigenetic regulation of cellular adhesion in cancer. *Carcinogenesis* **32**:1414–1418 (doi:10.1093/carcin/bgr120).

Khandige S, Shanbhogue VV, Chakrabarty S & Kapettu S (2011) Methylation markers: a potential force driving cancer diagnostics forward. *Oncol Res* **19**:105–110 (doi:10.3727/096504011X12935427587641).

Kristensen LS, Nielsen HM & Hansen LL (2009) Epigenetics and cancer treatment. *Eur J Pharmacol* **625**:131–142 (doi:10.1016/j.ejphar.2009.10.011).

Lopez-Serra P & Esteller M (2012) DNA methylation-associated silencing of tumor-suppressor microRNAs in cancer. *Oncogene* **31**:1609–1622 (doi:10.1038/onc.2011.354).

Mathews LA, Crea F & Farrar WL (2009) Epigenetic gene regulation in stem cells and correlation to cancer. *Differentiation* **78**:1–17 (doi:10.1016/j.diff.2009.04.002).

Muñoz P, Iliou MS & Esteller M (2012) Epigenetic alterations involved in cancer stem cell reprogramming. *Mol Oncol* **6**:620–636 (doi:10.1016/j.molonc.2012.10.006).

Qu Y, Dang S & Hou P (2013) Gene methylation in gastric cancer. *Clin Chim Acta* **424C**:53–65 (doi:10.1016/j.cca.2013.05.002).

Reddy KL & Feinberg AP (2013) Higher order chromatin organization in cancer. *Semin Cancer Biol* **23**:109–115 (doi:10.1016/j.semcancer.2012.12.001).

Rodríguez-Paredes M & Esteller M (2011) Cancer epigenetics reaches mainstream oncology. *Nat Med* **17**:330–339 (doi:10.1038/nm.2305).

Silahtaroglu A & Stenvang J (2010) MicroRNAs, epigenetics and disease. *Essays Biochem* **48**:165–185 (doi:10.1042/bse0480165).

Thompson LL, Guppy BJ, Sawchuk L et al. (2013) Regulation of chromatin structure via histone post-translational modification and the link to carcinogenesis. *Cancer Metastasis Rev* (doi:10.1007/s10555-013-9434-8).

Tollefsbol T (ed.) (2008) Cancer Epigenetics. CRC Press.

Verma M (2013) Cancer control and prevention: nutrition and epigenetics. *Curr Opin Clin Nutr Metab Care* **16**:376–384 (doi:10.1097/MCO.0b013e328361dc70).

Waldmann T & Schneider R (2013) Targeting histone modifications—epigenetics in cancer. *Curr Opin Cell Biol* **25**:184–189 (doi:10.1016/j.ceb.2013.01.001).

Watanabe Y & Maekawa M (2010) Methylation of DNA in cancer. *Adv Clin Chem* **52**:145–167 (doi:10.1016/S0065-2423(10)52006-7).

You JS & Jones PA (2012) Cancer genetics and epigenetics: two sides of the same coin? *Cancer Cell* **22**:9–20 (doi:10.1016/j.ccr.2012.06.008).

GLOSSARY

ACH *see* **active chromatin hub**

active chromatin hub (ACH)
A three-dimensional arrangement of chromatin into areas from which transcription can take place

active memory *see* **short-term memory**

AGC kinases *see* **cAMP-, cGMP-dependent protein kinase C-like kinases**

antisense repression method
The method by which artificial oligonucleotides bind to mRNA to prevent their translation

arteriosclerosis
Thickening and hardening of the walls of the arteries

atherosclerosis
A specific form of arteriosclerosis in which an artery wall thickens as a result of the accumulation of fatty materials such as cholesterol and triacylglycerols

ATP-dependent nucleosome remodeling factors
Complexes that move, eject, or restructure nucleosomes

axon
The long, threadlike part of a nerve cell along which impulses are conducted from the cell

axon terminal
Distal terminations of the branches of an axon

biotin holocarboxylase synthetase (HCS)
An enzyme with an essential role in biotin utilization in eukaryotic cells; also known as holocarboxylase synthetase

bivalent domain
An area of chromatin that contains both activating and repressing epigenetic modifications in the same area

bookmarking theory
A theory to explain the transmission of expression status of genes from mother to daughter cells

bromodomain
A protein domain that recognizes acetylated lysine residues such as those on the N-terminal tails of histones

Cajal bodies
Ubiquitous subnuclear organelles found in both plant and animal cells

cAMP-, cGMP-dependent protein kinase C-like kinases (AGC kinases)
Enzymes that phosphorylate target proteins only when the enzyme's regulatory subunits are bound to cyclic AMP or cyclic GMP

CDPs *see* **cysteine-dependent phosphatases**

cell cycle
The series of events that take place in a cell that lead to its division and duplication; also known as the cell-division cycle

CHD ATPases *see* **chromodomain and helicase-like domain (CHD) ATPases**

chromatin
The material of which the chromosomes of eukaryotes (organisms other than bacteria) are composed

chromocenter
An irregular mass of heterochromatin in some chromosomes

chromodomain
A protein structural domain of about 40–50 amino acid residues commonly found in proteins associated with the remodeling and manipulation of chromatin

chromodomain and helicase-like domain (CHD) ATPases
Chromatin-remodeling proteins

chromosomal instability (CIN)
A condition involving the unequal distribution of DNA to daughter cells upon mitosis, resulting in a failure to maintain euploidy

chromosome domain *see* **chromosome territory**

chromosome territory
The organization of chromosomes into specific locations in the nucleus depending on their transcriptional activity

CIMP *see* **CpG island methylator phenotype**

CIN *see* **chromosomal instability**

co-linear activation
The progressive activation of genes situated in clusters

conduction
The transmission of an impulse along a nerve fiber

CpG island methylator phenotype (CIMP)
Widespread CpG island methylation in the promoters of several genes in cancer cells

CpG islands
Genomic regions that contain a high frequency of CpG sites

cysteine-dependent phosphatases (CDPs)
Phosphatase enzymes with cysteine as a functional amino acid at the active site

dendrites
Branched projections of a neuron that act to conduct the electrochemical stimulation received from other neural cells to the cell body

DNA methyltransferases (DNA MTases)
Family of enzymes that catalyze the transfer of a methyl group to DNA

epigenetics
The study of mitotically and/or meiotically heritable changes in gene function that cannot be explained by changes in DNA sequence

fibrillar center
A substructural component of the nucleolus

foam cells
Lipid-rich macrophages that surround lesions in the artery walls

GCK *see* **human germinal-center kinase**

genome-wide association studies (GWAS)
An examination of many common genetic variants in different individuals to see whether any variant is associated with a trait

genomic imprinting
An epigenetic phenomenon by which certain genes can be expressed in a parent-of-origin-specific manner

genomic instability
The tendency to develop a high frequency of mutations within the genome of a cellular lineage

glomerulus
A cluster of nerve endings or blood vessels, especially the capillaries around the end of a kidney tubule, where waste products are filtered

GWAS *see* **genome-wide association studies**

HCS *see* **biotin holocarboxylase synthetase**

histone chaperones
Broadly defined as a group of proteins that bind histones and regulate nucleosome assembly

histone code hypothesis
A hypothesis that the transcription of genetic information encoded in DNA is in part regulated by chemical modifications to histone proteins

histone fold
The interaction motif involved in homodimerization of the core histones (H2A, H2B, H3, and H4) and their assembly into an octamer. It comprises three α-helices linked by two loops. The surface residues mediate octamer formation and interaction with the DNA that wraps around the octamer

histone tail bridging model
A model used to explain how the N-terminal tails of histones may allow interactions between nucleosomes

holocarboxylase synthetase (HCS) *see* **biotin holocarboxylase synthetase**

HPA axis *see* **hypothalamic–pituitary–adrenal axis**

human germinal-center kinase (GCK)
A subset of the Ste20 group of proteins

hypothalamic–pituitary–adrenal axis (HPA axis)
A complex set of interactions between the hypothalamus, pituitary, and adrenal glands that control reactions to stress and regulate many body processes, including digestion, the immune system, mood and emotions, sexuality, and energy storage and expenditure

ICM *see* **inner cell mass**

ICR *see* **imprinting control region**

imitation switch (ISWI) ATPases
ATP-dependent chromatin remodeling factors

imprinting control region (ICR)
A region of DNA within the locus of an imprinted gene that shows allele-specific methylation and controls expression of the imprinted genes

induced pluripotent stem cells (iPSCs)
A type of pluripotent stem cell artificially derived from a non-pluripotent cell—typically an adult somatic cell—by inducing a 'forced' expression of specific genes

inner cell mass (ICM)
The mass of cells inside the primordial embryo that will eventually give rise to the definitive structures of the fetus

interneurons
A neuron that transmits impulses between other neurons

iPSCs *see* **induced pluripotent stem cells**

ISWI ATPases *see* **imitation switch (ISWI) ATPases**

large organized chromatin K9 modifications (LOCKs)
Specific regions of chromatin that are frequently subjected to modification on lysine 9 of histone H3

lineage priming
Predisposition to develop into specific cell types

LOCKs *see* **large organized chromatin K9 modifications**

long-term potentiation
A long-lasting enhancement in signal transmission between two neurons that results from stimulating them synchronously

main cell body
The portion of a neuron that contains the nucleus but does not incorporate the dendrites or axon

MBD *see* **methyl-binding domain**

MECP1 *see* **methyl-CpG-binding domain protein 1**

MECP2 *see* **methyl-CpG-binding domain protein 2**

metallophosphatases (MMPs)
Phosphatase enzymes with active sites that have metal ions as part of their structure

methyl-binding domain (MBD)
Domain in a protein capable of recognizing and binding to methylated DNA

methyl-CpG-binding domain protein 1 (MECP1)
A protein capable of recognizing and binding to methylated DNA

methyl-CpG-binding domain protein 2 (MECP2)
A protein capable of recognizing and binding to methylated DNA

microangiopathy
An angiopathy affecting small blood vessels in the body

microRNAs (miRNAs)
Post-transcriptional regulators that bind to complementary sequences in the 3′ untranslated regions

microsatellite instability (MIN)
The condition of genetic hypermutability that results from impaired DNA mismatch repair

microsatellites
Repeating sequences of two to six base pairs of DNA; also known as simple sequence repeats (SSRs), **small tandem repeats (STRs) or variable number tandem repeats**

MIN *see* **microsatellite instability**

miRNAs *see* **microRNAs**

mitogen- and stress-activated protein kinase 1 (MSK1)
A member of a class of serine/threonine-specific protein kinases belonging to the CMGC (CDK/MAPK/GSK3/CLK) kinase group

mitogen- and stress-activated protein kinase 2 (MSK2)
A member of a class of serine/threonine-specific protein kinases belonging to the CMGC (CDK/MAPK/GSK3/CLK) kinase group

mixed-lineage leukemia 1 (MLL1)
A type of childhood leukemia in which a piece of chromosome 11 has been translocated

MLL1 *see* **mixed-lineage leukemia 1**

monoamine-deficiency hypothesis
A hypothesis stating that depression is caused by the underactivity in the brain of monoamines such as dopamine, serotonin, and norepinephrine

motor neurons
A nerve cell forming part of a pathway along which impulses pass from the brain or spinal cord to a muscle or gland

MSK1 *see* **mitogen- and stress-activated protein kinase 1**

MSK2 *see* **mitogen- and stress-activated protein kinase 2**

NCP *see* **nucleosome core particle**

NPS *see* **nucleosome positioning sequence**

nuclear architecture
The arrangement of the components of a cell's nucleus

nuclear bodies
Subnuclear foci of various transcription factors, hnRNP proteins, heat-shock factors, heterochromatin proteins, and even elements of the cleavage and polyadenylation machinery

nucleolar organizers
Regions containing the genes encoding ribosomal RNA

nucleolus
A distinct subnuclear structure that is the site of ribosome biogenesis

nucleoskeleton
A filamentous network within the nucleus that provides structural integrity

nucleosome core particle (NCP)
The histone octamer around which 14 turns of DNA are coiled

nucleosome exclusion element
DNA sequences or structures that inhibit the establishment of nucleosomes at those loci

nucleosome positioning sequence (NPS)
DNA sequence directing the formation of nucleosomes at specific locations on the genome

nucleosome sliding
The process by which DNA slides around a nucleosome spool. The effect is to displace the spool linearly along the DNA

nucleosomes
Structural units of a eukaryotic chromosome, consisting of a length of DNA coiled around a core of histones

oncogenes
Genes that in certain circumstances transform a cell into a tumor cell

p21-activated kinase (PAK1)
One of a family of serine/threonine protein kinases that bind to, and in some cases are stimulated by, activated forms of the small GTPases Cdc42 and Rac

PAD4 *see* **peptidyl arginine deaminase 4**

PAK1 *see* **p21-activated kinase**

parental conflict hypothesis
A hypothesis stating that the inequality between parental genomes due to imprinting is a result of the differing interests of each parent in terms of the evolutionary fitness of their genes

PcG *see* **polycomb-group proteins**

peptidyl arginine deaminase 4 (PAD4)
Enzyme responsible for the conversion of arginine residues to citrulline residues

perinucleolar heterochromatin
Silenced ribosomal RNA genes at the edge of the nucleolar membrane

permissive chromatin state
A state permitting access of the transcriptional machinery to chromatin

PEV *see* **position effect variegation**

pluripotent
The potential to differentiate into any of the three germ layers

PML bodies
Bodies of unknown function containing the promyelocytic leukemia protein (PML)

polycomb-group proteins (PcG)
Chromodomain proteins that transcriptionally repress genes by favoring the formation of closed chromatin structures

polycomb repressive complex 1 (PRC1)
Multiprotein complex that contains polycomb proteins and functions by repressing gene transcription

position effect variegation (PEV)
The effect on the expression of a gene when its location in a chromosome is changed

PRC1 *see* **polycomb repressive complex 1**

primary memory *see* **short-term memory**

PRMTs *see* **protein arginine methyltransferases**

protein arginine methyltransferases (PRMTs)
Enzymes that catalyze the methylation of arginine residues of proteins to yield N-monomethylarginine and N,N-dimethylarginine

protein lysine acetyl transferases
Enzymes that acetylate conserved lysine amino acids on histone proteins by transferring an acetyl group from acetyl-CoA to form ε-N-acetyllysine

protein tyrosine kinase (PTK)
A protein kinase that phosphorylates tyrosine residues

proto-oncogenes
A normal gene that has the potential to become an oncogene

PTK *see* **protein tyrosine kinase**

R *see* **restriction point**

restriction point (R)
A checkpoint in the G1 phase of the mitotic cell cycle

ribosomal S6 kinases (RSKs)
Enzymes catalyzing the phosphorylation of ribosomal protein S6

RNA–DNA interaction model
A model of post-translational gene silencing

RSKs *see* **ribosomal S6 kinases**

sensory neuron
A neuron conducting impulses inward to the brain or spinal cord

short-term memory
The capacity for holding a small amount of information in mind in an active, readily available state for a short period, *see also* **primary memory**

small tandem repeats *see* **microsatellites**

'statistical positioning' theory
A model in which nucleosomes are positioned primarily by steric exclusion and two-body interactions rather than intrinsic histone–DNA sequence preferences

Ste20 *see* **sterile 20 proteins**

sterile 20 proteins (Ste20)
Proteins in the mammalian Ste20 kinase (MST) signaling pathway with an important role in the regulation of apoptosis and cell cycle control

subconscious
The part of consciousness that is not currently in focal awareness

SWI/SNF (switch/sucrose nonfermentable) ATPases
Nucleosome remodeling complexes

synapse
A junction between two nerve cells, consisting of a minute gap across which impulses pass by diffusion of a neurotransmitter

totipotent
Totipotency is the ability of a single cell to divide and produce all of the differentiated cells in an organism

transcription factories
Discrete sites in a eukaryotic nucleus into which active gene transcription units can be clustered

transformation
Genetic alteration of a cell resulting from the direct uptake, incorporation, and expression of exogenous genetic material

tumor suppressor genes
Genes that protect a cell from one step on the path to cancer; also known as anti-oncogenes

two-hit hypothesis
A hypothesis that cancer is the result of accumulated mutations to a cell's DNA; also known as the Knudson hypothesis

uremia
A raised level in the blood of urea and other nitrogenous waste compounds that are normally eliminated by the kidneys

variable number tandem repeats *see* **microsatellites**

INDEX

Note

The index covers the main text but not the end-of-chapter bibliographies; the figure captions but not the content of the figures. The letters 'B', 'F', and 'T' after page references indicate that coverage on these page(s) is only in a box, figure, or table but boxes, figures, and tables are not distinguished on pages that include indexed text discussion.

Numeric and Greek-letter prefixes of purely positional significance have been ignored in sorting (so "5-methylcytosine" appears at "m" and "β-III tubulin" at "t"): those integral to a name are sorted as though spelled out (so "beta cells" at "b" and the "γ5 gene" at "g").